Mobile Phone Programming

Mobile Phone Programming
and its Application to Wireless Networking

Edited by

Frank H.P. Fitzek

Aalborg University
Denmark

and

Frank Reichert

Agder University College
Norway

 Springer

A C.I.P. Catalogue record for this book is available from the Library of Congress.

ISBN 978-1-4020-5968-1 (HB)
ISBN 978-1-4020-5969-8 (e-book)

Published by Springer,
P.O. Box 17, 3300 AA Dordrecht, The Netherlands.

www.springer.com

Printed on acid-free paper

To Sterica and Lilith.
— Frank H.P. Fitzek

I dedicate this book to Tim (WoW Level 70, mighty Undead Warrior), Max (Wow Level 70, fearless Tauren Hunter), and Aida (Reality Level 80++, loving Human Wife and Mother)
— Frank Reichert (Level 64)

Foreword

Saila Laitinen

Nokia

The technology evolution has been once and for all beyond comparison during the past decade or two. Any of us can nowadays do numerous things with numerous devices to help in everyday life. This applies not least to mobile phones. If we compare the feature set of a mobile phone model in 1995 with the latest smartphone models the most visible difference is of course in the user interface, the mp3 player, integrated camera, and the access to the mobile Internet.

This evolution is a result of multiple members in the entire mobile ecosystem: (1) device manufacturers have managed to bring richer devices into markets, (2) operators have invested in network enablers heavily, (3) developers have found mobile platforms fancy to innovate on, and last but not least (4) consumers have discovered the benefits of using the rich feature set of mobile devices and they have welcomed them into their lives.

This book gives a thorough picture of all options a developer can choose from when developing mobile applications or other content. It explains Python; that is especially suitable for scripting, supports procedural programming, and object-oriented programming. Java, an object-oriented programming language developed by Sun Microsystems. Symbian C++, a programming language tailored to mobile programming. Open C provides the familiar standard C function libraries in mobile development. Qtopia's greenphone is a Linux mobile development device open for innovation. Maemo Linux, an embedded Linux development environment for mobile innovation. Windows Mobile, a compact operating system combined with a suite of basic applications for mobile devices based on the Microsoft Win32 API.

In addition to the above, the book focuses on mobile application design for wireless communication systems. Besides the well-known cellular architecture also the peer-to-peer and cooperative networking is introduced. Concepts such as cross-layer design and the convergence of wireless sensors with mobile phones are introduced and examples are given. Furthermore the energy consumption of mobile phones is addressed and designing rules to develop energy-efficient mobile applications are presented. I truly see that once the wireless internet really goes live all around the world; there are only winners in this industry.

I highly recommend reading this book, start the real mobile innovation so that together we can make people all around the world connected with each other without compromising in our mobility.

Helsinki, Finland   *Saila Laitinen*
April 2007   *Head of Regional Operations, Forum Nokia*

Preface

The main source of motivation to write this book is the beauty of the mobile phones as an inspiring platform for research as well as teaching activities.

To prove an idea or a concept in the research domain, there are mainly three possibilities, namely analytical evaluation, simulative evaluation, and implementation on an appropriate platform. The problem with the analytical and the simulative evaluation is that analytical derivations become more and more complicated reducing the possible audience or that the chosen simulation environment is not fully transparent. Therefore we need to implement our proof of concept on omnipresent platforms to introduce it to a larger audience. The mobile phones are omnipresent entities and are suitable as proof-of-concept platforms as they have two advantages over any other platform; first they are not that cost-intensive and second mostly everybody owns one mobile phone by themselves or at least has had experience with it. As we will show throughout this book, the degree of freedom to implement whatever idea in the world of wireless networking is nearly infinite and sometime only restricted on purpose by the mobile phone manufacturer. Of course a mobile phone has some limitations in the memory capacity, the processing power, and the limited energy due to the battery-driven entity. But those limitations should not be understood as a disadvantage, but more as setting design rules for new solutions in the real world.

To use the mobile platform in student and master projects we started to introduce mobile phone programming within our teaching activities. This topic is mostly new to students and they come with heterogeneous knowledge of programming languages. Different courses cover Python, JAVA, and Symbian/C++. Within the book we have extended the group of programming languages and deployment environments with topics such as Windows Mobile, Maemo, Qtopia, and Open C. In addition to the programming teaching, we encourage students to use their new knowledge within future research projects and to make their own mobile applications.

The importance of mobile applications introducing new kinds of services has been and is still underestimated. With the introduction of the second generation networks, new services were provided to the customers namely mobile speech services, roaming between different countries, low data rate IP connection, and more advanced security. The introduction of the third generation did not bring de facto revolutionary new services, which may explain why this generation of mobile communication systems is not having the same success as its forerunner had when it was introduced. Thus services are the key for successful communication networks.

We sincerely hope that this book is a contribution to enable a new wave of exciting mobile applications that enrich our lives, and that gives students a great start ramp to launch their future careers.

Planet Earth   *Frank H.P. Fitzek*
April 2007   *Frank Reichert*

Contents

Part V Cooperative Communication

---

13 Cooperative Wireless Networking: An Introductory Overview

14 The Medium is the Message

15 Peer-to-Peer File Sharing for Mobile Devices

16 Energy Saving Aspects and Services
for Cooperative Wireless Networks

Part VI Cross-Layer Communication

17 Cross-Layer Protocol Design for Wireless Communication

18 Cross-Layer Example for Multimedia Services
over Bluetooth

Part VIII Power Consumption in Mobile Devices

List of Contributors

Anders Grauballe
Aalborg University
Department of Electronic Systems
Niels Jernes Vej 12
DK-9220 Aalborg, Denmark
agraubal@es.aau.dk

Andreas Häber
Agder University College
Faculty of Engineering and Science
Grooseveien 36
Servicebox 509
N-4898 Grimstad
Norway
andreas.haber@hia.no

Antti Saukko
Nokia
Itämerenkatu 11–13
FIN-00180 Helsinki
Finland
antti.saukko@nokia.com

Bertalan Forstner
Budapest University of Technology
and Economics
H-1111 Budapest, Goldmann György
tér 3. Hungary
bertalan.forstner@aut.bme.hu

Ben Krøyer
Aalborg University
Department of Electronic Systems
Niels Jernes Vej 12
DK-9220 Aalborg, Denmark
bk@es.aau.dk

Christian S. Jensen
Aalborg University
Department of Computer Science
Fredrik Bajers Vej 7E
DK-9220 Aalborg Øst
csj@cs.aau.dk

Clemens Gühmann
Technische Universität Berlin
Einsteinufer 17
10587 Berlin, Germany
clemens.guehmann@tu-berlin.de

Eero Penttinen
Nokia
Itämerenkatu 11–13
FIN-00180 Helsinki
Finland
eero.penttinen@nokia.com

Fadi Chehimi
Informatics
Infolab21
Lancaster University
Lancaster, LA1 4WA, UK
f.chehimi@lancaster.ac.uk

Frank H.P. Fitzek
Aalborg University
Department of Electronic Systems
Niels Jernes Vej 12
DK-9220 Aalborg, Denmark
ff@es.aau.dk

Frank Reichert
Agder University College
Faculty of Engineering and Science
Grooseveien 36
Servicebox 509
N-4898 Grimstad, Norway
frank.reichert@hia.no

Gerard Bosch i Creus
Nokia Research Center
Itämerenkatu 11–13
FIN-00180 Helsinki
Finland
gerard.bosch@nokia.com

Gergely Csúcs
Budapest University of Technology
and Economics
H-1111 Budapest, Goldmann György
tér 3. Hungary
gergely.csucs@aut.bme.hu

Gian Paolo Perrucci
Aalborg University
Department of Electronic Systems
Niels Jernes Vej 12
DK-9220 Aalborg, Denmark
gpp@es.aau.dk

Hassan Charaf
Budapest University of Technology
and Economics
H-1111 Budapest, Goldmann György
tér 3. Hungary
hassan@aut.bme.hu

Imre Kelényi
Budapest University of Technology
and Economics
H-1111 Budapest, Goldmann György
tér 3. Hungary
imre.kelenyi@aut.bme.hu

Janne Dahl Rasmussen
Aalborg University
Department of Electronic Systems
Niels Jernes Vej 12
DK-9220 Aalborg, Denmark
jannedr@es.aau.dk

Jeppe Jensen
Aalborg University
Department of Electronic Systems
Niels Jernes Vej 12
DK-9220 Aalborg, Denmark
jeppe85@es.aau.dk

Jürgen Scheible
University of Art and Design
Helsinki Media Lab
Hämeentie 135C
FI-00560 Helsinki, Finland
jscheib@uiah.fi

Kristian Torp
Aalborg University
Department of Computer Science
Fredrik Bajers Vej 7E
DK-9220 Aalborg Øst
torp@cs.aau.dk

Leonardo Militano
University Mediterranea of Reggio
Calabria
Faculty of Engineering, Department
DIMET
Via Graziella, Loc. Feo di Vito
89100 Reggio Calabria, Italy
leonardo.militano@unirc.it

Marcos Katz
VTT
Kaitoväylä 1, Oulu
P.O.Box 1100
FI-90571 Oulu
FINLAND

Martin Mauve
Heinrich-Heine-University
Universitätsstr. 1
D-40225 Düsseldorf, Germany
mauve@cs.uni-duesseldorf.de

Matti Sillanpää
Nokia Research Center
Itämerenkatu 11–13
FIN-00180 Helsinki
Finland
matti.jo.sillanpaa@nokia.com

Michael Stini
Heinrich-Heine-University
Universitätsstr. 1
D-40225 Düsseldorf, Germany
stini@cs.uni-duesseldorf.de

Mika Kuulusa
Nokia Technology Platforms
Itämerenkatu 11–13
FIN-00180 Helsinki
Finland
mika.kuulusa@nokia.com

Mikkel Gade Jensen
Aalborg University
Department of Electronic Systems
Niels Jernes Vej 12
DK-9220 Aalborg, Denmark
mgade@es.aau.dk

Morten L. Jørgensen
Aalborg University
Department of Electronic Systems
Niels Jernes Vej 12
DK-9220 Aalborg, Denmark
mljo@es.aau.dk

Morten V. Pedersen
Aalborg University
Department of Electronic Systems
Niels Jernes Vej 12
DK-9220 Aalborg, Denmark
mvpe@es.aau.dk

Olli Silvén
Department of Electrical
and Information Engineering
P.O.B. 4500
FI-90014 University of Oulu, Finland
Olli.Silven@ee.oulu.fi

Omer Rashid
Informatics
Infolab21
Lancaster University
Lancaster, LA1 4WA, UK
m.rashid@lancaster.ac.uk

Paul Coulton
Informatics
Infolab21
Lancaster University
Lancaster, LA1 4WA, UK
p.coulton@lancaster.ac.uk

Paul Gilbertson
Informatics
Infolab21
Lancaster University
Lancaster, LA1 4WA, UK
p.gilbertson@lancaster.ac.uk

Peter Østergaard
Aalborg University
Department of Electronic Systems
Niels Jernes Vej 12
DK-9220 Aalborg, Denmark
petero@es.aau.dk

Rico Wind
Aalborg University
Department of Computer Science
Fredrik Bajers Vej 7E
DK-9220 Aalborg Øst
rw@cs.aau.dk

Stephan Rein
Technische Universität Berlin
Einsteinufer 17
10587 Berlin, Germany
stephan.rein@tu-berlin.de

Tatiana Kozlova Madsen
Aalborg University
Department of Electronic Systems
Niels Jernes Vej 12
DK-9220 Aalborg, Denmark
tatiana@kom.aau.dk

Tero Rintaluoma
Hantro Products Oy
Kiviharjunlenkki 1
FI-90220 Oulu, Finland
Tero.Rintaluoma@hantro.com

Thomas Arildsen
Aalborg University
Department of Electronic Systems
Niels Jernes Vej 12

DK-9220 Aalborg, Denmark
tha@es.aau.dk

Ulrik W. Rasmussen
Aalborg University
Department of Electronic Systems
Niels Jernes Vej 12
DK-9220 Aalborg, Denmark
wilken@es.aau.dk

Will Bamford
Informatics
Infolab21
Lancaster University
Lancaster, LA1 4WA, UK
w.bamford@lancaster.ac.uk

Part I

Introduction

1

Introduction to Mobile Phone Programming

Frank H.P. Fitzek[1] and Frank Reichert[2]

[1] Aalborg University `ff@es.aau.dk`
[2] Agder University College `frank.reichert@hia.no`

1.1 Evolution of the Mobile Phones, Networks, and Services

1.1.1 Mobile Phone Family

Mobile phones got their name from their very first application, allowing people to make phone calls while being mobile. The communication architecture was dominated by *base stations* communicating with the mobile phones. Base stations were needed to enter the existing telephone networks and thereby allowing to communicate with fixed line communication partners. Furthermore, the base stations and the connected backbone allowed to communicate with other mobile users in different cells. In the beginning, referred to as *first generation* (1G) of mobile networks, there was no need for additional functionality, besides voice services, on top of the mobile phones. With the *second generation* (2G) , changing from analog mobile systems to digital ones, enhanced services were introduced such as the short message service (SMS) and data connections to the Internet. But another important change took place: the split of *network* and *service providers* such that we have four major players in the mobile world as given in Figure 1.1. Before that, the network provider had the monopoly to decide which services (in this case it was only the voice service) are available on the phones. Most of the European network providers were still fighting their monopoly, while in Japan DoCoMo launched their *i-mode* service. The idea was to maintain a platform where third-party service providers could offer their services to the mobile phone customers using the DoCoMo network. The business case was based on a fixed-ratio split between the service and the network provider. In contrast to the WAP services in Europe, which was a big financial disaster, the i-mode took off and was very successful. From that time on the mobile market has been divided into four different players, namely the customer as the main target of the business, the network providers enabling to convey bits toward the customer, the service provider, and the mobile phone manufacturer. These four entities

3

F.H.P. Fitzek and F. Reichert (eds.), Mobile Phone Programming and its Application to Wireless Networking, 3–20.
© 2007 *Springer.*

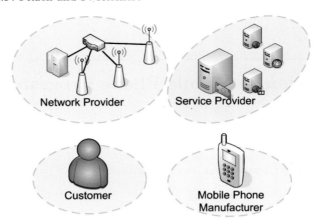

Figure 1.1. Interdependencies between the mobile phone customer, the network provider, the service provider, and the mobile phone manufacturer.

have some interdependencies. The most dominating one is the relationship between network providers and mobile phone manufacturers. To attract customers, the network providers bundle their contracts with a mobile phone, both being renewed within a given period. This ensures the mobile phone manufacturers sell a huge number of mobile phones. On the other side, the network operators also pull strings as to which services and applications are available on the mobile phone. Furthermore the network operator also controls which services are going over their network and makes the service provider dependent on the network operators' decision. This relationship should be taken into account making novel applications for mobile phones.

But the dominating position of the network provider is getting weaker as the monopoly of the cellular communication link is destroyed by wireless technologies such as wireless local area network (WLAN) and Bluetooth. These types of communication enable the creation of new type of services without tight coupling with the network operator. Mobile phone manufacturers are forced by the operators to keep this new freedom under control. Third-party developers seem to have a lot of freedom in the programming. But sometimes when it comes to the sensitive parts of the network operator, the mobile phone is not that flexible anymore. As given in one of the following chapters, voice-over IP (VoIP) just between two mobile phones can only be enabled with half-duplex. Another example for a possible restriction is the strict coupling of the IP protocol with the GPRS module. IP over Bluetooth or WLAN is either not supported or very difficult to use in some mobile phones. But these restrictions should not stop us from making interesting services, they cost only time.

The *third generation* (3G) of mobile networks did not bring radical changes in respect to the 2G system. Nevertheless, there is one important lesson learned so far with the new generation, i.e, that pure technology is not selling, but customers will ask for services. The main changes from 2G to 3G within protocol layers were that they were hardly visible for the customer and therefore 3G is not a selling keyword itself. Actually most of the network providers do not advertise their 3G network anymore as they did before and claim it is an evolutionary step such as it was with GPRS and EDGE. The search of new services was dominated by the term *killer application*. Many panels and conferences were dominated by the question what is the next application that will significantly change the wireless community. But the answer is not easy to find. DoCoMo tried to identify good services to start their i-mode service back in the 1990s by making public-opinion polls before the launch of i-mode. Comparing the wish list of the customers before with the actual use statistics after the launch of i-mode showed that there was absolutely no correlation. Thus even the customers do not know what they really want to have in advance. It is very difficult to design new applications and predict their success on the market beforehand. Many applications were generated over the past decades and only some of them were successful. To give the reader some designing rules for successful services, we distinguish between two sorts of services, namely the *personal* and the *community* services as given in Figure 1.2.

Personal services encompass all services, which are used by the customer on the phone with limited or no interaction with other customers. The community services are different in the interaction with other customers. These services are those made to interact with other people. Both, the personal and the community services, are separated into services with and without wireless network support. The wireless network support is realized by standard GSM connection, GPRS, EDGE, 3G data connections, Bluetooth, or WLAN. The services are colored in orange, for those existing on most phones already, in blue, for those that can be installed on the mobile phones, and in green, for those advocated in this book to rule the future of mobile phone services.

The personal services without network support include services such as calendars, cameras, downloading facilities to change the logo on the phone, and single-player games such as Tetris, pong, snake, etc. The logo services gave the network and service providers big revenues as they enabled people to customize their mobile phones. As customizing services were requested more and more by the customers, the next step was to introduce ring tones. As this service is not only noticed by the user, but by all people in the vicinity of the mobile phones, it is part of the community services. Ring tones are the ultimate way to customize the mobile phone and we refer to this as a way to increase the *coolness factor*. So the lesson learned within this paragraph is to enable users to express their way of life by customizing mobile phones. Personal services with network support include Web services and downloading tools (such as BitTorrent).

The most promising services are the community services with wireless network support. This relies on the fact that human beings are always interested in the lives of others. To proof this statement, the most successful services such as voice service and SMS services are within this group. But even new

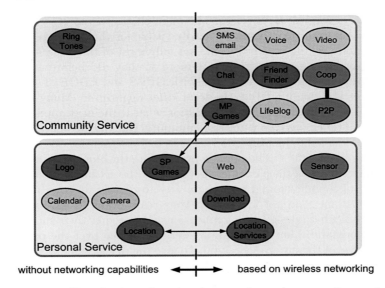

Figure 1.2. Classification of services into service and community services.

services, such as video services, friend finders, and multiplayer games will find their way to the market. Multiplayer games have already proven their potential for the wired networks, but even for the wireless world multi player games are the future making a big portion of the revenue in the near future. Thus, the lesson learned in this section is that successful services should interconnect people. A fact that is addressed by Nokia's sentence *Connecting people!* for the speech services, but it may be true also for the next generation of mobile services.

Paying attention to the two lessons learned in the last two sections, two new service classes, namely the *peer to peer* services and the *cooperative* services are introduced. These two services form the future of the upcoming *fourth generation* (4G) of mobile communication systems. But the focus is not restricted only to those services and we introduce *sensor* services, which are realized by mobile phones working with sensors in the vicinity of the mobile phone which may be introduced within the 4G or 5G systems.

1.1.2 The Flexible Mobile Phone

The mobile phone platform has become more and more flexible for third-party developers. The first programming language on the mobile phone was JAVA and now there are several ways to implement an application on the phone. As given in Figure 1.3 the applications run in the *application suite*. The *user interface framework* host all functionality that let us operate the mobile device. Qtopia is one example of that. The mobile phone manufacturer

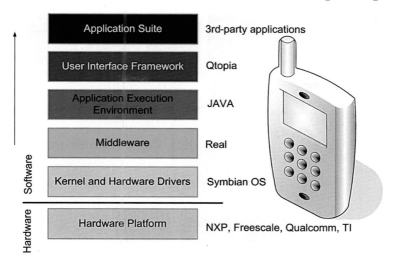

Figure 1.3. General architecture of the programmable mobile phone.

offers access to their platforms (with some restriction) using their *kernel and hardware drivers* such as the Symbian OS to access the *hardware platform*. So nearly anything that is in software can be changed on the mobile. The only static part of the phone is the *hardware platform*. Here and there some APIs can be found to change some parameters for a given wireless technology, but we are far from a composable or software-defined radio.

Beside the programming capabilities of the mobile phone, there is even more flexibility on the phone. As given in Figure 1.4 a mobile device has many different capabilities and functionalities. We have grouped those of a mobile (smart) phone into three classes, namely *user interfaces, communication interfaces*, and *built-in resources*. The user interfaces comprise typically the speaker, microphone, camera, display, built-in sensors, and keyboard capabilities. The built-in terminal resources include the battery, the central processing unit, and the data storage. From our point of view we highlight particularly the communication interfaces, which typically include cellular and short–range capabilities.

Instead of having one mobile device hosting all functionalities with the best possible quality, the concept presented later in our book considers using and sharing cleverly the capabilities of several, in principle different, mobile devices. Indeed, in such a heterogeneous scenario each terminal could feature a particular specialization, from a simple mobile phone for voice calls to advanced terminals with music and imaging devices. In the proposed wirelessly scalable architecture, a wireless grid of terminals is formed where the connecting links among cooperating terminals are implemented by the short-range air interfaces. In general terms, such a scalable concept is not totally new as Bill Joy, the visionary figure behind Jini Technology and cofunder of

Figure 1.4. A mobile device broken up into the several capabilities grouped into the user interface, communication interface, and the built-in resources.

Sun Microsystems argued already in 1999 for cooperative software:

A cell phone has a microphone. It has a numeric keypad. It has a display. It has a transmitter and receiver. That is what it is. It has a battery with a certain lifetime. It has a processor and some memory. If it is a European model it has a stored value card. It is a collection of things. It is some collection. The fact that it is all one bundle, well, that is pretty artificial – **Bill Joy**

Inspired by this developers may start to see the mobile phone as a collection of capabilities that can be reassembled on demand for a given purpose. Throughout the book we will give some examples for that.

But the flexibility of the mobile device is not limited to the combining of hardware components. Even the software can be used in a flexible way. One of the first ideas, which we will present in Part VI, is cross-layer design. Cross-layer protocol design is the idea of changing layers according to needs. Examples are given within this book. Some protocol layers may be even left out totally or grouped in a new way. The final step brings also flexibility to the lower levels of the protocol stack. While flexibility on commercial mobile devices is limited to multi-modality approaches, the future will be dominated by software-defined radio approaches. While multi-modality is only choosing among different available wireless technologies such as Bluetooth or WLAN, the software-defined radio can dynamically adjust the radio part toward the needed functionality. At the extreme, it could also change from short-range communication to cellular communication, exchanging the full protocol stack.

1.2 Wireless Technologies and Architectures

As advocated in the previous section the wireless network capabilities in mobile phones are the main enablers to invent new appealing applications. In this chapter different wireless technologies available on most mobile phones and underlying communication architectures are introduced. The technologies differ in the supported data rates, the communication protocols used, the communication coverage, and the energy consumption related to it. In general we distinguish between short-range and cellular communication systems. Historically the most important communication form is the cellular one. The 1Gs were equipped only with this technology as the main service was voice driven and a connection to the telephone operator network was the only needed wireless extension. Later on with the first GSM mobile phones short-range communication was introduced by infrared (IRDA). Later on further technologies such as Bluetooth and WLAN were also introduced into the device. Future short-range technologies are the ultra wide band and further evolution of the IEEE WLAN technology. In the following we revise shortly some of the communication technologies without claiming completeness. It is more a short revision focusing on the items we need in the later discussion of our programming parts.

1.2.1 Cellular Communication Systems

In the beginning the main service on mobile phones was voice. The first logical move was to enable connections of mobile users with users connected to fixed terminals or to use the core network to extend virtually a connection between two mobile users. As wireless radio communication is prone to path loss, a direct communication of mobile users would only be possible within a given range. But the cellular networks evolved from pure voice service support to data service support. The communication architecture is dominated by a point-to-point communication between a mobile phone and the so-called base station. The latter one can be referred to as the *bridge* between the wireless and the wired world. In the following, we introduce some examples and technologies for the European communication system GSM and its successors. To convey digital data from the mobile phone toward the core network, Circuit Switched Data (CSD), was introduced in the first 2G mobile phones, supporting a rate of 9.6 kbps. GSM is a time division multiple access (TDMA) system, where users get assigned one or a set of given time slots to communicate. To achieve the CSD data rate a single TDMA time slot is used. In this case the same capacity as a normal voice call is used. To increase the data rate the High Speed Circuit Switched Data (HSCSD) was introduced. In contrast to the CSD, different coding schemes were used. In case of a good channel condition where no error protection is used, a data rate of 14.4 kbps can be achieved. In case of a more unreliable wireless channel the full error correction is needed ending up in the same data rate as CSD. Furthermore

Table 1.1. Data rates for GPRS [kbit/s].

CS type	1	2	3	4	5	6	7	8
CS1	9.05	18.10	27.15	36.20	45.25	54.30	63.35	72.20
CS2	13.40	26.80	40.20	53.60	67.00	80.40	93.80	107.20
CS3	15.60	31.20	46.80	62.40	78.00	93.60	109.20	124.80
CS4	21.40	42.80	64.20	85.60	107.00	128.40	149.80	171.20

HSCSD allowed to bundle up to four of those time slots resulting in a maximum data rate of 57.6 kbps or 38.4 kbps for an error-free or an error-prone channel, respectively. General Packet Radio Service (GPRS) is an extension to CSD and HSCSD. GPRS as well as HSCSD is referred to as 2.5G technology and can be seen as a first step toward 3G networks. As HSCSD, GPRS uses bundling techniques to increase the data rate, but for GPRS bundling up to eight channels is supported. For GPRS four different coding schemes (CS) are available. If up to eight time slots can be bundled and four different CS are available, 32 different data rates are available as given in Table 1.1.

By bundling the unused TDMA time slots in the GSM cell it provides realistically data rates between 40 and 50 kbps (with a theoretical capacity of 171.2 kbps). GPRS is part of GSM since Release 97 and onwards. In contrast to the CSD and HSCSD, which were circuit-switched, GPRS is already packet-switched and allows the user to be *always on*. This means that GPRS can be, and is mostly, billed per data volume and not on a time basis as the circuit switched technologies. The available GPRS bandwidth is shared among all GPRS users in the cell, where voice services are prioritized, so no quality of service (QoS) guarantees can be made. The main application of GPRS is the support of email services and web browsing, which can end up in poor performance in fully loaded cells. The data rate supported by each mobile phone depends on the class, i.e., the Nokia N70 supports class 10 resulting in an uplink rate of 8–12 kbps and and a downlink rate of 32–48 kbps. Enhanced Data Rates for GSM Evolution (EDGE) or enhanced GPRS (EGPRS) will bring higher data rates, which can go up to 473.6 kbps in the very best case. The high data rate is achieved extending the four modes of Gaussian Minimum Shift Keying (GMSK) modulation used in GPRS with five modes of 8-phase shift keying (8-PSK) modulation. Moving to 3G networks the Universal Mobile Telecommunications System (UMTS) is introduced. The 3G technology is based on Wideband Code Division Multiple Access (W-CDMA) as the underlying standard. UMTS is standardized by 3GPP offering data rates of nearly 2 Mbps within one UMTS cell and 384 kbps per end user. The evolution of the cellular technology is shown in Figure 1.5.

In other regions of the world the technology is named differently but the provided services and data rates are mainly the same with respect to the granularity we need to discuss for the purpose of this book. The industry is continuing to evolve 3G networks. Mobile network providers are already installing system upgrades for using 3G High-Speed Downlink Packet Access (HSDPA)

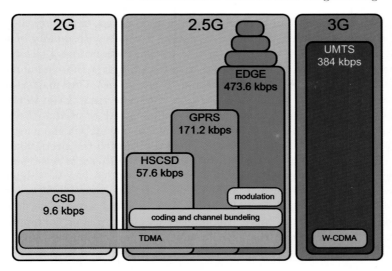

Figure 1.5. Evolution of the cellular communication systems.

and High-Speed Uplink Packet Access (HSUPA). HSDPA will achieve up to 28.8 Mbps for downloads and HSUPA up to 5.76 Mbps for uploads. This is already more than most of us have as fixed broadband access in our homes. But development does not stop there. The standardization path for 3G includes a *Long Term Evolution (LTE)* that will give users 100 Mbps peak rates for downloads and 50 Mbps for uploads. Manufacturers such as Ericsson, Nokia, and Motorola are discussing to introduce LTE into the market in 2009.

1.2.2 Short-Range Communication Systems

As candidates for the short-range communication we refer to wireless personal area networks (WPAN) and wireless local area networks (WLAN). As a candidate for WPAN and WLAN, Bluetooth and the IEEE802.11 family is named here, respectively. We mentioned before IRDA and even other technologies as ZigBee can be named. But here we refer to the technologies that have the biggest impact on mobile phone programming and which are found most likely on the mobile phones.

Bluetooth was introduced by Ericsson in 1999 as a cable replacement for mobile equipment to ease the data exchange between mobile phones, personal digital assistants, laptops, and personal computers. Later the initiative was taken over by IEEE and standardized in IEEE 802.15.1. Bluetooth was designed as low cost technology with low power consumption for short-range communication and can be found now on many mobile phones. The power consumption is addressed by three different power classes allowing transmission ranges of 1, 10, and 100 meters. Bluetooth started with version 1.0, which was still error prone those days and became obsolete with version 1.1 and 1.2.

Different transmission rates were possible. In contrast to all existing technologies, Bluetooth provided communication channels with quality of service (QoS) support even for voice. Bluetooth communication takes place between communication peers, where a master peer coordinates the communication in the group of peer slaves also referred to as *piconet*. One master and up to seven active peers can communicate at the same time. Even more peers can be assigned to the communication group, but they need to be parked and cannot participate actively. Slave peers communicate via the master peer with each other, thus no direct communication between the peers take place. Furthermore Bluetooth is able to form so called *scatternets* where multiple communication groups are pared via *bridges*. A bridge can be a master or a slave. Using multihop Bluetooth can increase the communication coverage and allow to use more than eight devices active at the same time. For communication two types of link exist, namely the synchronous communication oriented (SCO) link and the asynchronous communication-less (ACL) link. The SCO is mainly to support voice service, while the ACL is for any data traffic. Bluetooth is based on a slotted channel with a slot length of 0.625 ms. For multiple access TDMA/TDD is used. A frame consists of an uplink and a downlink phase with at least one slot of each. Each communication packet is acknowledged by the counterpart peer. To increase the throughput efficiency, one, three, or five slots can be grouped together for the uplink, the receiver will always use only one slot for the acknowledgement. The amount of payload data that can be stored within a group of slots depends on whether forward error correction is used or not. In Bluetooth terminology DM packet are those using FEC and DH packet are not. Thus for ACL six different packet types exist, namely DM1, DM3, DM5, DH1, DH3, and DH5. Table 1.2 gives the packet types with the related payload size and data rate. With Bluetooth version 2.0 data rates up to 3 Mbps are available. In the future Bluetooth will be merged with the ultra wide-band (UWB) technology. UWB allows up to 400 Mbit/s for short range communication consuming two up to three times the power of current Bluetooth technology as promised by the researchers working in this field.

Wireless local area network (WLAN) is standardized by the IEEE in the 802.11 family. First implementations of the IEEE WLAN 802.11 had one

Table 1.2. Bluetooth ACL packet types with symmetric (sym) and asymmetric (asym) data rates in [kbit/s].

Type	slots used	header[byte]	payload [byte]	FEC	sym	asym ↑	asym ↓
DM1	1	1	0-17	2/3	108.8	-	-
DH1	1	1	0-27	none	172.8	-	-
DM3	3	2	0-121	2/3	258.1	387.2	54.4
DH3	3	2	0-183	none	390.4	585.6	86.4
DM5	5	2	0-224	2/3	286.7	477.8	36.3
DH5	5	2	0-339	none	433.9	723.2	57.6

simple MAC layer and three different physical realizations with direct spectrum, frequency hopper, and infrared. Direct spectrum made the race as it costs less than the frequency hopper and was not dependent on the line of sight (LOS) as it was the case with infrared. The further development of 802.11b increased the data rate to 11 *Mbps* available on each of the three orthogonal channels. So far only the 2.4 *GHz* industrial science medical (ISM) spectrum was used. 802.11a used orthogonal frequency division multiplex (OFDM) as the access technology offering data rates up to 54 *Mbps* in the 5 *GHz*. The number of orthogonal channels available differs from region to region. For the USA the number of channels is twelve. The same data rate is also supported in the 2.4 *GHz* specified in IEEE 802.11g being backwards compatible with IEEE 802.11b. All different IEEE WLAN technologies named here have the same medium access (MAC) strategy, which is derived from the Ethernet world. Even though as point coordination function (PCF) was standardized by IEEE, only the distributed coordination function (DCF) found its way into the products. While PCF was intended to be used in access points, the DCF could be used for ad hoc type and infrastructure type of communication. The IEEE proposed carrier sense multiple access (CSMA) with collision avoidance (CA). Ethernet is using CSMA with the more efficient collision detection (CD). But due to the wireless nature, detection of collision is not possible for all participating communication partners and CSMA/CA is used. Any communication device will scan the wireless medium to make sure it is free for a given time. Once the medium was unused for a given time the device starts to transmit its packet. When the packet was received by the counterpart entity successfully it generates an acknowledgement to inform the sender that the packet has received successfully. An unsuccessful transmission can be due to some transmission errors or more likely a collision with another device sending at the same time. As both devices will not transmit their packet successfully when they convey their data at the same time, referred to as *collision*, they need to retransmit it. To avoid the same situation over and over again, the colliding stations chose randomly a time out of an interval specified as *collision window*, which will be waited until they try to retransmit. The size of the collision window will be doubled with each collision the device is involved in and reset to the initial value after the first successful transmission. By this strategy, also referred to as *exponential back-off*, the devices adapt to the current situation.

As the collisions reduce the efficiency on the wireless medium, potential collisions should be detected as soon as possible. So far the missing acknowledgment is the only way to detect it, but loosing the long data packets is never a good idea. Furthermore there exists the *hidden terminal problem*, which refers to the situation where two stations that cannot sense the existence of the other one directly (so they are not in each other's coverage), may transmit packets to a third station that is never able to receive the colliding packets. Therefore ready to send (RTS) and clear to send CTS) packets are introduced. Now devices that want to send will not send their long data packet

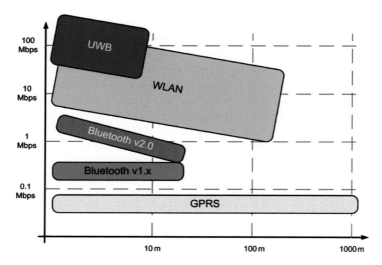

Figure 1.6. Comparison of Wireless Network Capabilities.

directly, but use the smaller RTS packet to indicate their willingness to send to a given destination. The called destination will answer this request by a CTS packet when it does not sense any ongoing communication. The RTS and the CTS packets also host the information as to how long they want to use the medium to inform the neighborhood about it. Now collision can still occur (between RTS packets or an RTS and ongoing data packet transmissions), but the overall collision time is reduced. The use of RTS/CTS packet introduces inefficiency only if the possibility of collision is small. In Figure 1.6, the physical data rates versus the coverage is given for some of the technologies used in the book.

1.3 Mobile Application Deployment

After months of hard work the new mobile application is tested, certified (e.g., by Symbian), and runs smoothly. The user trials were encouraging, and the new application should be a big success. Now a number of hard questions are waiting for answers:

- How to make target customers aware of the new application?
- How to charge for the service?
- How to prevent others from stealing and copying the idea?
- How to offer potentially millions of users a stable and reliable service?

The answers to these questions will be dependent on the application and the developer's preferences and vision for their business. Developers will always have the option to market and operate services on their own. However, powerful partners can rapidly reach out to millions of potential customers with well-established marketing and distribution channels. A further, serious obstacle for a successful market launch is the lack of trust that customers have in a new and unknown service provider. Therefore, before going all alone, a new service provider should consider the options offered by 3rd parties. All network providers (*operators*) are open to new ideas. New ideas will create new revenues and possibly attract new customers. Developers can either contact them directly or go through major handset manufacturers. Manufacturers are a good starting point as they have relations with many operators, application distributors, and thus potentially can give your application a wider audience. Usually manufacturers offer device emulators, development tools, verification and testing support, and marketing of solutions to operators and customers. Here are some examples:

- Forum Nokia – `http://developer.nokia.com/`
- Sony-Ericsson Developer World – `http://developer.sonyericsson.com/`
- Motorola Motodev – `http://developer.motorola.com/`
- Ericsson Mobility World – `http://www.ericsson.com/mobilityworld/`

Phone manufacturers can market your applications in many ways. The simplest way of cooperation is by offering application developers contacts to operators, or by publishing on a web shop. Nokia offers applications via *Nokia Catalogs* (`http://www.nokia.com/catalogs`). Many manufacturers have close relationships with web-based application resellers. Manufacturers use, e.g., Motricity (`http://www.motricity.com/`), Pocketgear (`http://www.pocketgear.com/`), and Handango (`http://www.handango.com/`) as partners to offer applications to hundreds of millions of potential customers worldwide.

The next levels of cooperation allow selected applications to be included in a phone's recommendation list or even be amongst the pre-loaded applications. Thus, every customer who buys a phone can either decide to download it or use the application right away. Obviously being part of a set of pre-loaded applications is very difficult to achieve.

All major operators offer partnership programs. A close relationship with operators can enrich applications significantly. Operators offer powerful application enablers via their partner portals that give developers access to, e.g., charging, security, device management, digital rights management, presence, push, messaging, and positioning services. For some applications this will be the only way to go. Examples of operator developer support sites are:

- Vodafone via.vodafone – `http://www.via.vodafone.com/`
- Cingular devCentral – `http://developer.cingular.com/`

Operators are large and powerful players. Before entering a relationship with an operator a developer should study the *Terms and Conditions* documents very, very carefully. For example, some operators may require 24/7

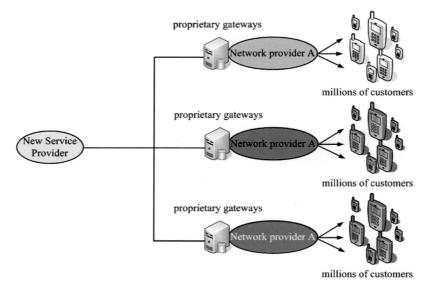

Figure 1.7. A service provider integrating business with several operators.

customer support to qualify, which is expensive to provide. Often operators will require exclusivity, meaning that the application can never or for a long time period not be marketed via other operators. In case the application shall reach millions of customers belonging to different network providers, service developers are faced with a major challenge. For a new service the task of interoperating with different global operators is likely overwhelming. Each operator has different legal documents and guidelines, and offers access to key enablers through own proprietary 3rd-party gateways (see Figure 1.7).

Service aggregators ("brokers") act as an intermediate between operators and service providers. Thereby a new service provider needs only to establish one legal relationship with the aggregator (see Figure 1.8). Aggregators like Ericsson IPX (http://www.ericsson.com/solutions/ipx/) offer charging, messaging, and other enablers for over 86 operators worldwide reaching out to 650 million customers. Aggregators are a way to getting up and running fast, but they will request their share of the potential revenue stream.

Deploying a new service involves creating a value network that interconnects different players, each contributing a specific component to the overall application (Figure 1.9). Besides network providers and aggregators, other parties may offer services such as charging (e.g., Paypal, MasterCard, Visa, CCBill, or operator), music files, news (e.g., Reuters) or financial information, data storage (e.g., Amazon S3), web hosting, user administration and authentication (e.g., by operators). In principle all functions not belonging to the core competence and core business of the new service company are outsourced. Therefore, the design of a new service needs to take into account

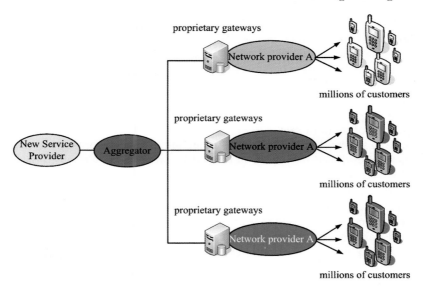

Figure 1.8. Using service aggregators to simplify interoperation with many operators.

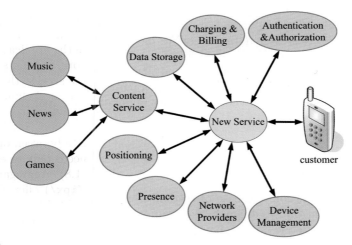

Figure 1.9. Example of a value network combing service components for a new service.

the deployment scenario. It will restrict the technical solution, and affect the business model.

Working together with 3rd parties will provide a new service provider with many benefits. For a small start-up with limited resources, it is nearly impossible to ramp up a service to ensure reliability, scalability, technical expertise in all technical subareas, great customer support, and fast time-to-market. Driving a business requires a business model that is suitable to support the operations. There are many ways to charge a user. Some basic methods are traffic volume based, distance based, time based, message based, flat-rate, one-time-fee, service-based charging, and more. Many of them cannot be realized without the cooperation of an operator. Service-based charging, for example, could charge customers only for each usage of a service without any fees for data transmission. A music download service would be a typical example of such a service, where the users expect only to pay for a song irrespective of its data size.

Once a service is launched it is important to provide customers with a channel for feedback. Feedback is critical in order to understand how users perceive the service, what problems are occurring, and what new features could be included in future versions. Most phone manufacturers and network providers offer pages for customer feedback, e.g., Motorola: `https://www.motorola.com/feedback.jsp`, or for where users can exchange advice and opinions, e.g., Cingular: `http://forums.cingular.com/cng/`. User discussions give hints what competitors are doing better, and whether users are satisfied with current performance and features. Finally, we recommend that you study carefully how to protect your inventions through early patent filings and nondisclosure agreements before you negotiate with other parties. Otherwise your competitors will copy your successful ideas. Or worse, you see your ideas patented by others. A great concern of many users is privacy and anonymity. Especially in Europe the lawmakers have been very strict in handling critical user data. Therefore data can only be stored in a pre-defined context, and only be used for that pre-defined purpose. For further reading and more details, we recommend *Mobile Media and Applications* by Christoffer Andersson et al., which provides more insights into the telecommunication industry as a whole, consumer understanding, and the complete development lifecycle.

1.4 Structure of the Book

For better illustration and ease of reading, this book is divided into several parts. Part II introduces different programming languages and deployment environments for mobile phones, namely *Python, Java, Symbian/C++, Open C, Qtopia, Maemo*, and *Windows Mobile*. The individual chapters are not a compendium of each programming language, but give the reader a good overview as to which programming language to choose. Furthermore the chapters create an understanding which is built upon in the later chapters. As the main focus of the book relies on wireless communication technologies those parts are more highlighted than others for each of the programming languages.

A mobile can be used in different application areas and communication environments as given in Figure 1.10.

Part III explains service discovery for infrastructure-based communication. Part IV focuses on the possibilities of peer-to-peer (P2P) networking. As P2P

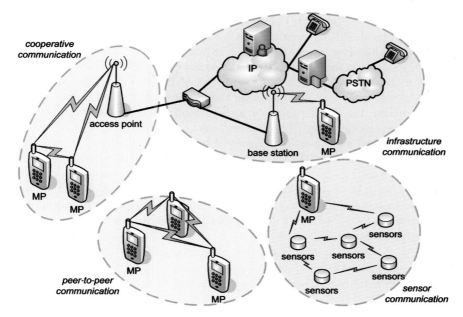

Figure 1.10. Overview of different Communication Forms.

networking has already found its role in the wired domain, the wireless domain will certainly adapt it, and new applications will make use of P2P links in many unexpected ways. As given in Figure 1.10 P2P networking allows mobile phones to communicate directly with each other. For this kind of communication short-range communication technologies are used. The first chapter of Part IV focuses on the need of wireless P2P networks, as they nicely map to the social behavior of humans to communicate and share items. In conjunction we develop a new concept for *Digital Ownership* as *Digital Right Management* (DRM) fails to address these new kind of network types. Part V contains four chapters with respect to wireless cooperative networking as introduced in [1]. The chapters scale in terms of complexity from very beginner-like example to more complex ones targeting cooperation between mobile phones. The *cooperative communication* as given in Figure 1.10 can also be referred to as *cellular controlled peer to peer* networking and in contrast to the pure P2P networking, the mobile phones connect simultaneously to a central access point (or base station in terms of cellular networks). BitTorrent is described as an example of such type of communication, which we advocate. Cross-layer design, introduced in Part VI, is a methodology to optimize network protocols exploiting flexibility of certain parts of the mobile phone. Some researchers claim that cross-layer design is too complex for mobile terminals and that the approach violates the layered protocol design. In a very first example we demonstrate that cross-layer design gives performance gains without increasing complexity. Part VII, investigates the possibility of merging the two interesting worlds of mobile phones and wireless sensors. Here we explain different architecture forms, show how to build own sensors, give examples how to use sensors, and demonstrate the sensors as context-aware supporters for the mobile phone. In the last part, Part VIII, we would like to pay attention to the energy consumption on mobile phones. As mobile phones are battery driven, the development of mobile applications has to take the limited resource into account. The chapters make the reader familiar with energy consumption and show how to measure it with an *external* and *internal* approach. The companion DVD contains all software such as SDKs and source code used within this book. The companion website of our book can be found at `http://mobiledevices.kom.aau.dk/MPbook`. The intention of the companion website is to offer upcoming examples for mobile phone programming. Even if we paid a lot of attention to avoid mistakes within the book, we will host an online errata document.

References

1. F.H.P. Fitzek and M. Katz, editors. *Cooperation in Wireless Networks: Principles and Applications – Real Egoistic Behavior is to Cooperate!* ISBN 1-4020-4710-X. Springer, April 2006.

## Part II

Mobile Phone Programming Languages

2

Python for Symbian Phones

Jürgen Scheible

University of Art and Design Helsinki, Media Lab `jscheib@uiah.fi`

2.1 Introduction

This chapter introduces *Python for S60* that brings the power and productivity of the programming language *Python* to the Symbian mobile platform-in particular to Nokia's S60 platform. *Python for S60* provides a scripting solution making use of Symbian C++ APIs. With its great capabilities and flexibility to rapidly prototype creative ideas and to make proof of concepts, it makes programming on the S60 platform fun. *Python for S60* is easy to learn and allows novice as well as experienced programmers, artists, and people from the web developer communities to quickly make scripting or standalone applications on S60 with completely free and open tools. Python is a dynamic object-oriented open source computer programming language that was created by Guido van Rossum. It can be used for many kinds of software development. *Python* runs on most common platforms, e.g., Windows, Mac OS, Linux, and through *Python for S60* it runs on Symbian OS. *Python for S60* has many Python Standard Library modules built-in but includes a number of additional mobile device specific modules, e.g., for SMS, telephone features, camera, and system info. Though Python's optional object-oriented programming (OOP) can be used in making *Python for S60* applications, it is left out here for didactical reasons in order to keep the structure of the code examples simple and easy to understand.

In recent years the processing power and memory capacity of smartphones have drastically advanced. This allows running an interpreted language like Python on such devices, simply by installing a Python execution environment. Wrapping complex low-level technical details behind simple interfaces, *Python for S60* allows developing fully working S60 applications even with just a few lines of code. Tapping into the rich set of smartphones features is therefore easy, features such as camera, Bluetooth, network access, sound recording and playing, sending and receiving SMS, sending MMS, text to speech, 2D graphics

F.H.P. Fitzek and F. Reichert (eds.), Mobile Phone Programming and its Application to Wireless Networking, 23–61.

as well as 3D graphics through OpenGL ES, file up-/download from the Internet, and many more. *Python for S60* is not sandboxed as some other nonnative programming frameworks that are built on mobile platforms. Nevertheless, security issues are handled through the Symbian Platform security framework available from Symbian OS9 onwards. *Python for S60* is provided by Nokia in the form of a ready-to-use install Python interpreter as well as open source code. It is available for free at `http://sourceforge.net/projects/pys60`. To extend its functionality by the developer community it is possible to create Python extensions in C++ based on Symbian API's. Further, so-called stand-alone applications can be created from Python scripts which are treated as any other S60 application from a user's point of view. With this chapter we aim to provide an efficient introduction to the *mobile phone specific* features of *Python for S60* including some basic Python language issues. We do this by looking at programming examples in the form of scripts which are fully working programs that run either on a Nokia S60 phone or on a S60 emulator. Note, we used *Python for S60* version 1.3.18 for the examples. To learn more about general features of Python and about the Python programming language itself we recommend to read some online tutorial that can be found from `http://www.python.org`. But you should be able to go through this chapter even without having any previous knowledge on Python or S60.

2.2 Python for S60 Installation and Resources

For developing Python for S60 scripts all you need to have is a simple text editor (even notepad works) on your computer and an S60 phone for instant testing without a build or compiling process in between. As an alternative for testing you can use an S60 platform device emulator on your PC, too. In order to test and run a Python script on your S60 phone, you first need to install the Python for S60 interpreter software to the phone, since it is not pre-installed. This installation provides the needed Python execution environment and a range of Standard and proprietary Python library modules. For testing Python for S60 scripts on your PC via an S60 platform device emulator some extra software has to be installed to your machine.

2.2.1 Software Installation on the Phone

1. Download the Python for S60 interpreter software to your computer from the SourceForge's PyS60 project website: `http://sourceforge.net/projects/pys60`. It is available for free as a Symbian install file (.SIS). Note, you need to choose the correct Python for S60 interpreter software file based on your phone model (e.g., whether your model is based on the 2nd or 3rd edition of the S60 developer platform).
2. Install the Python for S60 interpreter software to your phone via Bluetooth, USB cable, or Nokia PC Suite. After this, your phone is ready to

Figure 2.1. Phone menu.

run Python scripts. The Python icon should be visible on your phone's desktop or inside one of its subfolders (see Figure 2.1).

We cannot give detailed instructions here about the installation due to lack of space but also due to forthcoming developments of the Python for S60 interpreter software that might cause some changes in installation procedures, etc. To make sure you have the latest installation instructions available, check from Nokia's wiki site or relevant PyS60 online tutorials:

```
http://wiki.opensource.nokia.com/projects/Python_for_S60
http://www.mobilenin.com/pys60/menu.htm
http://mobiledevices.kom.aau.dk/index.php?id=909
```

2.2.2 Software Installation on the PC (for S60 Platform Device Emulator Use)

1. Download (free) and install to your PC the *S60 Platform SDK for Symbian OS* that includes the device emulator that allows applications to be run and tested on the PC. It can be found from Forum Nokia's website: `http://forum.nokia.com`. (For details about the installation, read the releasenote.txt file contained in the .zip file you download).
2. Download (free) and install the Python plug-in for the Symbian SDK. According to the phone model you want to target with your applications, you need a different plug-in to install on your PC. It can be found from the SourceForge's PyS60 project website `http://sourceforge.net/projects/pys60`.

2.3 Writing a Python Script

2.3.1 How Python Runs Programs

The Python interpreter reads the program you have written and carries out the code statements it contains. The simplest form of a Python program is just a text file containing Python statements which are executed. Such a text file is in our case the Python script that we are writing.

2.3.2 Three Basic Steps

Let us now look into how to write such a Python for S60 script and how to instantly run it on an S60 mobile phone or an emulator on your PC as an alternative. There are three basic steps to take:

1. Write a Python script with a text editor on your computer and save it with the .py file extension.
2. Move your *.py script file from your computer to your phone.
3. Start the Python interpreter application on the phone (or on the emulator) by clicking on the *Python* icon on the desktop of your phone (Figure 2.1) or emulator, then press *options* key and select *Run script* from the menu. From the appearing list choose the name of your script and press ok. Your script should now start up.

Detailed Explanation of Steps 1 and 2

1. You can use any simple or advanced text editor on Mac or PC or Linux to write and edit your Python script. Some developers have preferences regarding which editor to use. Useful editors are, e.g., ConTEXT or Pywin which are freely downloadable on the Internet, but also Notepad works. All you need to do is to write the Python code with the text-editor and save the file with the ending .py. After the code is typed, the file is ready to be executed on the mobile or an emulator without going through a build or compiling process. In order to run it, the file needs to be uploaded to the phone, or copied to a specific folder when using the emulator tool on your computer.
2. There are different possibilities available to move your Python script to your phone. Depending which phone model you have either of the following should work: a) using the Nokia PC suite (PC only), or a USB cable or Bluetooth for connecting the phone as an external hard drive to then copy paste your scripts from the PC to a folder you need to create on your phone named *Python* or b) via Bluetooth sending your *.py file from the computer (any OS) to your mobile where it will arrive in the message inbox. Open the received message from the inbox in order to install the Python script file. When asked to install your file as a script or a library, choose *Python script*. When using the emulator, all you need to do is to copy your script files to a specific folder on your computer.

Another alternative for developers to test and debug applications without using the emulator is to use the *Bluetooth console* which connects the mobile phone via Bluetooth to the computer's serial port to run the *Python for S60* interpreter in a kind of *remote* mode. Which means you can type Python commands from the Python prompt in a terminal application on the computer but the actual execution of the commands is done on the phone. To learn more about this alternative, refer to the Programming_with_Python document [1].

2.3.3 The First Script Example

Let us look at a simple script consisting of three lines of code that does the following:

1. Display a text input dialog on the screen. The instruction in the text input dialog should say: Type a word: (Figure 2.2(a))
2. Display a pop-up note stating the result from the previous text input dialog, after the user has typed in a word and has pressed *ok* (see Figure 2.2(b)).

Listing 2.1. Code Example First Script

```
import appuifw
data = appuifw.query(u"Type_a_word:", "text")
appuifw.note(u"The_typed_word_was:_" +data, "info")
```

(a) Text input dialog.   (b) Pop-up note.

Figure 2.2. The first script example.

Code Explanation

In the first line of Listing 2.1 we import the `appuifw` module that handles UI widgets like *text input dialogs* and *pop-up notes* etc. In Python, a module is a file that contains a collection of related functions grouped together. All modules that come with *Python for S60* got installed together with the Python execution environment. To address a function contained in a module, the syntax in Python is such that we first write the module name, then a dot, and then the function name. Like here in the second line where we create a single-field dialog using the `.query()` function of the `appuifw` module as `appuifw.query(label, type)`. For the first input parameter in brackets called `label` we put the text u"Type a word:". The u must be there to declare the text string as unicode because the phones' UI widgets cannot display text that is not declared as unicode. For the second input parameter in brackets called `type` we put "text". It declares the input field as text type dialog. The function `.query()` returns the string that the user types into the text input dialog and we assign that string to the variable `data` by using the equal sign `data = appuifw.query(u"Type a word:", "text")`. In the third line we create a pop-up note using the `.note()` function as `appuifw.note(label, type)`. As the first parameter `label` we put the text u"The typed word was: " + data. This is the text that will appear inside the pop-up note. In this case we add the content of the variable `data` to our string `label` by writing + data. As the second parameter `type` we put "info". It tells the pop-up note to put a green exclamation mark inside on it. Before we continue to look into further *Python for S60* issues and code examples let us get a short introduction to some Python syntax issues.

2.4 A Short Python Syntax Lesson

In Table 2.1 we introduce you to some basic Python syntax issues.

2.5 Overview of Python for S60 Modules

You might have heard of Python's philosophy of *batteries included* – having a rich and versatile standard library of modules available with your execution environment, which make many complex tasks practically effortless. In Python, a module is a file that contains a collection of related functions grouped together. As mentioned before *Python for S60* comes with many standard Python library modules which are described in the Python documentation at http://www.python.org/, but it includes also a whole range of proprietary modules that are mobile-phone-specific and do appear only in *Python for S60*. Table 2.2 is a list of most of the phone-specific *Python for S60* modules, giving you an overview of the functionalities and features that

Table 2.1. Basic Python syntax issues.

Variables	Lists
a=5 *b="hello"* In Python there is no need to declare data types of variables. Python finds out at runtime whether the variable is a string, integer, etc. Variable names can be arbitrarily long. They can contain both letters and numbers, but they have to begin with a letter. Note, case matters. Bert and bert are different variables.	In Python a list is an ordered set of values enclosed in square brackets, where each value is identified by an index. *names = [u"Symbian", u"PyS60", u"Mobile"]*

	String handling
	print "Hi %s you %s" % ("Jeff", geek) The resulting screen shows: *Hi Jeff you geek*

For loop	Dictionaries
foods = [u"cheese", u"sausage", u"milk"] *for x in foods:* *print x*	Dictionaries are written as comma-separated series of *key:value* pair inside curly braces. *{ 'bacon': 2, 'eggs': 3}*

For loop (cont.)	Tuples	If statement

This almost reads like English: "For x (every element) in (the list of) foods, print (the name of the) x." When executing this for loop, the resulting screen shows:
 cheese
 sausage
 milk
	Tuples are written as comma-separated series of values enclosed in parenthesis. Tuples are immutable (unchangeable) arrays.
	 (1,2) a two-item tuple
	 (1,2,3,4) a four-item tuple
		if x < 0:
		   # Action 1
		 elif x == 0:
		   # Action 2
		 elif x == 1:
		   # Action 3
		 else:
		   # Action 4

Double and single quotes
In Python Double and single quotes work the same. *"Python"* or *'Python'*

Python for S60 offers. During the rest of this chapter we will learn details about these modules and how use them.

2.6 Modules – and How to Program Python for S60 Scripts

In the following paragraphs we introduce several modules in order to show the basic principles of how write *Python for S60* scripts.

2.6.1 Messaging Module

The `messaging` module allows sending of SMS and MMS messages. In Listing 2.2 we show how ordinary short messages SMS are sent by using the `sms_send(recipient, message)` function with a telephone number as

Table 2.2. Python phone-specific modules.

Module name	Module description
appuifw	Offers an interface to S60 UI application framework, i.e., dialogs, notes, selection lists
audio	enables recording and playing of audio files and provides access to the device text-to-speech engine
messaging	Offers APIs to messaging services for sending SMS and MMS
inbox	Provides reading of incoming SMS messages incl. SMS content, arrival time, and sender's number. It allows also deleting of SMS messages
camera	Handles taking of photographs and starting and closing of the viewfinder
graphics	Provides access to 2D graphics primitives and image loading, saving, resizing, and transformation capabilities provided by the Symbian OS
sysinfo	Offers an API for checking the system information of a S60 mobile device such as battery level, IMEI code, signal strength, or memory space
glcanvas	Provides a UI control for displaying OpenGL ES graphics
gles	Provides Python bindings to OpenGL ES 3D/2D graphics
location	Offers an API to location services, i.e., reading the Cell-ID
socket	S60 extensions to standard Python socket module, Bluetooth protocol RFCOMM and OBEX is added, setting the default access point
telephone	Provides an API to telephone functionalities such as dial and hang-up
calendar	Offers an API to calendar services: reading, creating entries, setting alarms
contacts	Offers an API to address book services allowing the creation of contact information databases
e32	Offers Symbian OS related utilities that are not related to the UI and are not provided by the standard Python library modules
e32db	Provides an API for relational database manipulation with a restricted SQL syntax
keycapture	Offers an API for global capturing of key events
topwindow	Interface for creating windows that are shown on top of other applications

`recipient` parameter given in string format and a `message` parameter given as string in unicode format. Sending an MMS message with `message` as body text in unicode to a telephone number as `recipient` (string) is done with the function `mms_send(recipient, message, attachment)`. The parameter `attachment` is the full path to a file that should be attached to the message, for example an image file located in the gallery on the phone's memory card `"e:\\Images\\picture1.jpg"`. (Note, the # sign in the code marks a comment line).

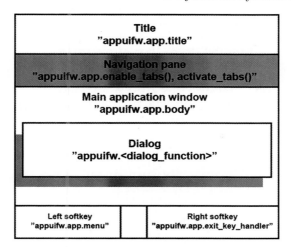

Figure 2.3. Python interpreter UI.

Listing 2.2. Code Example SMS and MMS

```
import messaging , appuifw

#Sending a SMS message to a number
messaging.sms_send('+3581234567', u'Hello!')
appuifw.note(u"SMS_sent", "info")

#Sending a MMS message to a number
messaging.mms_send('+3581234567', u'Hello!', attachment='e:\\
Images\\picture1.jpg')
appuifw.note(u"MMS_sent", "info")
```

2.6.2 Appuifw Module

The `appuifw` module establishes an interface to the S60 UI application framework, i.e., dialogs, notes, selection lists. The Python interpreter UI (Figure 2.3) has a common structure with almost any application UI on your S60 mobile phone. As UI elements you see on top of the screen the application title and below a row of navigation tabs which are usually not visible. The application body which is the large area underneath is used by different UI widgets such as dialogs and notes. On the bottom there are the left and right softkey. The left softkey activates the application menu and the right softkey exits a running Python script or application. The application UI is accessed through a special `app` object inside the `appuifw` module. Objects bind together variables and functions that manipulate them. It is easy to modify the UI elements of Python interpreter UI. Each of the elements, `title`, `body`, `menu`, and `exit_key_handler` are just special variables inside the `app` object to which you can assign values as to any other variable. For this you need to use their full name, i.e., `appuifw.app.title`.

Title

You need to give a string in unicode as title `appuifw.app.title= u'my␣ first␣script'` in order to make your application title appear on the top part of the screen when running your script.

Screen Types

There exist three different screen types such as: normal, large, full (Figure 2.4).

```
appuifw . app . screen=" full "
appuifw . app . screen=" normal"
appuifw . app . screen=" large "
```

Dialogs

Python for S60 offers a set of dialogs of which some are listed here and few explained in detail later:

1. note – pop-up notes
2. query – single-field input dialogues
3. multi query – two-field text input dialog
4. popup menu – simple menu
5. selection list – simple list with find pane
6. multi selection list – list to make multiple selections

(a) 'normal'

(b) 'large'

(c) 'full'

Figure 2.4. Screen types.

Dialog Query

The single field dialog is a query where the user needs to take action. The different types of query refer to the input value expected. The following queries exist: text, code, number, date, time, query (see Listing 2.3 and Figure 2.5).

Listing 2.3. Code Example Dialog Query

```
import appuifw

appuifw.query(u"Type_a_word:", "text")

appuifw.query(u"Type_a_number:", "number")

appuifw.query(u"Type_a_date:", "date")

appuifw.query(u"Type_a_time:", "time")

appuifw.query(u"Type_a_code:", "code")

appuifw.query(u"Do_you_like_PyS60:", "query")
```

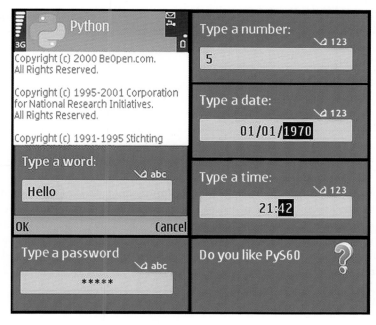

Figure 2.5. Dialog query.

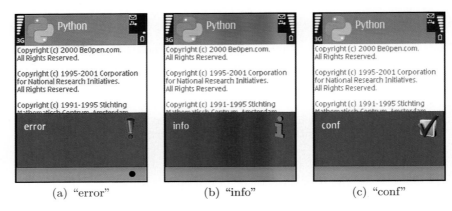

(a) "error"   (b) "info"   (c) "conf"

Figure 2.6. Dialog note.

Dialog Note

In contrast to the query, the dialog note just prompts information on the screen, without a user action. The following dialogs are supported: `error`, `info`, `conf` (see Listing 2.4 and Figure 2.6).

Listing 2.4. Code Example Note

```
   import appuifw
2
   appuifw.note(u"info", "info")

   appuifw.note(u"error", "error")
7  appuifw.note(u"conf", "conf")
```

Dialog Pop-up Menu

The pop-up menu function displays a pop-up menu style dialog. The parameters that need to be passed to it are `label` which is the text that appears on the top of the menu – given as a string in unicode – and `list` that represents the content of the menu given as a Python list. The function returns a number which is an index to the menu content's list, referring to the list element that was selected by the user when executing the script (see Listing 2.5 and Figure 2.7).

Listing 2.5. Code Example Pop-Up

```
   import appuifw

   mynewlist = [u"Symbian", u"PyS60", u"MobileArt"]
5  index = appuifw.popup_menu(mynewlist, u"Select:")

   if index == 0 :
```

Figure 2.7. Dialog pop-up menu.

```
            appuifw.note(u"Symbian, aha")
10  if index == 1 :
            appuifw.note(u"PyS60 - yeah")

    if index == 2 :
            appuifw.note(u"I love MobileArt")
```

Dialog Selection List

The `selection_list()` function executes a dialog that allows the users to select a list item (list to be defined). It has a built-in search field (find pane) at the bottom. The search field facilitates searching for items in long lists. The input parameter is 1 for search field on - 0 is for off (see Listing 2.6 and Figure 2.8).

Listing 2.6. Code Example Selectionlist

```
1   import appuifw

    mynewlist = [u"red", u"green", u"blue", u"brown"]

    index = appuifw.selection_list(mynewlist,1)

6   if index == 2:
        print "blue was selected"
    else:
        print "thanks for choosing"
```

Other Functions and Parameters UI Widgets

Figure 2.9 shows a few more available UI widgets: Form `appuifw.app.form ()`, Listbox `appuifw.Listbox()`, Tabs `appuifw.Listbox()`. Please check the

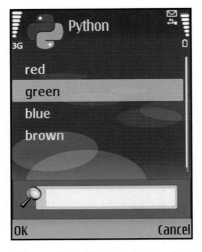

Figure 2.8. Dialog selection list.

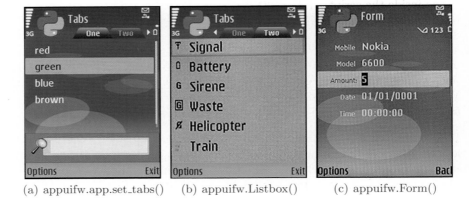

(a) appuifw.app.set_tabs()   (b) appuifw.Listbox()   (c) appuifw.Form()

Figure 2.9.

PyS60 Library Reference document for other functions and parameters regarding the native graphical user interface widgets offered through the `appuifw` module.

2.6.3 Application Basics

In contrast to simple scripts that execute line by line and drop out after the last line – as all previous examples do – Listing 2.7 shows the minimum requirements needed to program an application that runs until the user presses the exit key.

Listing 2.7. Code Example Basic Control

```
import appuifw, e32

def quit():
    print "Right_Softkey_pressed!"
    app_lock.signal()

appuifw.app.title = u"First_App"
appuifw.app.exit_key_handler = quit
appuifw.note(u"Application_running")

app_lock = e32.Ao_lock()
app_lock.wait()
```

We import the **appuifw** module and the **e32** module which offers special Symbian OS platform services. For shutting down the application, the user needs to press the exit key (right softkey). We need to tell in our script what to do (handle) when the exit key is pressed. We do this by assigning a so-called *callback function* (function name without parentheses) to the special UI variable **exit_key_handler** (Figure 2.3) with **appuifw.app.exit_key_handler =quit**. In this case we assign the function **quit** to it. A function which we define inside the script starting with **def quit():** and its following indented block of code. Note, regarding callback functions, whenever the Python API documentation asks you to provide a callback function in the code, do not add parentheses after the function name since Python will call the function on your behalf. To make sure that our script does not drop out after the last line, we need to stop the execution at the last line and wait for user actions. We achieve this with an object called AO lock which can put the script into a waiting mode. The AO lock is created by calling **e32.Ao_lock()** and assigning it to a variable with **app_lock = e32.Ao_lock()**. The script starts waiting for user actions when the **wait()** function is called in the last line with **app_lock.wait()**. The lock stops waiting when the lock's **signal()** function is called with **app_lock.signal()**. And this is typically done in the exit key handler function **quit()** in our case. No worries if this sounds complicated, you can copy-paste the three lines mentioning **app_lock** each time to your own code as such. In order to see why it is important to use the lock at the end of the script code, just leave the last line **app_lock.wait()** out from your code and run the script. The dialog *Application running* pops up but the application will not go into wait state and will finish instantly instead.

2.6.4 Audio Module: Sound Recording and Playing

In Listing 2.8 we want to introduce the **audio** module and show how to create an application menu. With our script we control a simple sound recorder and player via an application menu having three choices, namely: play, record, stop. The **audio** module which we import at the beginning of the script enables recording and playing of audio files and gives access to the text-to-speech engine of the device. It supports all formats that are supported by the device itself, typically: WAV, AMR, MIDI, MP3, AAC, and Real Audio.

Listing 2.8. Code Example Sound

```
import appuifw, e32, audio

def recording():
    global S
5   S=audio.Sound.open('e:\\boo.wav')
    S.record()
    print "Recording! To end it, stop from menu!"

def playing():
10  global S
    S=audio.Sound.open('e:\\boo.wav')
    S.play()
    print "Playing! To end it, stop from menu!"

15 def closing():
    global S
    S.stop()
    S.close()
    print "Stopped"
20
def quit():
    app_lock.signal()

appuifw.app.menu = [(u"play", playing),
25                  (u"record", recording),
                    (u"stop", closing)]

appuifw.app.title = u"Sound recorder"
appuifw.app.exit_key_handler = quit
30 app_lock = e32.Ao_lock()
app_lock.wait()
```

The line of code which creates an application menu is `appuifw.app.menu=[...]`. We need to assign a list to it that includes the menu entries. The list must consist of tuples like (u"play", playing) or (u"record", recording). The first value in the tuple is the text that appears inside the menu when it is shown on the screen and the second value is the name of the callback function that needs to be called when the user selects the entry. In our list of menu entries we have names of three callback functions `playing`, `recording`, `closing`. Each callback function needs to be defined within the script as you can see in the example beside, i.e., `def recording():`, `def playing():`. A function is defined starting with `def`, then the function name and colon followed by the block of code that the function should carry out when it is called. Let us look now at each of these functions. In the `recording()` function the line S=audio.Sound.open (filename) returns a newly initialized sound object S with the file named boo.wav opened. e:\\boo.wav refers to the location of the file on the phone, in this case e:\\ means memory card and the file boo.wav is located on the root. If the file does not exist yet, it will be automatically created.

Note that `filename` should be a full unicode path name and must also include the file extension. With `S.record()` the microphone of the phone is opened and sound is recorded into the file boo.wav. The recording needs to be stopped via the function `closing()` that includes `S.stop()` which stops the recording. With `S.close()` the opened audio file gets closed. Note, the functions `record()`, `stop()`, `close()` are built-in functions of the `audio` module's sound class. They are accessible through its instance `S` simply by putting the object name then a dot and the function name with parenthesis. Let us look shortly at the `playing()` function. There we need to create the sound object `S` too, and with `S.play()` the recorded material is played and can be heard through the phone's speaker. There is a range of other functions provided by the `audio` module. For example with the `.state()` function one can inquire if an audio file is being played or currently recorded. An interesting feature is also the recording of the phone's speech channel during an ongoing phone call. For further functions and parameters of the audio module check the PyS60 Library Reference documentation [2]. Note, when testing the Listing 2.8 the user must press the left softkey in order to open the application menu and then select from the application menu the entries in the order of first `record` then `stop`, and `play` and again `stop`, otherwise an error will emerge. Also be aware that the newly recorded sound is always appended to the previous one.

2.6.5 Variable Scope: Local, Global

Variables which are introduced or manipulated in a function are visible only inside that function (local). But they can be made visible to the global scope of a script by putting the keyword `global` in front of the name of the variable at the beginning of the functions code block. This means they are visible also to all other functions of that script. In the example in Figure 2.10, variable `b` is local to function `function_one()`. Outside this function it can not be used. But the keyword `global` in front of the variable `a` tells that it belongs to the global scope. Therefore variable `a` can be used also in function `function_two()`. If a variable is defined outside any function, it belongs

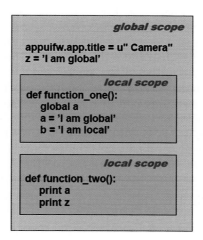

Figure 2.10. Variable scope.

automatically to the global scope, like variable z in the example. The output of `function_two()` is:

```
I am global
I am global
```

2.6.6 Debugging

There are three kinds of errors that can occur in a Python program: *syntax errors*, *runtime errors*, and *semantic errors*. Python may produce *syntax errors* when it translates the source code into byte code when starting up the program by the Python interpreter. A program can only be executed if is syntactically correct. If the translation process fails it returns usually an error message `SyntaxError: invalid syntax`, indicating where something is wrong with the syntax, including the line number of the code. If something goes wrong while the program is running, the Python interpreter produces a *runtime error*. Runtime error messages (`RuntimeError:`) usually include information about where in the code (line number) the error occurred and what is the problem. *Semantic errors* are problems with a program that compiles and runs but does not do the right thing.

There are several ways for debugging Python code, but only a few are mentioned here. One of them is the `print` statement. You can use it to print out values to the screen for example if a variable holds a correct value at a certain time. The print statement can also be placed at different parts of the code for printing a string to the screen or inside a log file in order to check if the script passes that point during runtime. Another approach is to use the `traceback.print_exc()` function through which all exceptions that the runtime throws can be printed and stored in a log file (note, you need to import the module `traceback` for this). Also the use of `try:` and `except:` statements can be useful for debugging. When these are used, the runtime executes first the block of code that follows the `try:` statement. If its code is faulty or causes some problem the script does not stop working, but instead the runtime executes the statements in the `except:` block. You can place for example, debugging information to the `except:` block. Another feature for debugging is the `e32.stdo` function. It takes as its argument the name of the file where C STDLIB's `stdout` and `stderr` are to be redirected. This makes it possible to capture the low-level error output when the Python interpreter has detected a fatal error and aborts. Details can be found in the PyS60 Reference Library document [2] and Programming with Python [1].

2.6.7 Inbox Module: Reading Incoming SMS

The `inbox` module provides reading of incoming SMS messages including SMS content, arrival time, and sender's number. It also allows deleting of SMS messages. Combined with the `audio` module's feature of accessing the phone's text-to-speech engine, your phone can speak out an incoming SMS and delete it instantly. Listing 2.9 shows one approach how this can be coded. First we import the **inbox** module and we use it in the lower part of the script to create an object i with i=inbox.Inbox(). With i.bind(read_sms), the inbox's **bind()** function is given a callback function named **read_sms** which is called when a new message arrives. It takes one parameter, the ID of the incoming message, **id**. After binding, the script stops execution with an AO lock. Once an SMS arrives, the function **read_sms()** is called a new instance of the inbox is created with i=inbox.Inbox() after that the content of the new SMS is read out and the message deleted. With audio.say(sms_text) the SMS text is spoken out.

Listing 2.9. Code Example Inbox

```
import inbox , e32 , audio

def read_sms ( id ) :
    i=inbox . Inbox ()
    sms_text = i . content ( id )
    i . delete ( id )
    audio . say ( sms_text )
    app_lock . signal ()

def quit () :
    app_lock . signal ()

i=inbox . Inbox ()
print "send_now_sms_to_this_phone"
appuifw . app . title = u" Say_SMS"
appuifw . app . exit_key_handler = quit
i . bind ( read_sms )
app_lock = e32 . Ao_lock ()
app_lock . wait ()
```

2.6.8 Graphics Module and the Use of Keyboard Keys

In Listing 2.10 we want show how to draw a red point on a blue background on the screen (Figure 2.11) and at the same time explain how to enable keyboard keys. In order to display 2D graphics to the screen, the application body appuifw.app.body (Figure 2.3) needs to be assigned a *canvas*. Canvas is a UI control that provides a drawable area on the screen and gives support for handling raw key events. For example, with c=appuifw.Canvas() we create an object named c and assign it with appuifw.app.body=c to the application body. Optional parameters can be passed to the Canvas() function that are callbacks to be called when specific events occur, e.g., redraw callback and event callback. They can be omitted as we do in our example. In order to learn how they are used, please refer to the PyS60 Reference Library document [2]. One of the possibilities to enable keyboard keys is to bind them with bind() to a callback function. In our example we bind the callback draw_point to the up arrow key which has the key code EKeyUpArrow. There exists a key code for each key on the keyboard. An overview of all key codes can be found in the PyS60 Reference Library document [2]. The binding is done

with `c.bind(key_codes.EKeyUpArrow, draw_point)`. The callback function
`draw_point` is called whenever the Up Arrow key is pressed. The `key_codes`
module needs to be imported at the beginning of our script. The basic idea
of using graphic elements within a script is to create first an image object,
manipulate it, and then draw it to the screen. For this we need to import the
`graphics` module.

Listing 2.10. Code Example Simple Graphics

```
import appuifw, e32, key_codes, graphics

def draw_point():
    img.clear(0x0000ff)
    img.point((120,100),0xff0000,width=70)
    c.blit(img)

def not_draw_point():
    img.clear(0x0000ff)
    c.blit(img)

def quit():
    app_lock.signal()

appuifw.app.screen='full'
c=appuifw.Canvas()
appuifw.app.body=c
s=c.size
img=graphics.Image.new(s)

c.bind(key_codes.EKeyUpArrow, draw_point)
c.bind(key_codes.EKeyDownArrow, not_draw_point)

appuifw.app.exit_key_handler=quit
draw_point()
app_lock = e32.Ao_lock()
app_lock.wait()
```

Figure 2.11. Drawing.

In Listing 2.10 we want to draw a red colored point to a blue background (Figure 2.11). First we create the image object during the start-up of the script with `img=graphics.Image.new(s)`, s specifying the size of the image. Since different phone models have different screen (canvas) sizes, the best thing to do is to inquire the canvas size in pixels with `s=c.size` (s gets width and height of the canvas as a tuple). In order to draw our red-colored point to the blue screen all we need to do is to manipulate the `img` object. We do this within the `draw_point()` function. First we turn the entire `img` object to blue with `img.clear(0x0000ff)` (color code for blue is: `0x0000ff`). This creates the background color of the screen. Then we draw the red point on top of it using `img.point((120,100),0xff0000,width=70)`. The first parameter gives the x- and y-coordinates of the red point on the screen as tuple `(120,100)`, the second parameter specifies the color of the point (red: `0xff0000`) and finally the last parameter specifies the width of the point in pixels. After that we draw of the image object `img` on the canvas with `c.blit(img)`. Once the user presses the Down Arrow key the `not_draw_point()` function is called which draws only the blue background but not the red point. Pressing the Up Arrow key brings again the red point to the screen by calling again the `draw_point()` function. Besides the graphics element `point` there exist also `rectangle`, `line`, `polygon`, `ellipse`, `pieslice`, `text`, and more. For `text` there is a font specification format available. The `graphics` module offers also several image manipulation methods such as resizing, flipping, rotating, and saving. There can be multiple Image objects used at the same time. An Image object can be created, for example, also from existing images located in your phone gallery or from graphics you specifically make for your application. You can use `img=graphics.Image.open ('e:\\Images\\picture.jpg')` to create Image objects from such sources. This can be useful if you want to create your own GUI with graphics object and images. As mentioned before there exist also other possibilities to enable keyboard keys than the one seen in Listing 2.10. Please refer to PyS60 Reference Library document [2] to learn more about it.

2.6.9 Image Composition Application: MobileArtBlog

The images in Figure 2.12 are created with a *Python for S60* script using the `graphics` module. It is a simple image composition tool that allows a user to compose and draw images by using the camera and the navigation keys. The tool was created by Jürgen Scheible, the author of this chapter, for his MobileArtBlog application. He says: *I use this 'toy' during my travels when I get stimulated by things I see and experience in different cities, places and situations, trying to capture the moment and turn it into a memorable 'art piece' (mobile art).* A taken photo can be placed multiple times on the canvas and its size can be changed. It can also be used to leave color traces by moving the photo on the canvas via the navigation keys. The ready images are directly posted to his MobileArtBlog website over the mobile data network together

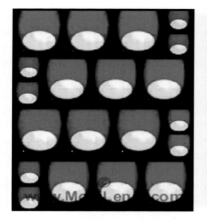

Figure 2.12. MobileArtBlog.

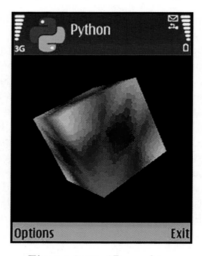

Figure 2.13. 3D graphics.

with some meta data like names of places as well as GPS coordinates that are read via Bluetooth from a GPS device. A collection of images can be seen at http://www.mobileartblog.org/ including their originating positions in a Google map view.

2.6.10 3D Graphics: gles and glcanvas Modules

The modules `gles` and `glcanvas` enable *Python for S60* to display 3D graphics for example as shown in Figure 2.13. The `gles` module provides Python bindings to OpenGL ES 2D/3D graphics C API, the `glcanvas` module provides a UI control for displaying OpenGL ES graphics. *Python for S60* supports OpenGL ES version 1.0 from Series 60 version 2.6 onwards, and OpenGL ES version 1.1, which requires S60 version 3.0 or newer. OpenGL ES is a standard defined by Khronos Group (http://www.khronos.org). Due to lack of space we cannot give details here, but please check the PyS60 Library Reference documentation [2] on how to use the 3D graphics functionalities.

2.6.11 Camera Module

Since different S60 phones have different types of built-in cameras, the `camera` module of *Python for S60* makes it possible to inquire certain characteristics of them. You can get the available hardware information with functions such as: `cameras_available()`, `image_modes()`, `image_sizes()`, `flash_modes()`, `max_zoom()`, `exposure_modes()`, `white_balance_modes()`, and so on. An important feature of the `camera` module is the viewfinder. The function `start_finder()` starts the camera viewfinder and `stop_finder()` stops it. The most important functionality that one expects from the camera module though is the `take_photo()` function. Simply by writing, i.e., `image = camera.take_photo(size = (640,480))` and `image.save('e:\\ Images\\picture.jpg')` a photo is taken and saved to the image gallery of the phone. The `take_photo()` function takes a whole range of input parameters: `take_photo(mode, size, flash, zoom, exposure, white_balance, position)`. To automatically open and close the viewfinder via a Python script can bring significant usability advantages for example, by providing the user the opportunity with one click to take a photo and send it away automatically via the mobile data network, wifi or via Bluetooth. What *Python for S60* cannot do (at the time of writing this book) is to record video.

In Listing 2.11 we show how to use the viewfinder, take a photo, and store it to the image gallery of the phone. You might be already familiar with a number of lines of this code, so we will explain only the new issues. We need to import the modules `camera` and `graphics`. Further we have to create a canvas and assign it to the application body since we need to draw the viewfinder output as well as the newly taken photo to the screen. When we start up the application we call the function `view_finder()` that activates the viewfinder with `camera.start_finder(finder_cb)` whereas `finder_cb` is the name of the callback function (defined in the upper part) that draws the viewfinder imagery to the canvas with `c.blit(im)`. In order to take a photo the user needs to press the 'Select' key which calls the callback function `make_photo()`. This stops first the viewfinder with `camera.stop_finder()`. Then we shoot a photo of the size 640 by 480 pixels with `image = camera.take_photo(size = (640,480))` whereas the function returns the photo as an Image object that we assign to the variable `image`. By simply writing `image.save('e: \\Images\\picture.jpg')` we store the new photo to the phone's gallery. The string `'e:\\Images\\picture.jpg'` represents the path of the file location on the phone as mentioned already in previous examples.

Listing 2.11. Code Example Camera

```
import camera, appuifw, graphics, key_codes

def finder_cb(im):
    c.blit(im)

def view_finder():
    camera.start_finder(finder_cb)
    c.bind(key_codes.EKeySelect, make_photo)

def make_photo():
    camera.stop_finder()
    image = camera.take_photo(size = (640,480))
    s = c.size
    c.blit(image,target=(0,0,s[0],(s[0]/4*3)),scale=1)
    image.save('e:\\Images\\picture.jpg')
    appuifw.app.menu = [(u"Back", view_finder)]

def quit():
    app_lock.signal()
    appuifw.app.set_exit()

appuifw.app.body = c =appuifw.Canvas()
appuifw.app.title = u"Camera_App"
appuifw.app.exit_key_handler = quit
view_finder()
app_lock = e32.Ao_lock()
app_lock.wait()
```

With `c.blit(image,target=(0,0,s[0],(s[0]/4*3)),scale=1)` the newly taken photo is drawn to screen area whereas `(0,0,s[0],(s[0]/4*3))` are the coordinates that specify the position of the drawing. The coordinates are interpreted as the top-left and bottom-right corners of the area. Since different phone models have different canvas sizes, the photo would appear at different positions with different sizes on each screen. Therefore we inquire the canvas size in pixels so it works on each model the same way. This is what we do: with `s = c.size` we receive a tuple (e.g., `(240,234)`) with values referring to width and height of the canvas. Using `s[0]` we read out the first value – the width – of the tuple. Since the proportions of the drawable photo should be 4:3 we can calculate the height with `s[0]/4*3`. To go back for taking another photo, the user can press *options* and select 'back'. This opens again the viewfinder. To learn about further parameters and functions of the `camera` module please refer to the PyS60 Library Reference documentation [2].

2.6.12 Video Download and Playing – the Use of Content Handler

In Listing 2.12 we explain how to download a file from the Internet. The script downloads a video of the format *.3gp that is stored on a server behind a URL. In order to do that, we need to import the `urllib` module, which is a standard Python module that provides a high-level interface for fetching data across the World Wide Web. We define a URL with `url = "http://w..."` which is the video file's URL in the Internet. Next we define a path and filename for storing the downloaded video to the phone writing `tempfile = 'e:\\Videos\\video.3gp'`. Simply by using the function `urlretrieve(url,tempfile)` the video is downloaded and stored to the phone. As mentioned earlier, *Python for S60* can neither record nor play video. But it offers a so called 'content handler' that uses the phone's native players to play various contents.

With `content_handler = appuifw.Content_handler()` we create an object of the content handler and with `content\_handler.open(tempfile)` we hand over the video file to be played. The content handler opens then automatically the phone's native video player and plays the video. Note, when you run the script, your phone prompts you to select an access point for connecting to the Internet in order to be able to download the video.

Listing 2.12. Code Example Video Download

```
import appuifw, urllib

url = "http://www.../vid001.3gp"
tempfile = 'e:\\Videos\\video.3gp'
urllib.urlretrieve(url, tempfile)
content_handler=appuifw.Content_handler()
content_handler.open(tempfile)
```

2.6.13 File Upload to a URL

In Listing 2.13 we explain how to upload a file via HTTP from the phone to a URL using a small Php script on the server side. We upload an image that is located in the image gallery on the phone at `'e:\\Images\\picture.jpg'`. With `f = open('e:\\Images\\picture.jpg','rb')` we open the image file in 'read binary' mode `'rb'`. After that we store the image content into a variable named `chunk` by writing `chunk = f.read()`. The next step is to establish an HTTP connection and upload our image file via a POST request to the server. This is done with the help of the standard Python module `httplib`. You can find detailed information on the `httplib` module in the Python documentation at `http://www.python.org`.

For coding the HTTP POST request we need to define first the `headers`. It is a mapping of typical HTTP headers done according to RFC822 standard. Next we create an object of an HTTP connection with `conn = httplib.HTTPConnection("www....com")`.

An HTTPConnection object represents one transaction with an HTTP server. It should be instantiated by passing a `host` and optional `port` number to it. If no port number is passed, the port is extracted from the host string if it has the form `host:port`, otherwise the default HTTP port (80) is used. Here we use the default port and as host we type: `"www....com"`. You need to put here the address of your own server to which the image should be uploaded. Next we send a request to the server using the HTTP POST request method that includes as parameters the selector `url`, `body`, and `headers`. We pass the above defined `headers` to the `headers` parameter. The `body` parameter is a string of data and we pass the variable `chunk` to it which holds the image content. The selector `url` is in our case a Php file that 'receives' the uploaded content on the server. The Php file is located on the root of the server and is named `image_upload.php` (you could place it to any location on the server). See the content of the Php file below (Listing 2.14). Due to lack of space we will not explain details. Here are a few things regarding your server: Make sure it has Php running and you have set the *receiving* directory's write permissions correctly, e.g., to octal to: 0777 (in our case the *receiving* directory is named 'pictures', which is a subdirectory of the one where the Php script itself is located).

Listing 2.13. Code Example Upload

```
import httplib, appuifw

def upload_to_url():
    f = open('e:\\Images\\picture.jpg','rb')
5   chunk = f.read()
    f.close()

    headers = {"Content-type": "application/octet-stream","Accept"
        :"text/plain"}
10  conn = httplib.HTTPConnection("www....com")
    conn.request("POST", "/image_upload.php", chunk, headers)
    conn.close()
    appuifw.note(u"Photo sent")

15 upload_to_url()
```

Listing 2.14. Code Example Upload in php

```
<?php
// read the incoming image data handed over
// from the Python S60 phone
$data = file_get_contents('php://input');
5 // write the file to the server into a sub directory
// (named videos) of this php script's directory
$filepathname = "pictures/picture.jpg";
$handle = fopen($filepathname, 'wb');
fputs($handle, $data, strlen($data));
10 fclose($handle);
?>
```

2.6.14 Mobile Photo Blogging Application

Simply by combining the code of the camera application (Listing 2.11) with the code of the file upload to URL example (Listing 2.13) you can make a ready-to-use mobile photo blogging application (see Listing 2.25). As you can see from this, often we can (more or less) *copy paste* different Python code snippets into a new script and easily create a new kind of application!

2.6.15 File Handling and Data Storing

In Table 2.3 we show how basic file handling and data storing can be done:

Table 2.3. File handling and data storing.

List content of a directory:	Example of storing settings in a
import e32, os	**dictionary as 'key:value' pair:**
# define the directory	*#store settings:*
imagedir = 'c:\\nokia\images'	*value1 = 'man'*
# read the directory	*value2 = 3.15*
files=map(unicode,os.listdir(imagedir))	*config={}*
Read image content to variable:	*config['var1']= value1*
# open the image file	*config['var2']= value2*
f=open('e:\\images\\test.jpg', 'rb')	*f = open(u'e:\\mysettings.txt', 'wt')*
pic = f.read()	*f.write(repr(config))*
f.close()	*f.close()*
Write text into a file:	*# read settings:*
f = open('c:\\writetest.txt', 'w')	*f=open('e:\\mysettings.txt', 'rt')*
f.write('hello, this works')	*content = f.read()*
f.close()	*config = eval(content)*
Conversion string / object:	*f.close()*
eval() converts from string to object	*value1=config.get('var1','')*
repr() converts any object to string	*value2=config.get('var2','')*
	print value1,value2

2.6.16 Socket Module: Bluetooth, TCP/IP, and SSL

The `socket` module is an extended version of the Python standard `socket` module. It offers Bluetooth protocol RFCOMM and OBEX, but also TCP/IP and SSL sockets can be created. It allows also setting a default access point for connecting to the Internet. In the next few code examples we first look at Bluetooth OBEX and later at RFCOMM. After that we show a simple example of a TCP/IP socket. We do not cover SSL sockets here. Python can also handle FTP but *Python for S60* does not include a `ftp` module. The standard Python `ftp` module can be separately installed as a *library* to the phone though.

Sending a Photo from Phone to Phone via Bluetooth OBEX

In this Bluetooth OBEX example (Listing 2.15) we show how to take a picture and send it to a phone using the Bluetooth OBEX protocol.

Listing 2.15. Code Example Obex

```
import camera,e32,socket,appuifw

def start():
    image= camera.take_photo()
    appuifw.app.body=c=appuifw.Canvas()
    c.blit(image,scale=1)
    file=(u'e:\\Images\\pic001.jpg')
    image.save(file)
    device=socket.bt_obex_discover()
```

```
        address=device [0]
        channel=device [1][ u ’OBEX_Object_Push’]
        socket . bt_obex_send_file ( address , channel , file )
13      appuifw . note (u” Picture_sent” ,” info ” )

    def quit ( ) :
        app_lock . signal ( )
        appuifw . app . set_exit ( )
18
    app_lock = e32 . Ao_lock ( )
    appuifw . app . title = u” Bluetooth_photo”
    appuifw . app . menu = [( u” Start” , start ) ,(u” Exit” , quit )]
    app_lock . wait ( )
```

Since you might be familiar with most of the lines of code, we explain here only the Bluetooth-related ones. First of all when dealing with Bluetooth we have to import the `socket` module. With `socket.bt_obex_discover()` the phone scans for other Bluetooth devices around and opens the phone's native Bluetooth selection list on the screen showing the nick names of all devices found. After selecting the nickname of our target device where we want to send our photo to, the variable `device` holds a tuple. It contains the inquired data from the scanning process regarding the Bluetooth service that the chosen target device offers. The inquired data is not seen anywhere during the script execution, but we show it here for didactical reasons: `(00:16:bc:41:bf:7f,{'OBEX Object Push' :9})` whereas the first value inside the parenthesis is the Bluetooth address of our target device in HEX format. The second value is a dictionary that contains the Bluetooth service offered as a 'key:value' pair. Behind the colon is the Bluetooth communication channel through which the service is offered. With `address=device[0]` we read the Bluetooth address `00:16:bc:41:bf:7f` and store it in the variable `address`. With `channel=device[1][u'OBEX Object Push']` we read the value of the dictionary's *key:value* pair which in this case is the communication channel on which the 'OBEX Object Push' service is offered. We store it in the variable `channel`. These two parameters `address` and `channel` plus the path of the file to be sent `file=(u'e:\\Images\\pic001.jpg')` are needed for pushing the image to our target phone. Note, the path must be given as a Unicode string. We do this with `socket.bt_obex_send_file( address, channel, file)`. The user of the target device is then triggered a pop-up menu asking *Receive message via Bluetooth from ....* After she presses *yes* the photo is transferred. There is one more new line of code `appuifw.app.set_exit()` inside the `quit()` function. It exits the Python interpreter completely.

Sending Data from Phone to Phone via Bluetooth RFCOMM

Bluetooth sockets can be set up for sending and receiving data using the Bluetooth protocol RFCOMM. This can be done from phone to phone, or phone to computer, or from phone to microcontroller that has a Bluetooth chip. For example, the arduino board (`http://www.arduino.cc/`) is a microcontroller board with a Bluetooth extension chip. This might be of interest to projects dealing with social interaction through wearables, physical computing, and mobile devices. Furthermore the sensor introduced in Chapter 21 is doing the same.

To establish an RFCOMM connection from phone to phone over Bluetooth two different scripts are needed. One phone runs the *server side* script and the other phone runs the *client side* script. Since we cannot go into detail due to lack of space we show here some code example of a simple chat application (Listing 2.16, 2.17) including comments. Make sure you have Bluetooth switched on at both phones. All we do is setting up a Bluetooth RFCOMM connection between two phones and each side sends and receives a string. With these two scripts a multiuser game can easily be created, too.

Listing 2.16. Code Example Chat Server with RFCOMM

```
# 'Server side' script
# the server side must perform the sequence socket(),
     bt_rfcomm_get_available_server_channel(),
# bind(), listen(), bt_advertise_service(), set_security(), accept
     ()
# Note: the 'server side' script must be started first and then the
     'client side' script
5
import socket, appuifw

def start_server_connection():
    # create socket using AF_BT for Bluetooth socket
10  server_socket = socket.socket(socket.AF_BT, socket.SOCK_STREAM)
    # Returns an available RFCOMM server channel for server_socket.
    channel = socket.bt_rfcomm_get_available_server_channel(
        server_socket)
    # binds the socket to the desired channel
    server_socket.bind(("", channel))
15  # the server will listen for incoming requests
    server_socket.listen(1)
    print "Wait..."
    # advertising the BT RFCOMM service via name (Unicode) 'mychat'
        on local
    # channel that is bound to socket
20  socket.bt_advertise_service( u"mychat", server_socket, True,
        socket.RFCOMM)
    # sets the security level of the given bound socket to AUTH (
        authentication);
    # when listening to a BT socket on the phone, you must set the
        security level
    socket.set_security(server_socket, socket.AUTH)
    # when client side script on other phone responds, the accept()
        function returns
25  # a new socket (different than the one for listening) for send
        () and recv()
    (new_server_socket, peer_addr)= server_socket.accept()
    run = 1
    while run:
        try:
30          # trigger a dialog to type in a word
            send_data = appuifw.query(u"Type_a_word:", "text")
            if send_data is not None:
                # send the typed word to the other phone (client
                    side script)
                new_server_socket.send(send_data)
35              print 'Sent:_', repr(send_data)
            else:
                run = 0
```

```
                            new_server_socket.close()
                        # go into receiving mode to receive data from the
                            client side script
40                      data = new_server_socket.recv(1024)
                        print 'Received:_', repr(data)
                        appuifw.note(u"" +data, "info")
                    except:
                        # if something went wrong, exit the interpter
45                      run = 0
                        new_server_socket.close()
                        appuifw.note(u"Connection_lost", "info")
                        appuifw.app.set_exit()

50  start_server_connection()
```

Listing 2.17. Code Example Chat Client with RFCOMM

```
    # 'Client side' script
    import socket,appuifw

    def start_client_connection():
5       # create socket using AF_BT for Bluetooth socket
        client_socket=socket.socket(socket.AF_BT,socket.SOCK_STREAM)
        #Performs the BT device discovery and the discovery of RFCOMM
            class services
        # on the chosen device. Returns BT device address, dictionary
            of services
        addr,services=socket.bt_discover()
10      # getting the channel of specific service 'mychat'
        channel=socket.bt_discover(addr)[1][u'mychat']
        address=(addr,channel)
        # connect to the server side script
        client_socket.connect(address)
15      print "Connected."
        run = 1
        while run:
            try:
                # go into receiving mode to receive data from the
                    server side script
20              data = client_socket.recv(1024)
                print 'Received:', repr(data)
                appuifw.note(u"" +data, "info")
                # trigger a dialog to type in a word
                send_data = appuifw.query(u"Type_a_word:", "text")
25              if send_data is not None:
                    # send the typed word to the other phone (server
                        side script)
                    client_socket.send(send_data)
                    print 'Sent:_', repr(send_data)
                else:
30                  # if the user has pressed cancel, exit the
                        interpreter
                    run = 0
                    client_socket.close()
                    appuifw.note(u"Connection_lost", "info")
                    appuifw.app.set_exit()
35          except:
                # if something went wrong, exit the interpreter
                run = 0
                client_socket.close()
                appuifw.note(u"Connection_lost", "info")
40              appuifw.app.set_exit()

    start_client_connection()
```

TCP/IP Socket

With the following scripts (Listing 2.18, 2.19) you can connect your phone via a TCP/IP socket to a server in the Internet or to a local machine in an ad hoc network. A `host` and `port` is needed for the AF_INET address family used, where `host` is a string representing either a hostname in Internet domain notation and `port` is an integral port number. As in the Bluetooth socket RFCOMM example we need also here two different scripts, one for the server and the other one for the phone as the client script.

Listing 2.18. Code Example TCP/IP Client

```
     import socket

3    HOST = '192.22.1.100'      # The remote host
     PORT = 12012               # The same port as used by the server
     print "define_socket"
     s = socket.socket(socket.AF_INET, socket.SOCK_STREAM)
     print "trying_to_connect_to_socket"
8    s.connect((HOST, PORT))
     print "connected"
     s.send('Hello,_world')
     print "data_send"
     data = s.recv(1024)
13   s.close()
     print 'Received', 'data'
```

Listing 2.19. Code Example TCP/IP Server

```
1    import socket

     host = ''                  # local host
     port = 12012               # port
     print "define_the_socket"
6    s = socket.socket(socket.AF_INET, socket.SOCK_STREAM)
     print "bind_the_socket"
     s.bind((host, port))
     s.listen(1)
     print "waiting_of_the_client_to_connect"
11   conn, addr = s.accept()
     print 'Connected_by', addr
     while 1:
         data = conn.recv(1024)
         if not data: break
16       print 'Received:_', data
         send_data = 'This_works'
         conn.send(send_data)
         print 'Sent:_', send_data
     conn.close()
```

Setting a Default Access Point

The `socket` module offers also the functionality of setting a default access point for connecting to the Internet. This omits the task of manually selecting an access point each time when you open a Python script that connects to the Internet. The following script (Listing 2.20) can be used to set and clear a default access point.

Listing 2.20. Code Example Access Point Setting

```
import socket

def set_accesspoint():
    apid = socket.select_access_point()
    f = open('e:\\apid.txt','w')
    f.write(repr(apid))
    f.close()
    app_lock.signal()

def clear_accesspoint():
    f = open('e:\\apid.txt','w')
    f.write(repr(None))
    f.close()
    app_lock.signal()

appuifw.app.menu = [(u"set_access_point", set_accesspoint),
    (u"unset", clear_accesspoint)]
app_lock = e32.Ao_lock()
app_lock.wait()
```

First the `socket` module must be imported. Let us look at the function `set_accesspoint()`. With `apid = socket.select_access_point()` a pop-up menu is opened on the screen that lists all available access points that are defined on the phone. When the user selects an access point from the menu, an id is returned to the variable `apid`. The id is then written as a string to a file named `apid.txt` with `f.write(repr(apid))`. For clearing the access point simply `"None"` needs be stored to the `apid.txt` file inside the `clear_accesspoint()` function.

Listing 2.21. Code Example Access Point Reading

```
import appuifw, urllib

def download():
    f=open('e:\\apid.txt','rb')
    setting = f.read()
    apid = eval(setting)
    f.close()
    if not apid == None :
        apo = socket.access_point(apid)
        socket.set_default_access_point(apo)
    url = "http://www.../vid001.3gp"
    tempfile = 'e:\\Videos\\video.3gp'
    urllib.urlretrieve(url, tempfile)
```

The code in Listing 2.21 starting from the line `f=open('e:\\apid.txt', 'rb')` till `socket.set_default_access_point(apo)` can be used to prepare your script to automatically connect to the Internet using the set default access point. When starting the script it reads the stored id of the access point from the file `apid.txt` into the variable `apid`. With `apo=socket.access_point (apid)` and `socket.set_default_access_point(apo)` the script is then prepared. The script does not use the default access point though, in case the `apid` is set to `None`, for example, due to clearance of the default access point. In this case the script will ask you to select an access point manually each time. In the given script the line `urllib.urlretrieve(url,tempfile)` invokes the

Internet connection, therefore the described code needs to be placed before that line.

Tapping into Web Service APIs

Companies such as Google or Yahoo and other large Internet companies are offering more and more API's to their web services, for example, for photo sharing, shopping, social-event calendars, Search, Maps, etc. By combining these web services with mobile technology, completely new types of applications and services can be created. By using *Python for S60* it is easy and fast to create applications that connect to such services via JSON, REST, etc. This allows, for example, to combine several features of different services to be jointly used in one single mobile application. More info's on API's of web services can be found from `http://code.google.com/apis/` or `http://developer.yahoo.com/`.

2.6.17 sysinfo Module

The `sysinfo` module offers an API for checking the system information of the mobile device. The following functions as shown in Table 2.4 are available.

Here is a code example:

```
import sysinfo
sysinfo.battery()
```

2.6.18 Location Module

The `location` module offers APIs to location information-related services. It has one function: `gsm_location()` which retrieves GSM location information: Mobile Country Code (MCC), Mobile Network Code (MNC), Location Area Code (LAC), and Cell ID. A location area normally consists of several base stations. It is the area where the mobile terminal can move without notifying the network about its exact position. The Mobile Country Code and the Mobile Network Code form together a unique identification number of the network into which the phone is logged. Note, for 3rd edition devices the `location` module requires so-called capabilities ReadDeviceData, ReadUserData, and Location. These refer to the Symbian platform security framework. Depending on the certificate level of the *Python for S60* interpreter that you have installed, these capabilities might not be assigned, meaning they will not work. In order to get them to work you need a different certificate level. Find more on this in Section 2.8. The Listing 2.22 shows how to retrieve the GSM location information: Mobile Country Code, Mobile Network Code, Location Area Code, Cell ID. The `location`, `appuifw`, and `e32` modules are used.

Table 2.4. sysinfo module.

battery()	Current battery level ranging from 0 to 7 is returned. If using an emulator, value 0 is always returned.
imei()	IMEI code of the device as a Unicode string is returned. If using an emulator, the hardcoded string u'000000000000000' is returned.
active_profile()	Current active profile as a string is returned, which can be one of the following: 'general', 'silent', 'meeting', 'outdoor', 'pager', 'offline', 'drive', or 'user <profile value>'.
display_pixels()	Width and height of the display in pixels is returned.
display_twips()	Width and height of the display is returned in twips which are screen-independent units to ensure that the proportion of screen elements are the same on all display systems. A twip is defined as 1/1440 of an inch, or 1/567 of a centimeter.
free_drivespace()	Amount of free space left on the drives in bytes is returned.
max_ramdrive_size()	Maximum size of the RAM drive on the device is returned.
total_ram()	Amount of RAM memory on the device is returned.
free_ram()	Amount of free RAM memory available on the device is returned.
total_rom()	Amount of read-only ROM memory on the device is returned.
ring_type()	Current ringing type as a string is returned, which can be one of the following: 'normal', 'ascending', 'ring_once', 'beep', or 'silent'.
os_version()	Operating system version number of the device as integers is returned.
signal_bars()	Current network signal strength ranging from 0 to 7 is returned, with 0 meaning no signal and 7 meaning a strong signal. If using an emulator, value 0 is always returned.
signal dbm()	Current network signal strength in dBm is returned. If using an emulator the returned value is always 0.
sw_version()	Software version as a Unicode string is returned. If using an emulator, it returns the hardcoded string u'emulator'.

The output can be seen in Figure 2.14. The function gsm_location() returns the location information values as a tuple (mcc, mnc, lac, cellid) and we then printed out each value separately.

Listing 2.22. Code Example Location

```
import appuifw , e32 , location

def gsm_location ( ) :
    (mcc, mnc, lac , cellid ) = location . gsm_location ( )
    print u"MCC:_" + unicode (mcc)
    print u"MNC:_" + unicode (mnc)
    print u"LAC:_" + unicode ( lac )
    print u" Cell_id :_" + unicode ( cellid )
```

Figure 2.14. Location.

```
     def quit():
         app_lock.signal()
         appuifw.app.set_exit()
13
     app_lock = e32.Ao_lock()
     appuifw.app.title = u"Location"
     appuifw.app.exit_key_handler = quit
     appuifw.app.menu = [(u"get_location", gsm_location)]
18   app_lock.wait()
```

2.6.19 Telephone Module

The `telephone` module provides an API to telephone functionalities such as dialing a number and hanging-up. Here is some example code (Listing 2.23):

Listing 2.23. Code Example Telephone

```
    import telephone
2
    telephone.dial('+3581234567')
    telephone.hang_up()
```

There exist only two functions in this module:`dial()`, `hang_up()`. A telephone number must be passed as a parameter to the function `dial()` as a string.

2.6.20 Contact Module

The `contact` module offers an API to address book services allowing the creation of contact information databases to gather and change information of the contact database entries. Listing 2.24 shows how to retrieve a telephone number of a contact and call it. Three modules are used: `appuifw`, `contacts`, and `telephone`.

Listing 2.24. Code Example Contacts

```
import contacts, appuifw, telephone

db=contacts.open()
entry =db.find(appuifw.query(u'Insert a name','text'))
L=[]
for item in entry:
    L.append(item.title)
if len(L)>0:
    index = appuifw.selection_list(choices=L , search_field=0)
    num=entry[index].find('mobile_number')[0].value
    telephone.dial(num)
else:
    appuifw.note(u'No matches','error')

appuifw.app.set_exit()
```

2.7 Creating Stand-Alone Applications

A stand-alone application lets you start your *Python application* directly from an own icon on the phone's desktop and not having to open first the Python interpreter anymore and pressing *run script* and so on. A stand-alone application is done by creating a Symbian install file *.sis from your Python script using a utility called py2sis. Note, the created stand-alone application will run only if *Python for S60* is already installed on the target device. Though there are ways round to avoid this by including the *Python for S60* interpreter inside your *.sis package as well. Detailed information on the steps on creating a stand-alone application as well as making your own icons can be found from the PyS60 tutorial written by the author of this chapter at `http://www.mobilenin.com/pys60/info_standalone_application.htm`. The following section describes only roughly the major steps on how to create a stand-alone application.

1. Download and install the S60 SDK for Symbian OS on your Windows machine. It can be found from `http://forum.nokia.com/`. Make sure the utilities `makesis` and `uidcrc` are in your system path.
2. Install the Python plug-in. It comes with the *Python for S60* SDK package provided at `http://sourceforge.net/projects/pys60`. It includes the py2sis utility.
3. Open a *Command Prompt* window.
4. Use the command:
   `py2sis <src> [sisfile] [--uid=0x1234567]`
   `<src>` is the path of your Python script
   `[sisfile]` is the name of your newly created *.sis file

[--uid=0x1234567]: A UID can be chosen from range 0x01000000 to 0x0fffffff (for more information about UID, see the help documentation of the S60 SDK).

Example:

py2sis test.py myfirst.sis --uid=0x0FFFFFFF

IMPORTANT: Whether you can install your ready *.sis file immediately to your phone or not depends on your phone model. If it is a 2nd edition phone or older, you can install the *.sis file instantly. If it is a 3rd edition phone, you must undergo an additional process of creating a *certificate* for your application. This is due to the Symbian platform security framework which handles access to sensitive APIs through a *capability model*. In the next section we explain a few more details.

2.8 Symbian Platform Security and Python for S60

The Symbian Platform Security is a security framework that limits what applications can 'do' on devices based on Symbian OS v9 or later. For S60 devices, Platform Security is included starting from S60 3rd edition. The way Platform Security affects a developer is a complex topic and therefore hard to summarize both briefly and precisely. Nevertheless we give you here some rough introduction. As long as a Python application does not require capabilities beyond the so called user-grantable set (including capabilities known as LocalServices, NetworkServices, UserEnvironment, ReadUserData, and WriteUserData) things are fairly simple: You can package your Python application into a self-signed SIS file that is then installable to all phones that allow the installation of untrusted applications, or if you want to distribute your Python application as a plain .py script for advanced users, they can simply install a self-signed *Python for S60* interpreter and run your script in that environment. Fortunately most of the current *Python for S60* functionality can be accessed using the user-grantable capability set. You should check the latest details from Nokia's wiki: http://wiki.opensource.nokia.com/projects/ Python_for_S60 or the PyS60 Library Reference documentation [2].

Examples of functionality that are not accessible with the user-grantable capability set include:

- Access to Cell ID information (function location.gsm_location()). It requires capabilities ReadUserData, ReadDeviceData, and Location.
- Access to global key event capture (module keycapture). It requires capability SwEvent.
- Setting the device time (function e32.set_home_time()). It requires capability WriteDeviceData.

If you require access to capabilities beyond the user-grantable set, things get too complicated to explain in the space available here. Consult the PyS60

Library Reference documentation [2] or Symbian Signed documentation at `http://www.symbiansigned.com/` for details.

2.9 Creating Python Extensions in C++

Python for S60 supports dynamically loadable native extension modules (.PYDs), that can be used to add access to APIs not available through the existing Python modules or increase the computational performance of selected parts of your Python application. Through suitable native extension modules, a *Python for S60* application can have the same level of access to platform functionality as a native Symbian C++ application.

To write an extension for Python for S60, you need:

- A PC with Windows
- The S60 C++ SDK
- The *Python for S60* plugin for the S60 C++ SDK

With these tools you can write .PYD extensions and wrap them into .SIS packages either on their own, or bundled as a part of a Python application packaged as a .SIS file. For details on how to write Python extensions, check the PyS60 Library Reference documentation [2] or Nokia's wiki: `http://wiki.opensource.nokia.com/projects/Python_for_S60`.

2.10 Further Reading Sources on Python for S60

For those interested in getting deeper knowledge in programming *Python for S60* a complete book on the topic is available from fall 2007, published by Symbian Press/Wiley, written by the same author as this chapter with two additional authors Ville Tuulos and Jukka Laurila. The book covers the entire scope of *Python for S60* in detail and focuses on rapid prototyping. It provides many creative exercise examples as kick and inspiration.

2.11 Code Appendix

Listing 2.25. Appendix Code Example

```
# Mobile photo blogging application:

import camera, appuifw, graphics, httplib

def finder_cb(im):
    c.blit(im)

def make_photo():
    camera.stop_finder()
    image = camera.take_photo(size = (640,480))
```

```
       s = c.size
       c.blit(image,target=(0,0,s[0],(s[0]/4*3)),scale=1)
       # here we call the uploading function
14     image.save('e:\\Images\\picture.jpg')
       upload_to_url()

   def upload_to_url():
       f = open('e:\\Images\\picture.jpg',"rb")
19     chunk = f.read()
       f.close()
       headers = {"Content-type": "application/octet-stream",
                   "Accept": "text/plain"}
       conn = httplib.HTTPConnection("www....com")
       conn.request("POST", "/image_upload.php", chunk, headers)
24     conn.close()
       appuifw.note(u"Photo_sent")
       # after the upload is done, the application will exit
       app_lock.signal()

29 def quit():
       app_lock.signal()

   appuifw.app.body = c =appuifw.Canvas()
   appuifw.app.menu = [(u"Capture_+_upload", make_photo)]
34 camera.start_finder(finder_cb)
   appuifw.app.title = u"Camera_+_Upload"
   appuifw.app.exit_key_handler = quit
   app_lock = e32.Ao_lock()
   app_lock.wait()
```

2.12 Links

- http://sourceforge.net/projects/pys60
- http://www.python.org/
- http://wiki.opensource.nokia.com/projects/Python_for_S60
- http://forum.nokia.com/
- http://www.mobilenin.com/pys60/menu.htm
- http://mobiledevices.kom.aau.dk/index.php?id=909
- http://www.symbiansigned.com/
- http://www.mobilenin.com/pys60/info_standalone_application.htm
- http://www.arduino.cc/
- http://www.mobileartblog.org/

References

1. Programming with Python. World Wide Web. Available at: http://www.forum.nokia.com/.
2. Pys60 library reference documentation. World Wide Web. Available at: http://sourceforge.net/projects/pys60.

3

Java 2 Micro Edition

Gian Paolo Perrucci[1] and Andreas Häber[2]

[1] Aalborg University `gpp@es.aau.dk`
[2] Agder University College `andreas.haber@hia.no`

3.1 Java Family Overview

Java technology is an object-oriented programming environment created by Sun Microsystems. The main goal was to offer an environment to develop platform-independent applications following the slogan *write once, run anywhere*. Today Java is one of the most popular software developing environment. It offers the opportunity to create enterprise applications for server and desktop computers as well as applications for small and mobile devices. To get familiar with J2ME technology for mobile application development, let us start with exploring the overall Java family first. As shown in Figure 3.1, Java has been split up in 4 distinct editions [3]:

J2EE is mainly intended for building distributed enterprise applications with emphasis on server-side development and web applications.
J2SE is intended for conventional desktop applications development.
J2ME is a subset of J2SE intended for embedded devices which cannot support a full J2SE implementation.
Java Card offers an environment for developing applications running on smart cards.

This split has been made to offer a solution for different needs and circumstances. It would be inefficient, for example, to use a full J2EE implementation, which includes packages targeted for an enterprise application, if a desktop machine is used. Also, a full implementation of J2SE cannot be supported by a small capabilities device, which has fewer resources than a desktop machine.

---

[3] Nowadays it is popular to refer to the different Java editions as: Java ME, Java SE, Java EE. But in this book we will use the terms: J2ME, J2SE, and J2EE since they are still widely used.

F.H.P. Fitzek and F. Reichert (eds.), Mobile Phone Programming and its Application to Wireless Networking, 63–93.
© 2007 *Springer.*

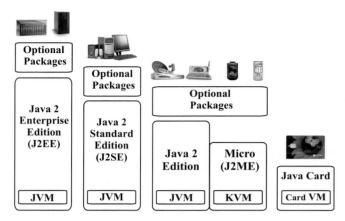

Figure 3.1. Java family architecture.

3.2 J2ME Platform

J2ME is part of the Java family (Figure 3.1), intended to build applications running on battery driven platforms such as mobile phones, PDAs, and other embedded devices. Since all these devices have different capabilities in terms of memory, processing time, hardware, and display, they have been grouped together according to their purposes and functionalities into device groups. Those with similar capabilities are grouped in configurations, which define a lowest common denominator of features for each device group. Within each configuration, different profiles are created in order to further classify the device type. J2ME is not a new language, but it has adapted the existing Java technology for running on embedded devices, removing parts of J2SE that they cannot support and adding new features that they need. Nevertheless, the J2ME platform encloses a large range of devices types: TV set-top box, Internet TVs, PDA s, pagers, mobile phones, etc. All these devices have different capabilities such as user interface, memory budget, and processing power. At the moment two configurations exists in the J2ME platform:

- **CLDC**, for devices with:
  - Very simple User Interface
  - Low level memory budget (160 Kb to 512 Kb) for Java
  - Wireless communication, targeting low bandwidth
  - 16 or 32 bit processors
  - Limited power, often battery operating
  - Examples: mobile phones, pagers, entry level PDAs.
  These devices will here be known as *less capable devices*.
- **CDC**, for devices with:
  - Large range of User Interface capabilities
  - Memory budget in the range of 2–16 MB for Java

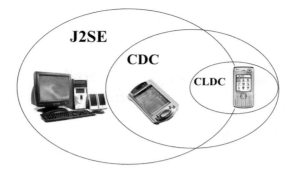

Figure 3.2. J2ME inheritance.

- Connectivity to some type of network
- 16- or 32- bit processors
- Examples: Internet TVs, TV set-top boxes, PDA.

In the following we will call devices running this configuration as *more capable devices*.

The relationship between CDC, CLDC, and J2SE is shown in Figure 3.2: all the main functionalities of CDC and CLDC are inherited from J2SE. Moreover, all the features implemented in the CLDC are implemented in the CDC as well, in order to achieve upward compatibility between the two J2ME configurations. The J2ME platform runs on the top of the operating system of the Java-enabled mobile phones and consists of two different versions: one for *more capable devices* and one for *less capable devices* as shown in Figure 3.3. Each of them includes:

Configuration: This is the basic set of class libraries and the VM which interprets and runs the applications. The VM is either a full JVM or a subset of this. For CLDC and CDC the class libraries are mostly a subset of the J2SE API in addition to APIs specific for J2ME such as the GCF(see Section 3.2.3).

Profiles: The profiles complement the configuration with more high level APIs specific for a device group. APIs provided by a profile normally consists of application life-cycle model, user interface, storage, specific network support (see Section 3.2.3), and more. Notice that profiles can be layered on top of each other, as will be seen with CDC in Section 3.2.2.
The configuration together with one or more profiles creates a complete J2ME runtime environment.

Optional Packages: With optional packages the runtime environment is extended with APIs targeting specific requirements, such as database connectivity, communication methods, graphics, etc.
An optional package can be configuration independent (i.e., support both CLDC and CDC) or it can be targeted at one specific configuration.

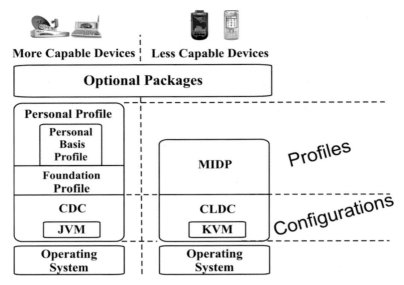

Figure 3.3. J2ME platform.

Unfortunately optional packages can make deployment of the application hard, because you might end up with a long list of optional packages required to support your application. We will later, in Section 3.2.5, explain the so-called roadmap specifications which remedies this deployment headache.

3.2.1 Less Capable Devices: Targeting CLDC

As stated earlier, J2ME can target two different groups of devices. In this section the configuration and the profile supported by *less capable devices* are described.

Connected Limited Device Configuration

CLDC defines a set of libraries and a VM for resource constrained devices. At the moment two versions of CLDC are available:

- CLDC 1.0 is the first release of the CLDC specification.
- CLDC 1.1 is a backward compatible version of the CLDC 1.0 with some enhancements, such as support for floating-point math and weak references.

The CLDC provides a Java runtime environment, but the set of core classes is usually quite small and needs to be enhanced with additional classes. In fact the CLDC does not define any APIs to manage the user interface and does not implement any I/O models. Moreover, it does not define how applications are managed during their life cycle. For this reason, the CLDC is usually coupled with a profile such as the MIDP.

Mobile Information Device Profile

To fill in the missing functionality of the CLDC and to support specific features of the device, a profile is added on the top of the configuration. The MIDP is a collection of APIs and libraries aimed to offer some more functionalities required by mobile applications, such as user interface, networking, and a persistent data storage support. Currently there are two versions of MIDP implemented on real devices:

MIDP 1.0 is the first version of the profile and it includes basic user interface and network security support.

MIDP 2.0 is a backward compatible version of the MIDP 1.0, which adds new features: an enhanced user interface, multimedia, game, and connectivity functionalities.

The 3rd generation of the MIDP is currently being specified as JSR271 [15] (in Section 3.2.5 we explain the Java standardization process). This specification is based on the MIDP 2.0, and provides backward compatibility with it so that MIDlets written for MIDP 2.0 can execute in MIDP 3.0[4] environments. An application developed for the MIDP profile is referred as a MIDlet. It does not use a static main method as an entry point, like traditional desktop applications. Its entry point is a class that extends the `javax.microedition.midlet.MIDlet` class. The MIDlet class defines abstract methods that are called when the state of the MIDlet is changing. For instance, whenever the MIDlet starts up, its startApp method is called. In Listing 3.1 you can see an example of the source code of a MIDlet. One or more MIDlets, together with all the other files and classes needed by the application, are packaged into a so-called MIDlet suite. It consists of a standard JAR file and a JAD file. Later in this chapter we will see that the JAR file is actually the one needed to be sent to the phone for installing the application.

3.2.2 More Capable Devices: Targeting CDC

Here we present the configuration and profiles which can be supported by the *more capable devices*.

Connected Device Configuration

The CDC is similar to CLDC (see Section 3.2.1), but more powerful. It has full support for the Java language and the JVM specification. In addition it

---

[4] Since it is not finalized yet, we will not discuss MIDP 3.0 further in this book but the specifications will be available on: www.jpc.org

includes more APIs from J2SE than CLDC. In fact, CDC is fully compatible with CLDC with regards to APIs supported. Most compatibility issues arise at the profile-level, because of different user interface APIs supported. If cross-platform support is important for your application then make sure that you design the presentation layer different from control, communication, and/or data access layers. With only the user interface different it becomes much easier to achieve cross-platform support, but testing is of course a key to success.

Foundation Profile

The FP [13], together with the CDC, forms a Java application environment for resource-constrained devices. In particular the Foundation Profile adds APIs for communication, security, compression, and utility functionality such as timers and generic event handling. Notice that the Foundation Profile *does not* include any APIs for creating graphical user interfaces.

Personal Basic Profile

In the PBP [12] APIs for basic graphic support are added along with support for JavaBean and the Xlet programming model[5]. The graphics support of PBP is derived from the J2SE 1.4 API specification and therefore applications written for PBP v1.0 can run on J2SE 1.4 as well.

Personal Profile

PP [11] is, like PBP, layered on top of the Foundation Profile. The difference between the PBP and PP is that in the Personal Profile more heavyweight widgets are supported in addition to execution of Java web applets. The first version of PP [5] provided an upgrade path for applications created for the Personal Java Application Environment Specification [18] to migrate to J2ME. In addition, just as for the Personal Basic Profile, applications using the java.* APIs in the Personal Profile will run on J2SE.

3.2.3 Networking Support

Mobile phones offer a great connectivity allowing users to be in contact with the outside world from anywhere at anytime. This is not limited to voice calls or simple text messages (e.g., SMS) anymore, but is extended to wireless connectivity as well. J2ME provides networking interfaces for wireless application development to take advantage of the connectivity that these devices offer. J2ME has to support a variety of mobile devices with different networking

---

[5] Xlet comes from JavaTV to support applications running on digital TV receivers

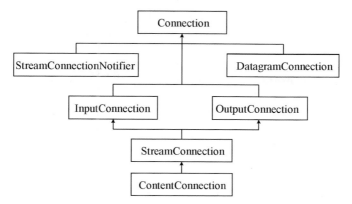

Figure 3.4. GCF.

capabilities and I/O requirements, and this task is fulfilled by the GCF. The main concept of the GCF is to define a network architecture as general as possible, but still leave full access to protocol-specific features. GCF provides a single set of abstractions that the programmer can use to handle multiple protocols, instead of having different abstractions for different types of communication. Both CLDC and CDC uses GCF, although there are some differences in which protocols they support. In Figure 3.4 the framework, or interface hierarchy, is shown.

On top of the GCF configurations, profiles and optional packages add support for specific protocols. Some examples of connection types and which specification they appear in, are listed below.

- HttpConnection (MIDP, FP)
- UDPDatagramConnection (MIDP, FP)
- SipConnection (Optional Package (JSR180))

The entry point to the GCF for applications is the Connector class, found in the `javax.microedition.io` package, which creates Connection-objects (i.e., objects implementing the Connection interface or one of its derivatives). It has several methods for opening connections. Common for all of them is that they include a string argument, *name* in the specifications, which specifies the connection type. The format of this string follows the URI syntax defined in RFC2396 [1]:

```
Connector.open(''<protocol>:<address>;<parameters>'');
```

Protocol: The protocol to be used for this connection. In the underlying implementation of GCF this protocol will be used to find the correct Connection-class to be used (called *late binding*).

Address: Network address of the endpoint for this connection. Connection-classes which support server functionality let you specify only the port the connection should be bound to, for example *socket://:80* to create a server socket connection bound to port 80 at the local host.

Parameters: Optionally the connection allows you to specify extra parameters to be used. For example the SipConnection allows you to add the transport parameter to specify the underlying transport mechanism. Parameters are specified as a list of one or more *name = value* pairs which are prefixed by a semicolon.

An example of opening a connection is shown below:

```
SipConnection sc = (SipConnection)Connector.open
  (''sip:+47-555-1234567;postd = pp22@foo.com;user=phone'');
```

Here the protocol part is *sip:*, the address part is *+47-555-1234567* and two parameters are specified with values, namely *postd* and *user*. See also the example in Section 3.4.3 where we show how to use GCF to download a document from any URL.

3.2.4 Optional Packages

As explained earlier optional packages add additional APIs to the platform, or they may be included as a part of a configuration and/or profile. Here we provide a quick overview for some optional packages frequently used.

JSR075 – PDA Optional Packages for the J2ME Platform [6]
This JSR specifies two independent optional packages: FileConnection API and PIM APIs. The FileConnection API (in the package `javax. microedition.io.file`) gives access to the device's file system by extending the GCF (see Section 3.2.3).
The PIM API (in the package `javax.microedition.pim`) allows applications to access personal information stored on the device, such as contact list, event list, and todo list.

JSR082 – Java APIs for Bluetooth Wireless Technology [7]
Defines APIs to use Bluetooth on the device. The APIs extend the GCF with several protocol schemes, such as *btspp:* (Serial Port Profile), *btgoep:* (Generic Object Exchange Profile), and *btl2cap:* (Logical Link Control and Adaption Protocol). Device and service discovery, service registration, device management, and communication is supported.

JSR120 – WMA [8] Based on the GCF this set of APIs defines support for using wireless messaging protocols, such as SMS. The specification is generic and can be extended by device manufacturers to support the protocols a device supports. Such extensions are called *adapter specifications* and JSR120 includes three such adapter specifications: GSM SMS Adapter, GSM Cell Broadcast Adapter, and CDMA IS-637 SMS Adapter.

JSR135 – MMAPI for J2ME [9] This optional package makes it possible to use a device's multimedia capabilities, such as playing and recording of audio and video and taking pictures. The API is flexible and the features an application can use depends of course on the capabilities of the device running the application. Therefore it is very important that an application first checks the multimedia capabilities (found in system properties) before using these APIs. In addition to being flexible it is also extensible, e.g., it is possible to support protocols for streaming of audio/video not included on the device. See Section 3.4.2 for an example which shows how MMAPI is used. We recommend the book *Pro Java ME MMAPI: Mobile Media API for Java Micro Edition* [4] for further information about MMAPI.

3.2.5 Java Technology Standardization

All the Java technology which will be discussed here has been specified through the JCP. Here a short introduction to what this process is and what it means for you as a J2ME developer. It was introduced in 1998 as an open, participating process to develop and revise Java technology. The whole process is overseen by the EC, which consists of 16 JCP Members and a nonvoting Chair. There are two ECs: one with responsibility for the J2SE and J2EE technology specifications and another which is responsible for J2ME technology. It is the EC's responsibility to select JSRs to be developed and approve when specifications are ready for public review, release, etc. The outcome of the process usually consists of three artifacts:

Specification: This is the API defined and is the most important outcome of the process. For an end-user it says how to use this technology, e.g., what classes to use to support Bluetooth in an application.
On the other hand the API must be implemented by someone. For example, the provider of the Java environment on a mobile phone must follow the API in this specification to provide Bluetooth support.

TCK: Provides tests which an implementation of this specification must pass.

RI: This is an implementation of the specification to let users start working with it. When using a RI you should be aware that they are not optimized for performance and any other quality, except that they have passed the TCK for this specific JSR. But it is sufficient to start learning the new technology and perhaps use it for prototyping.

To start the process JCP member(s) submit a JSR, which proposes either development of a new specification or a significant revision of an existing specification. Development of a specification is divided up into four phases: initiation, early draft, release, and maintenance. Depending on the particular specification, drafts of it might be available for reviewing by the public before it is released. A quick overview of these four phases:

Initiation Phase: The JSR is submitted to the JCP by JCP community member(s) and may be approved for development by the responsible EC. But it will take some time before this JSR is supported by devices.

Draft Phase: The EG has completed the first draft of the specification for review, these may not be available for non-JCP members.

Public Review: Based on feedback gathered from the early draft the specification is updated and re-released. This draft should be publicly available.

Proposed Final Draft: The final draft awaits completion of its associated RI and TCK. This draft may be changed/updated based on the experience gathered through development of the RI and/or TCK.

Final Release: The EC has voted to approve the specification and it is released as final. Depending on the nature of the specification it can take some time before it will be available on mobile devices. If you have developed an application which depends on this specification you need to consider your deployment options.

All Java technology discussed earlier in this chapter, including configurations and profiles, has passed through the JCP. In the remainder of this section we will look at a special kind of JSRs for J2ME technology which are called *Roadmap Specifications*.

Roadmap Specifications

In addition to the JSR s which specify a particular technology there are also a few specifications which provides a roadmap for technologies to be supported on devices. As written earlier about optional packages (see Section 3.2.4 on page 70) they make it harder to deploy the application because the application only runs on devices with a specific optional package available. In particular this makes it hard for customers to know if their device supports an application or not. In addition the specifications leaves some properties open to the implementor to decide. By mandating values to these properties (e.g., which character encodings are available, and protocols which must be supported) it becomes more predictable to develop applications for devices compliant to a roadmap specification.

Java Technology for the Wireless Industry

The JTWI [10] was the first roadmap specification available for J2ME. Figure 3.5 shows the JSRs which JTWI includes. In addition to just specifying that these technologies must be included in JTWI-compliant platforms, the JTWI provides clarifications and mandates requirements to properties of these specifications as well. Such requirements are for example minimum number of running threads and timers, character encodings, time zone names, messaging adapters (for WMA), and for MMAPI protocol and playback types supported.

Mobile Service Architecture

After many years of JTWI being the roadmap for J2ME mobile phones, the MSA [14] was finalized in December 2006 as an evolution to JTWI. It adds many more optional packages, as shown in Figure 3.6, as well as more

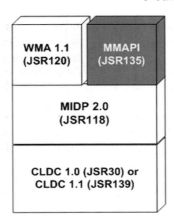

Figure 3.5. Technologies included in JTWI. MMAPI is optional.

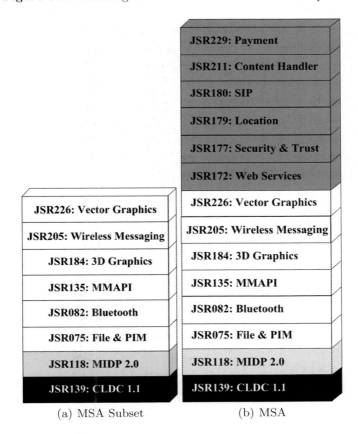

(a) MSA Subset   (b) MSA

Figure 3.6. MSA platform types.

clarifications and requirements. It contains two platform definitions: MSA and MSA Subset. In addition to the MSA and MSA Subset there is also the MSA Advanced specification (JSR249), but that specification is *not* currently final (current JCP stage is "Expert Group Formation"). Figure 3.6 shows the technologies which MSA consists of. What is missing in that figure is that CDC can be used as the configuration as well. However, the specification is short on details about this option. As can be seen the MSA provides two rich platforms to develop mobile applications on, and it will be interesting to see when the first device supporting MSA is released. Before any MSA-compliant hardware is released the Sun Java Wireless Toolkit 2.5 [17] supports MSA and can be used for development for either the whole platform or single technologies included in it.

3.3 Development Environment

For J2ME development you need a Java compiler (see Section 3.3.1), a J2ME runtime and an editor to write code in. This is called the development environment and is explained here starting with the lowest layer and moving upward the tool chain. However, the programming productivity can be greatly increased using one of the several Java development tools present on the market today. They offer an IDE for easy editing, compiling, building, and debugging Java programs. The most popular IDEs and tools for J2ME are: Eclipse [3], Carbide.j [22], and NetBeans [20]. They provide an environment to test and debug applications on a phone emulator before installing it on a real device. The development environment explained in the following sections will be based on this:

- Java SE Development Kit (JDK) 5.0_06 or later [16]
- NetBeans IDE [20]
- NetBeans Mobility Pack [21] for CLDC and/or CDC

The installation of the files must follow the order in the above list. Notice that there is an option to download NetBeans IDE bundled with the JDK if you do not have that already. If a JDK is already installed on your system then the NetBeans IDE installer will find it during installation, or allow you to provide the path for it. The screenshots in the following sections have been taken from Microsoft Windows XP SP2 using NetBeans IDE 5.5.

3.3.1 Java Development Kit

The compiler and the interpreter offered by JDK can be used to compile Java applications for both CLDC and CDC. However, it must produce binaries (i.e., .class files) which are binary compatible with version 1.2 of the JVM specification. The compiler is a command line tool which is named *javac*.

To adhere to the binary compatibility requirements the *-source* and *-target* arguments are used, with the value 1.2 for both. You can use the same code as shown in Listing 3.1 for CLDC or Listing 3.2 for CDC. For example:

```
javac −source 1.2 −target 1.2 HelloWorld.java
```

The result should be a *HelloWorld.class* file which is in the application. Next this must be packaged into a JAR file. This is done by using the *jar.exe* tool, like this:

```
jar cf HelloWorld.jar HelloWorld.class
```

Here the *c* option specifies that we want to create a new archive and the *f* option that the filename is specified (HelloWorld.jar). At the end of the command, the name of the file(s) to be added to the archive comes. The result should be a HelloWorld.jar. Use *jar –help* to see all options for the jar tool. The .jar file should now be deployed to the phone where you want to run it. This is explained in Section 3.3.3.

3.3.2 NetBeans IDE

NetBeans is an IDE for developing Java server and client applications. Assuming that the installation of the software was successfull, when NetBeans is started the main window appears as shown in the Figure 3.7.

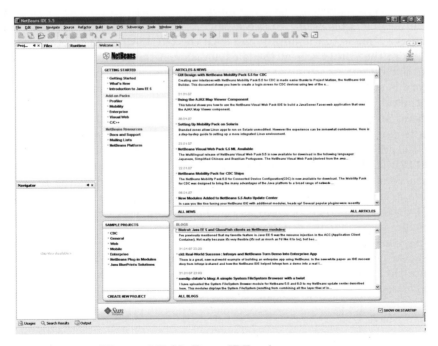

Figure 3.7. NetBeans IDE welcome screen.

NetBeans Mobility Pack

To extend the functionality of NetBeans for J2ME applications development you must install the plug-in called NetBeans Mobility Pack. It comes in two different flavors:

- NetBeans Mobility Pack for CLDC
- NetBeans Mobility Pack for CDC

A mobile phone emulator is integrated in the Mobility Pack, which has the purpose to test the applications on the PC. Device manufacturers provide their own mobility extension platforms in order to have a more robust testing environment for a specific device. These must be installed separately after the NetBeans Mobility Pack has been installed. You will find links to all supported device manufacturers' SDKs at NetBeans website [19].

Adding a New Library to NetBeans

Most libraries come as part of the platform you are using, which is specified in the *Java Platform Manager*. Additional libraries can be added through the *Library Manager* which you find on the Tools menu. This makes it easier to use libraries because the NetBeans IDE will use this information to provide you documentation support (JavaDoc), to see the source code, etc. In addition NetBeans will ensure that the library is available on the classpath when executing the application. To add a new library click the *New Library...* button. Give a name to the library and click the *OK* button. The library should appear in the library tree at the left-hand side of the Library Manager dialog box. But do not exit from this dialog yet, an important information for the library is still missing! Library information consist of three items: classpath information, source information, and javadoc information. Add the relevant archives and/or folders to these lists and then NetBeans has enough information to use the library later. We will now go through the steps necessary to add the JSR172 XML Parsing API. In the *New Library* dialog enter the name **JSR172-XML Parsing**. For the classpath you can use the *j2me − ws.jar* which comes with the Sun Java Wireless Toolkit (located in the lib folder where it is installed). The Javadoc for JSR172 can be downloaded from the JCP-website, and then be used as the Javadoc for this library. The *Library Manager* dialog should now look similar to Figure 3.8.

As you see, it is trivial to add a new library to the Library Manager and doing so gives you some nice benefits. The hardest part is to acquire and locate the necessary files. Note, NetBeans does not allow you to modify the *bootclasspath* from the Library Manager. This limits the kinds of libraries which can be added.

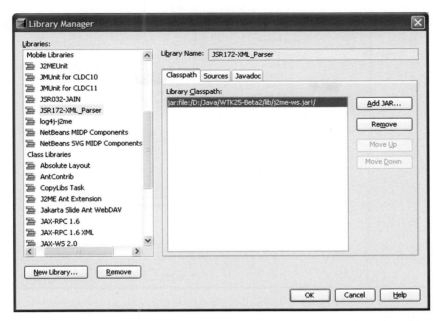

Figure 3.8. Library Manager dialog with JSR172 XML Parsing API library selected.

3.3.3 Developing MIDP Applications for CLDC

In the following section we will go through all the steps for creating a "Hello World" MIDP application for CLDC.

Creating a New Project

A project contains information about programs files and IDE environment settings. To create a new project click on File->New Project... (Ctrl+Shift+N). Then a window like the one in the Figure 3.9 appears.

After choosing Mobile Application, click on "Next" button. The window shown in the Figure 3.10 pops up and information about the project name and its location on the file system are required to be inserted:

Name and location: Enter a name and location for the project as of your choice. In later sections, we will refer to this application as *HelloWorld*.

Main Project checkbox: Checking the "Main Project" checkbox will make it easier to build, run, and debug this project because Netbeans provides several hotkeys for the Main Project. The "Main Project" is displayed with boldface in the project-listing.

By clicking on the "Next" button, another window shows up where the Emulator Platform can be chosen if other third-party emulators are installed.

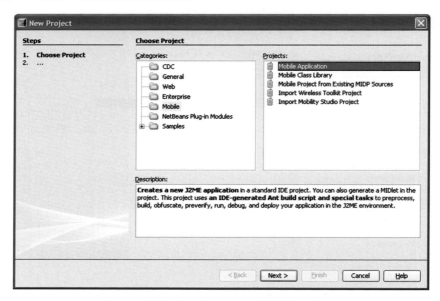

Figure 3.9. Creating a new MIDP project.

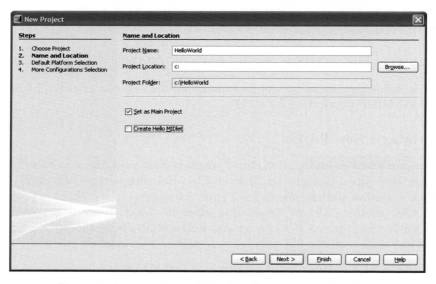

Figure 3.10. Naming and locating the new MIDP project.

Otherwise the default J2ME Wireless toolkit 2.2 is used. Moreover the device configuration and profile can be set to better emulate the target device. By pressing the "Finish" button, the new project creation is done. Next step is to add a MIDlet to the project.

Add Code to the Project

The MIDlet class is the core of the application. The steps for creating a new MIDlet class are shown below. By right clicking on the "<default package>"->New->MIDlet... in the Project tab, a new window appears, as the one shown in the Figure 3.11. After you have inserted all the information about the class name and you have clicked on "Finish" button, the main window of NetBeans displays the skeleton of the HelloWorld MIDlet as shown in the Figure 3.12.

In this example the code listed in Listing 3.1 will be used.

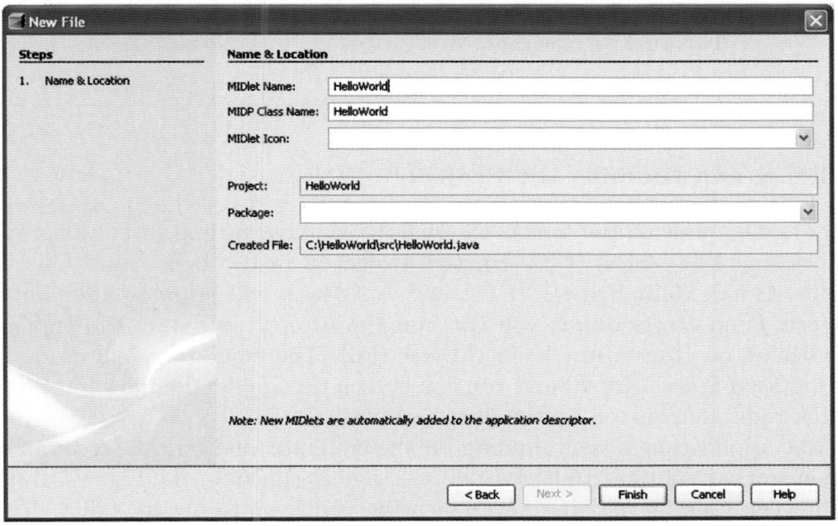

Figure 3.11. Naming and locating the new MIDlet.

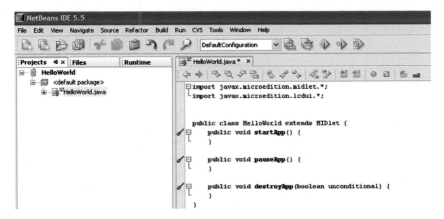

Figure 3.12. An example of the skeleton of a MIDlet.

Listing 3.1. Hello World Example

```
   import javax.microedition.lcdui.*;
   import javax.microedition.midlet.*;

4  public class HelloWorld extends MIDlet {
       Display display;
       Form form;
       public void destroyApp(boolean unconditional) {
           notifyDestroyed();
9      }
       public void pauseApp() {}
       public void startApp() {
           display = Display.getDisplay(this);
           form = new Form("Helloworld_example");
14         form.append("Hello_World!");
           display.setCurrent(form);
       }
   }
```

Building and Running the Project

After having inserted the code in the HelloWorld.java file and saved the project by pressing File->Save (Ctrl+S), the project is ready to be built. Click on Build->Build Main Project (F11) and NetBeans will start to compile the project. If no errors occur, you can run the project and start the emulator by clicking on Run->Run Main Project (F6). The emulator pops up on the screen (see Figure 3.13(a)) and you can launch the application now by clicking on the right soft button on the phone emulator.

The application is now running on the emulator and on the screen of the phone you can see the "Hello World" message as shown in the Figure 3.13(b). Before you start developing an application you have to decide which device your application is target for. It will help you to choose the proper emulator to use. For this purpose NetBeans can integrate other third-party emulators which look like real devices on the market. In fact the screen size and graphical interface of the default Wireless Toolkit emulator do not look like any specific mobile phone.

MIDlet on the Phone

When the project is built, NetBeans creates the MIDlet Suite for the project. In the *HelloWorld/dist* folder, two new files are created: HelloWorld.jad and Helloworld.jar as shown in the Figure 3.14. The latter one is the archive that contains all the files needed by the application to work. Therefore, for installing the MIDlet on the phone, this file needs to be transferred to it. In some cases: if the MIDlet is signed or it has additional attributes, the .jad file needs to be sent to the phone as well. There are two ways of sending the application on a real device:

(a) The names of available MI-Dlets are listed, but the project is not running; pressing the "Launch" key, the application starts.

(b) HelloWorld example running.

Figure 3.13. Emulator screenshot.

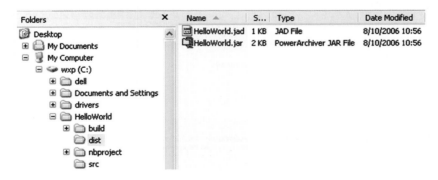

Figure 3.14. Location of the Jar and JAD files in the Windows file system.

- Using the software that the vendor of the handset provides, such as Nokia PC Suite [23] for Nokia phones or Sony Ericsson PC Suite for Smartphones [2].
- Using Infrared or Bluetooth technology.

As soon as the .jar file is sent to the phone, the installation process starts by opening the received file.

3.3.4 Developing CDC Applications

In the following section we will go through all the steps for creating a "Hello World" CDC application.

Creating a New Project

Now that every component of the environment has been correctly set up it is time to create a sample CDC application. Click on File->New Project... (Ctrl+Shift+N) in Netbeans. Then a window like the one in the Figure 3.15 appears. On the left-hand side select the "CDC" folder. Then select "CDC Application" in the list at the right-hand side. Now click on the "Next" button to proceed to the next step of the wizard. The window should look similar to Figure 3.16.

Click "Next" after selecting the appropriate platform for your project. In the following we will assume that the default "Sun Java Toolkit 1.0 for Connected Device Configuration" platform with the "DefaultColorPhone" device is used. A window like the one in the Figure 3.17 should now appear where you enter information about this application. The fields: *Name and location* and *Main Project checkbox* have been described in Section 3.3.3; the other ones are explained below.

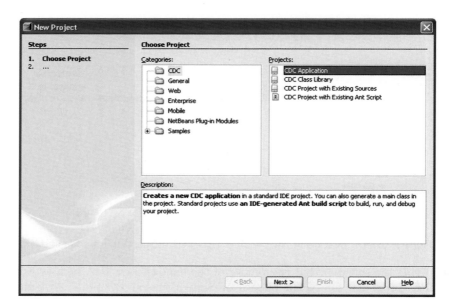

Figure 3.15. New CDC project.

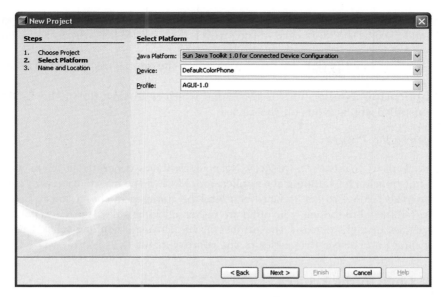

Figure 3.16. New project - platform selection.

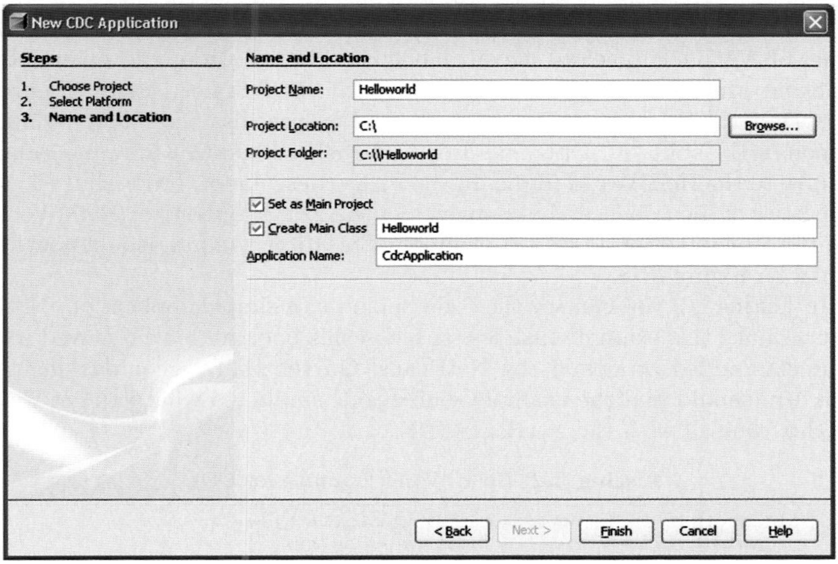

Figure 3.17. New CDC application.

Note, depending on the platform selected this dialog may ask you for other information than what we describe below. This information is specific for the platform you are working with.

Main Class: Enter the name of a class, including the package, to be created and used as the Main class of the project. This is the class where execution will start at.

For this walkthrough use "HelloWorld" as the main class (no package) as shown in Figure 3.17.

Application Name: This is the name which will show up in the list of installed applications on the device.

Generate the Project

Click "Finish" to create the project. Netbeans will now generate the Main class and information for building the application. Most of the information provided during the "New Project"-wizard can later be changed in the "Project Properties" dialog box which you open by either clicking File->"project name" Properties or right clicking the project in the Projects window (click Ctrl-1 to display) and select Properties in the context menu.

Add Code to the Project

Now we are ready to add some functionality to the project. The Main class generated by Netbeans is an AWT Frame (i.e., it extends the java.awt.Frame class). This means that it is a top-level window. An AWT Frame is a container of AWT components such as label, checkbox, button, and many more. Netbeans provides two different views for such AWT Frame classes: "Source" and "Design". Select the "Design" view and ensure that the Palette window is open (ctrl +shift+8). Drag and drop a "Label" component from the Palette window to the HelloWorld frame. In the Properties window (ctrl+shift+7) the properties of the Label can be changed. Set the *Text* property to "HelloWorld" and the *Alignment* property to "Center". The editor window should now look similar to Figure 3.18.

In Listing 3.2 you can see the code behind this simple application. Notice that to make this example take less space in this book we have removed some comments added by default by NetBeans. On the DVD accompanying this book you should find the example source code similar to what you get if you try this yourself with the NetBeans IDE.

Listing 3.2. Hello World Example for CDC

```
public class HelloWorld extends java.awt.Frame {
    public HelloWorld() { initComponents(); }

    private void initComponents() {
        label1 = new java.awt.Label();

        addWindowListener(new java.awt.event.WindowAdapter() {
            public void windowClosing(
                java.awt.event.WindowEvent evt)
            {
                exitForm(evt);
            }
```

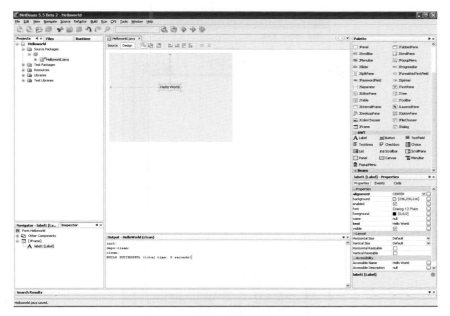

Figure 3.18. HelloWorldCDC with label.

```
13    });

      label1.setAlignment(java.awt.Label.CENTER);
      label1.setText("HelloWorld");
      add(label1, java.awt.BorderLayout.CENTER);
18
      pack();
   }

   private void exitForm(java.awt.event.WindowEvent evt) {
23       System.exit(0);
   }

   public static void main(String args[]) {
      java.awt.EventQueue.invokeLater(new Runnable() {
28       public void run() { new HelloWorld().setVisible(true);
         }
      });
   }

   private java.awt.Label label1;
33 }
```

Building and Running the Project

Now we can finally run the application and see the results. If the project is set
as the *main project* click *Run->Run Main Project* or hit the F6 key. Else right
click the project in the *Projects window* and select *Run Project*. First Netbeans

Figure 3.19. Emulator running HelloWorld in CDC.

will build the project and if everything goes well then it will start the emulator associated with the project. The result should look similar to Figure 3.19.

The application just displays the *HelloWorld* label added above. You can terminate the emulator from Netbeans by clicking *Build->Stop Build/Run*. Another option is to terminate the emulator application through the operating system, for example, by pressing *Alt+F4* on Microsoft Windows.

3.4 Application Examples

In the following sections, some examples of J2ME application are shown. However, they include just the basic code to fulfill some tasks (such as sending an SMS, playing an audio, file, downloading, and reading the content of a file), but they do not have any GUI or exception handlers. Their purpose is to give the reader an entry point to get started with J2ME applications development. The code for these and other unpublished examples can be found on the DVD.

3.4.1 Wireless Messaging API: Sending SMS

The Messaging API is based on GCF and designed to handle both binary and text messages. In the following example a simple application to send a text message is presented. It consists of one file containing the MIDlet responsible for the data sending (see Listing 3.3).

Listing 3.3. SMS Send MIDLet

```
   import javax.microedition.midlet.*;
 2 import javax.microedition.io.*;
   import javax.wireless.messaging.*;

   public class SMS extends MIDlet
   {
 7   String text="SMS_exampple_test";
     String address="sms://+123456789";
     MessageConnection clientConn;
     public void startApp()
     {
12     try{
            clientConn =
               (MessageConnection)Connector.open(address);
            TextMessage Tmsg = (TextMessage)clientConn.newMessage
               (MessageConnection.TEXT_MESSAGE);
17          Tmsg.setPayloadText(text);
            clientConn.send(Tmsg);
         }
       catch (Exception EX){
            System.out.println("Error:"+EX);
22       }
       }
     public void pauseApp() {}
     public void destroyApp(boolean unconditional) {}
   }
```

The String in line 8 specifies the string passed to the Connector created in lines 13–14. The URL has to have the following format:

``sms://``+" address ``+":"+`` port"

where *address* is the phone number of the recipient (which can include the international code as well) and *port* is the port number on the recipient phone. Examples of valid URL connection strings are:

- "sms://+123456789" : to send a message to the specified phone number
- "sms://+123456789:5000" : to send a message to a port (5000 in this case) on the specified phone number
- "sms//:5000" : to send a message on the local device on the specified port

The Listing 3.4 shows the code needed to handle the receiving of an SMS on a specific port.

Listing 3.4. SMS Receiver Code

```
   try
   {
       String address ="sms://:5000";
4      MessageConnection conn=(MessageConnection)Connector.open(
           address);
       Message Msg = null;
       while (true)
       {
           Msg = conn.receive();
9          if (Msg instanceof TextMessage)
           {
               TextMessage Tmsg = (TextMessage)Msg;
               String ReceivedText = Tmsg.getPayloadText();
               System.out.println("Received:_"+ReceivedText);
14         }
       }
   }
   catch (Exception EX)
   {
19     System.out.println("Error:_"+EX);
   }
```

The line 8 is the key of the code:

```
Msg = conn.receive();
```

it makes the application listen on the port specified in the string *address* at line 3. It is done by calling the method *receive()* on the Connector (*conn*) created at line 4.
To retrieve the text from the message the method *getPayloadText()* is called on the TextMessage object (line 12):

```
String ReceivedText = Tmsg.getPayloadText();
```

When developing an application for both sending and receiving SMS, it is better to have two different threads for the sending and the receiving parts. More information can be found in the WMA documentation [8].

3.4.2 Mobile Multimedia API: Audio Player

The code in the Listing 3.5, is an example of the MMAPI usage. It shows how to create a simple audio player using a wav file. Nevertheless, mobile phones can support other audio file formats, such as:

- Wave
- AU
- MP3
- MIDI
- Tone

Listing 3.5. Audio Player

```
import java.io.InputStream;
import javax.microedition.midlet.*;
import javax.microedition.lcdui.*;
import javax.microedition.media.*;
import java.io.IOException;

public class AudioPlayer extends MIDlet {
    public void startApp() {
        Player p;
        InputStream resource;
        try {
            resource=getClass().getResourceAsStream("/files/test.
                wav");
            p = Manager.createPlayer(resource,"audio/x-wav");
            p.start();
        }
        catch(IOException ioe) {
            System.out.println("error: "+ioe);
        }
        catch(MediaException EX) {
            System.out.println("error: "+EX);
        }
    }
    public void pauseApp() {
    }
    public void destroyApp(boolean unconditional) {
    }
}
```

In line 12 the path of the audio file is specified to create an InputStream:

```
resource = getClass().getResourceAsStream("/files/test.wav");
```

In the line 13, the player is created:

```
p = Manager.createPlayer(resource,''audio/x-wav");
```

The first parameter represents the InputStream related to the audio file; the second one represents the type of audio format;
If you want to play a different audio file you will need to change the file name in line 12 and the second parameter in line 13 (if the file has a different format). Possible choices for the second parameter are:

- *Wave audio files*: "audio/x-wav"
- *AU audio files*: "audio/basic"
- *MP3 audio files*: "audio/mpeg"
- *MIDI files*: "audio/midi"
- *Tone sequences:*: "audio/x-tone-seq"

More features can be implemented to extend the functionality of this simple player using the methods offered by the MMAPI. More info can be found in the MMAPI documentation [9].

3.4.3 Communication: Downloading Documents with GCF

Here we will show you how to use the GCF to download information using the HTTP (see Listing 3.6). In the next example, Section 3.4.4, the download method is reused to parse RSS documents downloaded (see Listing 3.7).

Listing 3.6. GCF Connection example.

```java
import java.io.*;
import javax.microedition.io.Connector;
public final class GCFDownload{
   public static void main(String[] args){
      InputStream is = null;
      try {
         is = getDocument(args[0]);
         int c;
         while((c = is.read()) != -1){
            System.out.print((char)c);
         }
      } catch (IOException e) {
         System.err.println("Error occured while receiving document")
            ;
         e.printStackTrace();
      } finally {
         if (is != null)
            try{
               is.close();
            } catch (IOException e) { e.printStackTrace(); }
      }
   }
   public static InputStream getDocument(String url)throws
         IOException{
      return Connector.openInputStream(url);
   }
}
```

3.4.4 XML Parsing: Parsing RSS

RSS [24] is a content syndication format encoded in XML. Here we will show how to parse one of these and display the contents using the J2ME XML Parsing API specified in JSR172, and is a subset of the SAX API. One example RSS document is the feed for the Agder Mobility Lab weblog (http://agdermobilitylab.blogspot.com/):

http://agdermobilitylab.blogspot.com/feeds/posts/default?alt=rss

Running this example on CDC the main method will be invoked, and the URL to be downloaded must be provided as the first argument. On CLDC the classes can be used from a MIDlet, which is not shown here. First a SAX-ParserFactory is created, which is then used to create a SAXParser. Using the parse method of the SAXParser we provide it with the InputStream acquired from GCF 's Connector. Since the XML parser gets a plain InputStream object instance this means that it will not notice anything about the communication protocol (e.g., HTTP) at all, due to the GCF. The XML parsing logic is handled by the subclass RSSHandler, which overrides some of the methods of the base class, DefaultHandler. Obviously more intelligent parsing can be done here; this is a very trivial example!

Listing 3.7. RSSViewer.java

```java
import java.io.*;
import javax.microedition.io.Connector;
import javax.xml.parsers.*;
import org.xml.sax.helpers.DefaultHandler;
import org.xml.sax.*;
public final class RSSViewer {
  public static void main(String[] args) { new RSSViewer(args[0]);
        }
  public RSSViewer(String rssUrl) {
    try {
      SAXParserFactory pFactory = SAXParserFactory.newInstance();
      SAXParser parser = pFactory.newSAXParser();
      parser.parse(getDocument(rssUrl), new RSSHandler());
    } catch (ParserConfigurationException e) {
      System.err.println("Error initiating XML parser");
      e.printStackTrace();
    } catch (SAXException e) {
      System.err.println("Parsing error");
      e.printStackTrace();
    } catch (IOException e) {
      System.err.println("Error occured while receiving document")
          ;
      e.printStackTrace();
    } finally {}
  }
  public static InputStream getDocument(String url)throws
      IOException{
    return Connector.openInputStream(url);
  }
  private final class RSSHandler extends DefaultHandler {
    boolean isChannel = false, isTitle = false, isItem = false,
            isDesc = false, isLink = false;
    public void startElement(
      String uri, String localName, String qName, Attributes attr)
      {
        if (qName.equals("channel")) { isChannel = true; }
        else if (qName.equals("title")) { isTitle = true; }
        else if (qName.equals("item")) { isItem = true; }
        else if (qName.equals("description")) { isDesc = true; }
        else if (qName.equals("link")) { isLink = true; }
    }
    public void characters(char[] ch, int start, int length){
      String text = new String(ch, start, length);

      if (isItem && isTitle){System.out.println("\r\n\tTitle: "+
          text);}
      else if(isItem && isDesc){System.out.println("\t" + text);}
      else if(isItem && isLink){System.out.println("\tLink: "+text
          );}
      else if(isChannel && isTitle){System.out.println(text);}
      else if(isChannel && isLink){System.out.println("Link: "+
          text);}
      else if(isChannel && isDesc){System.out.println(text+"\r\n")
          ;}
    }
    public void endElement(String uri,String localName,String
        qName){
      if (qName.equals("channel")) { isChannel = false; }
      else if (qName.equals("title")) { isTitle = false; }
      else if (qName.equals("item")) { isItem = false; }
      else if (qName.equals("description")) { isDesc = false; }
      else if (qName.equals("link")) { isLink = false; }
    }
  }
}
```

References

1. T. Berners-Lee, R. Fielding, and L. Masinter. Uniform Resource Identifiers (URI): Generic Syntax. RFC 2396, IETF, August 1998.
2. Sony Ericsson. Sony ericsson pc suite. Available at http://www.sonyericsson.com/.
3. Eclipse Foundation. Eclipse ide. Available at http://www.eclipse.org/downloads/.
4. Vikram Goyal. *Pro Java ME MMAPI: Mobile Media API for Java Micro Edition.* Apress, May 2006.
5. JSR 062 Expert Group. Personal Profile 1.0. Java Specification Request 062, Java Community Process, March 2006. Available at http://jcp.org/en/jsr/detail?id=062.
6. JSR 075 Expert Group. PDA Optional Packages for the J2ME Platform. Java Specification Request 075, Java Community Process, June 2004. Available at http://jcp.org/en/jsr/detail?id=075.
7. JSR 082 Expert Group. Java APIs for Bluetooth Wireless Technology. Java Specification Request 082, Java Community Process, June 2006. Available at http://jcp.org/en/jsr/detail?id=082.
8. JSR 120 Expert Group. Wireless Messaging API. Java Specification Request 120, Java Community Process, April 2003. Available at http://jcp.org/en/jsr/detail?id=120.
9. JSR 135 Expert Group. Mobile Media API. Java Specification Request 135, Java Community Process, June 2006. Available at http://jcp.org/en/jsr/detail?id=135.
10. JSR 185 Expert Group. Java Technology for the Wireless industry. Java Specification Request 185, Java Community Process, January 2006. Available at http://jcp.org/en/jsr/detail?id=185.
11. JSR 216 Expert Group. Personal Profile 1.1. Java Specification Request 216, Java Community Process, August 2006. Available at http://jcp.org/en/jsr/detail?id=216.
12. JSR 217 Expert Group. Personal Basis Profile 1.1. Java Specification Request 217, Java Community Process, August 2006. Available at http://jcp.org/en/jsr/detail?id=217.
13. JSR 219 Expert Group. Foundation Profile 1.1. Java Specification Request 219, Java Community Process, August 2004. Available at http://jcp.org/en/jsr/detail?id=219.
14. JSR 248 Expert Group. Mobile Service Architecture. Java Specification Request 248, Java Community Process, December 2006. Available at http://jcp.org/en/jsr/detail?id=248.
15. JSR 271 Expert Group. Mobile Information Device Profile 3.0. Java Specification Request 271, Java Community Process, December 2006. Available at http://jcp.org/en/jsr/detail?id=271.
16. Sun Microsystems. Java se development kit. Available at http://java.sun.com/javase/downloads/index.jsp.
17. Sun Microsystems. Sun java wireless toolkit 2.5. Available at http://java.sun.com/products/sjwtoolkit/.
18. Sun Microsystems. Personaljava application environment specification, January 1999. Available at http://java.sun.com/products/personaljava/spec-1-1-1/pjae-spec.pdf.

19. NetBeans. Netbeans emulators. Available at http://www.netbeans.org/kb/50/midpemulators.html.
20. NetBeans. Netbeans ide. Available at http://www.netbeans.info/downloads/index.php.
21. NetBeans. Netbeans mobility pack. Available at http://www.netbeans.org/products/mobility/.
22. Nokia. Carbide.j. Available at `http://forum.nokia.com/main/resources/tools\_and\_sdks/carbide/index.html`.
23. Nokia. Nokia pc suite. Available at http://www.nokia.com/pcsuite.
24. D. Winer. Rss 2.0 specification. Technical report, January 2005. Available at http://blogs.law.harvard.edu/tech/rss.

4

Symbian/C++

Morten V. Pedersen and Frank H.P. Fitzek

Aalborg University {mvpe|ff}@es.aau.dk

4.1 Introduction

The purpose of this chapter is to give the reader an overview of Symbian OS application development. These first sections will serve as an entry-point for new developers by giving an introduction to the development environment, tools, and Symbian C++ programming language. The final sections will go through a number of the networking capabilities supported by Symbian OS, accompanied by code examples. Several researchers claim that Symbian C++ is difficult to learn, and the particular steep learning curve is typically a show-stopper for many new developers, students, and other adopters. Often making it less suitable than, e.g., Java or Python for quick prototyping. However, this downside can in many cases be balanced by the fact that we through Symbian C++ obtain the full access to the devices capabilities and the speed advantage of native complied applications. Additionally Symbian OS has the advantage of a huge penetration of the smart phone market (55% in Q2 2006). Such a big penetration means that no serious mobile developer can disregard the Symbian platform. Throughout the years a number of different Symbian OS versions have been released and adopted by phone manufactures. In the following sections we will strive at being version-independent. However, in some cases where the information or code examples apply to only one particular OS version this will be noted. When writing applications be aware, that Symbian OS v9 introduced a complete binary break from the previous versions. This means that applications written previous to version 9 will need to be recompiled and partly rewritten to be able to run on version 9 phones and vice versa. In addition to the binary break, Symbian introduces an enhanced platform security model, which will not be covered in this chapter. Readers should refer to [2] for extensive information about Symbian OS platform security.

   The following section will introduce the reader to the tools comprising a Symbian OS development environment.

F.H.P. Fitzek and F. Reichert (eds.), Mobile Phone Programming and its Application to Wireless Networking, 95–138.
© 2007 *Springer.*

4.2 Setting up the Environment

In order to get started we need a couple of tools to be installed on our PC. These tools include an Integrated Development Environment (IDE) for building our applications and a suitable Software Development Kit (SDK). The SDK contains API documentation, example code, and a number of development tools, including an emulator for testing and debugging our application before actually deploying it on the target phone. The recommended IDE by Symbian and Nokia is currently the Eclipse–based Carbide.c++. The Carbide.c++ IDE is available in four different versions:

Carbide.c++ Express – Free version for noncommercial developers
Carbide.c++ Developer – Additional capabilities for commercial developers
Carbide.c++ Pro – For advanced commercial developers
Carbide.c++ OEM – For devices manufactures, etc.

We will give an overview of the free Carbide.c++ Express IDE, as it provides all the basic functionalities we need in order to get started (the current version at the time of writing the chapter is v1.1, which can be found on the DVD). Several other IDEs for Symbian development exist, these will however not be covered in this chapter. For running the Carbide.c++ Express edition, it is recommended that you have a fairly fast development PC. The following configuration is recommended: Windows XP (SP2) or Windows 2000 (SP4), 1800 MHz processor, 1024 MB RAM, and enough free hard drive space for the IDE and a SDK – typically around 650 MB. The choice of which SDK to install depends on the target device. As you may have noticed, the UI and input capabilities of Symbian OS-based devices can vary from phone to phone. Some phones provide pen-based input while others are designed for one-hand use with a numeric keypad. This flexibility means that we have to use the SDK matching the UI capabilities of our target device. Currently the majority of Symbian phones use the S60 UI from Nokia. S60 is available in a number of revisions; these are shown in the list below.

- S60 3rd Edition Feature Pack 1 – Symbian OS v9.2
- S60 3rd Edition – Symbian OS v9.1
- S60 2nd Edition Feature Pack 3 – Symbian OS v8.1
- S60 2nd Edition Feature Pack 2 – Symbian OS v8.0a
- S60 2nd Edition Feature Pack 1 – Symbian OS v7.0s enhanced
- S60 2nd Edition – Symbian OS v7.0s
- S60 1st Edition – Symbian OS v6.1

In addition to the S60 UI platform, other Symbian-based phones use the UIQ UI platform. UIQ is also available in a number of revisions:

- UIQ 2.0 – Symbian OS v7.0
- UIQ 2.1 – Symbian OS v7.0
- UIQ 3.0 – Symbian OS v9.1

Finally, a number of SDKs exist for phones not using the two above-mentioned UIs:

- Series 80 platform 2.0 – Symbian OS v7.0s
- Nokia 7710 SDK – Symbian OS v7.0

Depending on your target device you need to find the suitable SDK. Information about this can be found at `http://developer.symbian.com/main/tools/sdks/`. The page contains a list of the various SDKs and which phones they support. You can typically find this information also on the device manufactures' homepage.

4.2.1 Additional Installation Requirements

Besides installing the SDK and IDE some additional tools are either required or optional to make the development environment fully functional.

Perl (`www.activestate.com`) You need to install Perl as several of the build scripts in the SDK rely on this.

Java (`www.java.com`) In order to utilize the phone emulator fully, you also need a working installation of Java Runtime Environment.

Nokia PC Suite (`www.nokia.com`) A final tool that will come in handy is the PC Suite from Nokia. Installing the PC Suite will enable us to easily transfer our application to the actual phone.

As a general installation tip: Do not use directory paths with spaces for the SDK or Carbide.c++ workspace, as this has been known to cause problems when building projects.

4.2.2 Overview of the SDK

If installed into the default path, all SDKs share a common directory structure as shown in Figure 4.1, allowing several kits to be grouped under a common root using the following structure *Symbian**Symbian version*\ *UI version*\.

The directories shown contain useful information and tools needed for application development.

Epoc32 : Contains cross compiler, emulator and system headers, and libraries.
Examples : Code examples, supplied by Symbian.
Series60Doc : The SDK documentation files, including API descriptions.
Series60Ex : Code examples, typically UI platform specific.
Series60Tools : Development tools.

The Epoc32 folder containing the emulator also contains the drive mapping used when building for the emulator. Note that, on the target device the z : drive is used for the ROM and contains the operating system files and standard applications. The c : drive is use for RAM/flash and provides disk space for

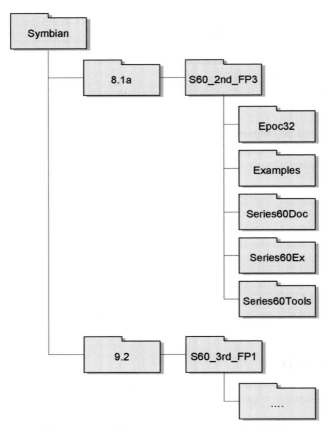

Figure 4.1. Directory structure of the SDK.

new applications and memory for running processes. Building for the emulator the drives are mapped into the following directories:

c: is mapped to \Epoc32\winscw\c

z: is mapped to \Epoc32\release\winscw\$variant$\z[1]

When building for the emulator the build tools assume that we are building system applications therefore placing them under z :. However, any files we create while running our applications in the emulator will be visible under the c : directory. Another important folder is the documentation folder, which contains various help files. Here we will also find the Symbian Developer Library as a compile HTML help file (.chm) containing searchable guides and API documentation see Figure 4.2. The SDK used in the following section giving

---

[1] Variant refers to the build variant, and should be replaced with udeb for debug builds and urel for release builds.

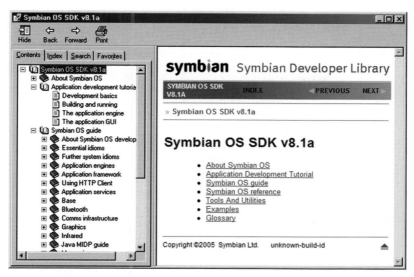

Figure 4.2. SDK documentation help file, containing essential information about the Symbian APIs and essential information about Symbian OS architecture, development tools, and programming idioms.

an overview of the Carbide.c++ IDE is S60 Feature Pack 3 which can be found on the DVD.

4.2.3 Overview of the Carbide.c++ Express IDE 1.1

This section provides the reader with an overview of the Eclipse-based Carbide.c++ IDE 1.1. We will go through the basic features of the IDE while building a *hello world* application. The section is written so that the reader may follow the examples stepwise on his own PC. After launching the Carbide.c++ IDE we will be greeted with a welcome screen as seen in Figure 4.3. The welcome screen contains shortcuts giving us different possibilities to explore the features of Carbide.c++, e.g., through tutorials or the platform overview documentation.

The reader should go through this documentation at some stage as it provides more detailed explanation of how to use the advanced features of the IDE. The actual development and project management is done in the workbench window, which can be accessed by pressing the *switch to workbench* icon in the top-right corner. You can always return to the welcome screen by using the *help→welcome menu* option. The workbench window shown in Figure 4.4 can be broken down into several important elements:

- C/C++ Project view: This shows the folder structure of our current projects and allows us to navigate the files.
- The Editor: Here we can edit our source files.

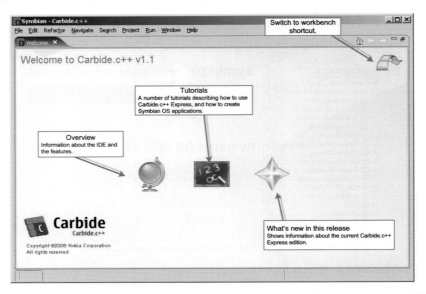

Figure 4.3. Carbide.c++ 1.1 welcome screen, containing shortcuts to, e.g., IDE overview, tutorials, and the workbench.

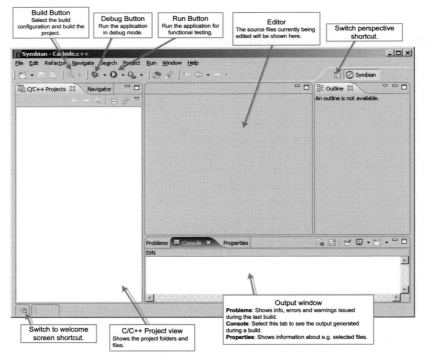

Figure 4.4. Carbide.c++ 1.1 workbench containing the tools needed to manage and edit our projects.

- The Output window: The output window actually contains three different views.
  - Problems: The problems encountered during a build of a project.
  - Console: Output from the build process (compiler, linker, etc.).
  - Properties: Shows various properties for a selected item.
- The tool bar among others containing:
  - The Build button allows us to quickly build our projects.
  - The Debug button launches application in debug mode.
  - The Run button launches the application for functional testing.

This collection of views/windows is in Eclipse terminology called a perspective. So more precisely we are now looking at the Symbian perspective. There are other very useful perspectives such as the Debug perspective that contains the windows/views useful during a debug session (e.g., allowing us to inspect variables and breakpoints). We can switch perspectives using the *Window→Open Perspective* menu item. We introduce the Debug perspective in Section 4.2.3. If you are already familiar with the Eclipse IDE, e.g., Java development, you might have already recognized many of the elements. Also, a large part of the operational aspects are the same. Now that we have gotten the IDE up and running, lets move on and create our first application using the Carbide.c++ application wizards.

Building a Helloworld Application

The application wizard enables us to quickly create a project and get the application skeleton in place. So we can start adding our own code to the project and building applications. However, before we get started the first thing we will do is to disable the *Automatic Build* feature of the IDE. If this feature is not disabled, the IDE will automatically build our project when, e.g., saving changes to files. This can lead to quite an excessive amount of project builds. The feature can be disabled by clicking *Project→Build Automatically* and ensure that the feature is unchecked. To create a new application project select the *File→New* from the menu bar. We can now choose between a number of different project types. In our case we will select the *C++ Application for S60 Project*, see Figure 4.5. The application wizard will now ask you to fill

Figure 4.5. Carbide.c++ 1.1 workbench containing the tools needed to manage and edit our projects.

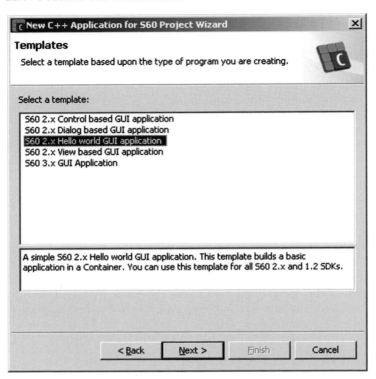

Figure 4.6. A number of different templates are available for different project types. Using a suitable template will generate a ready to use code skeleton for our applications.

in a project name and choose the application template to use. Here we will use HelloWorld as project name and select the *S60 2.x Helloworld GUI application* template, shown in Figure 4.6. The application wizard as given in Figure 4.7 will now show a dialog containing a list of SDKs and build configurations that we can use for our project. The list depends on the SDKs that we have installed and in our case that means the Series 60 2nd Edition Feature Pack 3 (abbreviation S60_2nd_FP3) will be listed. The wizard also suggests three different build configurations:

- S60 2.8 Emulator Debug: This build configuration is used when building our application for execution in the emulator on our PC. This is very useful in the development phase.
- S60 2.8 Phone (ARMI) Release: This build configuration is used to create binaries suitable for execution on the phone.
- S60 2.8 Phone (Thumb) Release: Another build configuration suitable for the phone.

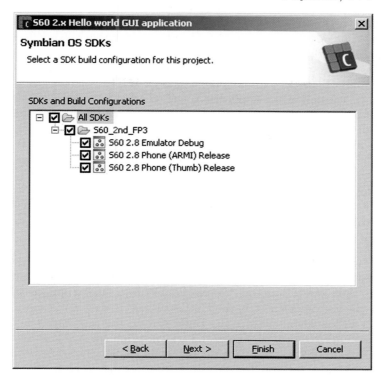

Figure 4.7. The application wizard will suggest build configurations available for the selected project template.

We will not discuss whether you should use ARMI or Thumb when building for the phone, but as a general rule ARMI will generate the most compatible executable. We will leave the selection unchanged and accept all three build configurations. You can press *Finish* now as we will not make any other changes to the template configuration. If you are curious, you can press *Next* to see the other possible options, but leave them unchanged for now.

Congratulations you have now successfully created your first project using the Carbide.c++ IDE. Let us explore our newly created project a bit to see which files and folders the wizard has created for us. In the C/C++ Projects view previously shown in Figure 4.4, we can see that a number of folders and files have been created. In the following we take a quick look at the different files and their purposes. For additional information about these consult the SDK documentation.

\src folder: Contains all the source (.cpp) files for our project.
\inc folder: Contains all the header files (.h) and some additional include files namely the .hrh and .loc files(the .loc file can also be called .rls). The .hrh file contains command ids (enumeration) which are assigned to menu

items (the id allows us to identify which menu button is pressed by the user). The .loc file can be used for localization so that the application can support multiple languages.

\data folder: Contains the resources files (.rss) used in our application. A resource file is used for defining a number of UI elements such as menus, dialogs, caption of the application, and so on.

\group folder: Contains the component definition (bld.inf) file and the project definition (.mmp) file. These two files describe our project, they are used, e.g., when building the project from commandline.

\aif folder: Contains application icon files (.bmp), and additional resources files containing information about the application.

\sis folder: Contains the installation information file (.pkg) which is used for creating the application installation file (.sis) containing our final application. The .sis file can be sent to our phone in order to install our application.

To edit a file, simply double-click it and it will be opened in the editor window. Try it out and let us move on and build the application. We can start a build in three ways: from the *Project→Build Project* menu options, or simply by right-clicking the project in the C/C++ Projects view and choosing *Build Project*, and finally by clicking the Build icon in the tool bar. The building workspace dialog shown in Figure 4.8 will now appear and our hello world application will be built using the default build configuration. As we have not changed anything, the application will be built for execution in the emulator.

To test that everything works we have two options: we can run the application for functional testing or we can run the application in debug mode. We will just run the application for functional testing without debugging facilities (breakpoints will not work here). The next section will show

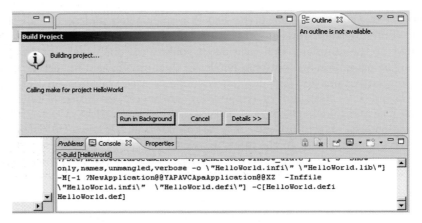

Figure 4.8. Carbide.c++ 1.1 build dialog showing the current build progress of our project.

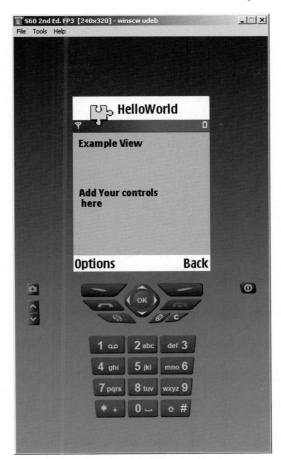

Figure 4.9. Device emulator allowing testing and debugging to be performed without the use of an actual device.

you how to debug the application using breakpoints. In order to run the application, we can either right-click our project and selecting *Run As→Run Symbian OS Application* option or click the Run icon in the tool bar. This will start the application in the emulator using a default run configuration. Starting the emulator can take quite some time, so have patience. Once the emulator is launched, locate our application and open it, the result can be seen in Figure 4.9.

You can read more about using the emulator in the SDK documentation and in the Carbide.c++ help in the *Help→Help Contents* menu option. Now exit the application and close the emulator and let us try the Carbide.c++ debugging facilities.

Debugging the Helloworld Application

A very useful feature of the Carbide.c++ Express IDE is that it allows us to debug our application by setting breakpoints at executable code and, e.g., inspect variables to diagnose problems. Let us try the debugger by setting a breakpoint in one of the source files. Open the `HelloWorldAppUi.cpp` file in the editor by locating it in the \src folder by double-clicking it. We now need to choose a suitable location for a breakpoint, e.g., in the `HandleCommandL()` function. This function will be called whenever the user presses a menu item in our application. As shown in Figure 4.10 we can set the breakpoint by right-clicking the bar in the left side of the editor window and selecting the *Toggle breakpoint* option.

We now need to run the application in debug mode. This can be done easily by either right-clicking the project folder in the C/C++ Project View and selecting *Debug As→Debug Symbian OS Application* or by pressing the Debug icon in the tool bar. This will create a default debug configuration, open the emulator, and automatically switch us to the debug perspective. We placed the breakpoint so that it would be activated if we select the *Test* menu option in our application. As soon as the breakpoint is activated, the Carbide.c++ debug perspective will be activated.

You can now start, e.g., inspecting variables and stepping through the code execution. Additional instructions on how to use the debug perspective can be found in the Carbide.c++ help. We can resume execution by pressing the resume button shown in Figure 4.11. We know now how to build the application for the emulator. Let us change build configuration and build for the target device.

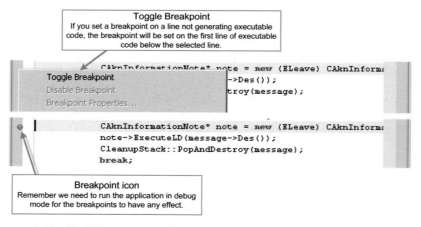

Figure 4.10. Carbide.c++ 1.1 allows us to set breakpoints in our source code, especially useful when tracking down errors in our applications.

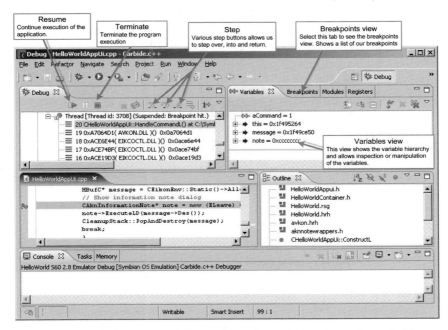

Figure 4.11. Carbide.c++ 1.1 debug perspective contains the tools needed to, e.g., step through execution of the source code and inspect variable values.

Building for a Target Device

In order to deploy the application on an actual phone we only need to change the build configuration. This can be done in several ways, we will do it by selecting *Project→Active Build Configuration→2 S60 2.8 Phone (ARMI) Release* option. The new active build configuration is now marked as shown in Figure 4.12. As when building for the emulator select, e.g., the *Project→Build Project* menu option, this will create a new file in the \sis folder, namely the `HelloWorld.sis` (Symbian installation file). The easiest way to install the application on a phone is to use the Nokia PC suite application installer and a Bluetooth dongle. Make sure your PC Suite is connected to your phone and then simply double-click the .sis file in the C/C++ Project view and the PC Suite application installer will be activated and the dialog shown in Figure 4.13 allow you to begin the installation on the phone. Complete the installation on the phone. You can find and run the application from the phone menu, where you should be able so see the newly installed application.

4.2.4 Summary

The following section is step two in the Symbian OS introduction, while we in the this section concentrated on getting the development tools installed and

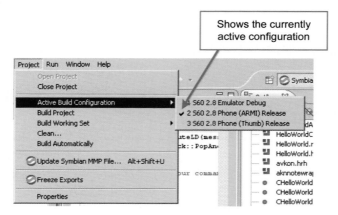

Figure 4.12. Using the build configuration selection submenu allows us to quickly switch between building for the emulator and device.

Figure 4.13. Nokia PC suite Application Installer dialog box. Using PC suite makes it faster to install and test our applications on the device.

ready, the next section will focus on the actual programming related issues, which a new developer will face when beginning Symbian OS application development.

4.3 Symbian OS in a Nutshell

In order to create and deploy successful Symbian applications there are a number of key concepts that the application developer needs to familiarize him/herself with. These concepts are often referred to as the *Symbian Fundamentals* comprising the following topics:

- Naming Conventions
- Exception Handling and Resource Management
- Descriptors
- Active Objects

The purpose of this section is not to explain every corner of each item, but rather give a short description of the subjects necessary to understand and use the programming examples given later. For a more elaborate explanation of these subjects the reader is referred to [5] [1] and the Symbian SDK.

4.3.1 Naming Conventions

A set of specific naming conventions are introduced in Symbian C++, the purpose of these are primarily to ease resource-management and exception handling. This is done by allowing specific information regarding the characteristics and behavior of a particular class or function, to be reflected directly in its name. In the following overview the different naming rules are presented (refer also to Table 4.1).

Static Classes

Classes consisting purely of static functions; these classes have no prefix letter and cannot be instantiated. Examples: `Math`, `User`, and `MyStaticClass`. Today static classes are not used anymore and our static functions should instead be wrapped in a namespace.

T Classes

A class prefixed with a T is a simple class that owns no heap memory, and therefore does not require any explicit destructor. One should be careful where to instantiate a T class, because even though it is simple it can consume a rather large amount of memory, example:

```
1  class  TCoordinate
       {
       public:
            TInt  iX;
5           TInt  iY;
       };
```

All basic types also have T prefix (see `e32def.h` and `e32std.h`):

Table 4.1. Naming convention used for basic C++ types.

Type	Symbian C++ replacement of	Comment
TAny	void (`TAny*` ≡ `void*`)	Shold only be used as `TAny*`
TInt	Integer: `int`	Minimum 32 bits
TInt8/16/32	Integer: `int`	8/16/32 bit versions
TUint	Unsigned integer: `unsigned int`	Minimum 32 bits
TUint8/16/32	Unsigned integer: `unsigned int`	8/16/32 bit versions
TReal	Floating point: `double`	64 bit
TReal32/64	Floating point: `float` or `double`	32/64 bit versions
TChar	Charater: `char`	32 bit
TText	Charater	16 bit
TText8/16	Charater	8/16 bit versions
TBool	Boolean: `bool`	32 bit. Use the enum `ETrue`/ `EFalse`

C Classes

A C-class derives directly from **CBase** or another C-class. C-classes are always constructed on the heap. The prefix C means "Cleanup", and indicates that we should be careful to avoid memory leaks. You never derive from more than one C class. Example:

```
CMyExample : public CBase
    {
    TInt iNumber;
    //...
    public:
        ~CMyExample();
    };
```

R Classes

The R-classes are special classes that are used as a handle to a resource. Normally they are instantiated on the stack or wrapped in a C-class on the heap. They typically must be opened and closed in some way (e.g. through a call to **Connect()** and **Close()**) before they can be used. Do not forget to close an open R-class, otherwise memory leaks can occur. Example:

```
CFileAccess : public CBase
    {
    // R-class, through which we connect to the file server
    RFs iFileServer;
    ...
    public:
        ~CFileAccess()
            {
            // Remember to close open handles
            iFileServer.Close();
            }
    };
```

M Classes

A M class defines a set of abstract interfaces – consisting of purely virtual functions. M classes are the only classes in the Symbian OS that can be used in connection with multiple inheritance. Example of a M-class:

```
class MNotify
    {
    virtual void HandleEvent(TInt aEvent) = 0;
    }
```

Structs

An ordinary C/C++ struct, without any member functions should be prefixed with a uppercase S. Not very often seen in Symbian OS, but here are a few examples from **eikdialg.h**:

```
struct  SEikControlInfo;
struct  SEikRange;
struct  SEikDegreesMinutesDirection;
```

Normally replaced by a T-class.

Member Variables

Member variables are prefixed with a lower-case i (for instance). Example:

```
CMyExample  :  public  CBase
    {
    TInt  iNumber;
    TChar  iLetter;
    //...
    };
```

Automatic Variables

The only rule applying to automatic variables are that they start with a lower-case letter.

Function Names

- Should start with an upper-case letter, e.g., `ShowPath(TFileName&aPath)`.
- Functions that might leave should be named with a trailing upper-case L, e.g., `NewL()`.
- Functions that pushes pointers on the cleanup stack should be named with a trailing upper-case C, e.g., `NewLC()`.
- Functions with a trailing D indicates the deletion of the calling object, e.g., ExecuteLD().

We will take a closer look at leaves and the cleanup stack in the following sections.

Function Arguments

Function arguments are prefixed with a lower-case a (for argument). Example:

```
TInt  CFileConnect::ShowPath(TFileName  &aPath)
    {
    return  iFs.DefaultPath(aPath);
    }
```

Constant Data

Constant data are prefixed with a upper-case K. Example:

```
const  TInt  KNumber = 1982;
```

Namespaces

Namespaces are prefixed with a upper-case N. Namespaces should be used to group functions that do not belong anywhere else. Example:

```
namespace NMyFunctions
    {
    // ...
    }
```

Enumerations

Enumerations are types, and therefore prefixed with a upper-case T. The enumerations members are prefixed with a upper-case E. Example:

```
enum TGames = {EMonopoly, ETetris, EChess = 0x05};
```

4.3.2 Exception Handling and Resource Management

This section describes an important aspect of Symbian OS programming, namely handling exceptions and resources. In Symbian we have several mechanisms that will help us write stable and memory-leak free code. The following will explain the usage of these, and present some example code and common pitfalls. Exceptions are used runtime to signal that some piece of code could not perform its assigned task, e.g., due to lack of resources. This could be low memory or an unavailable communication port. Exception handling in Symbian C++ employs a more light-weight system than the standard C++ exceptions, namely the leave and trap mechanisms.[2] Essentially working with leaves and traps are very similar to the try, throw, and catch statements of C++. Thinking of a leave as equivalent to a throw, and a trap as equivalent to the catch and try statement, is a good start when understanding Symbian exception handling.

Using Leaving Functions

A leave can occur in two ways, either by calling another leaving function or by explicitly calling the system function User::Leave(), the leave will carry an error code that allows its cause to be determined. An error code is an integer value with a predefine meaning, Symbian defines a wide range of different error codes, e.g., KErrNotFound = −1, KErrGeneral = −2, KErrCancel = −3, etc. see the e32err.h file in \Epoc32\include folder. In order for leaving functions to be easily identified, they should be suffixed "L", so appropriate measures

---

[2] Support for C++ exceptions was added in Symbian OS v9, but should only be used directly when porting C++ code to the platform.

to handle the leave safely can be taken by the calling code. Program execution stops at the point of the leave, and returns immediately to the trap harness from within which the function was called. We will look more at traps shortly.

The following is an example of how a leaving function could be called to trap a possible leave:

```
1  TRAPD( err , DoMagicTrickL () ) ;
   if ( err != KErrNone)
      {
      // Something went wrong
5     }
```

Allocating Memory

When allocating memory using the `operator new()` a possibility exists that the operation will fail under low-memory conditions. In that case we will have to check this using an if-statement as shown below.

```
1  CSomeObject *ptr = new CSomeObject ;
   If ( ptr == 0)
      {
      User :: Leave (KErrNoMemory) ;
5     }
```

This would become quite tedious to do, so fortunately Symbian provides us with a global overload of the `operator new()` that takes a parameter `ELeave`, which will leave with `KErrNoMemory` if it fails to allocate the required memory. So the code example above becomes:

```
1  CSomeObject *ptr = new (ELeave) CSomeObject ;
```

Using the Trap Harness

As mentioned when a leave occurs the control immediately returns the trap harness from within which the function call was made. Symbian provides us with two similar trap harnesses, i.e., `TRAP` and `TRAPD`. To call a function `DoMagicTrickL()` under a trap harness, we use the `TRAPD` macro as shown

```
1  TRAPD( err , DoMagicTrickL ()) ;
   if ( err != KErrNone)
      {
      // Some exception occurred
5     }
```

Note, that any leaving function called within `DoMagicTrickL()` will also be called under the trap harness. Additionally it is possible to nest trap harnesses so that exception handling can be performed at several levels. However, since the use of trap harnesses has an impact on both the executable size and the runtime performance we should try to minimize the use of traps. Using TRAP is very similar to the TRAPD macro, the difference is that using the TRAP macro we must ourselves define the error variable. The example code from above then becomes:

```
1  TInt  err;
   TRAP(err,  DoMagicTrickL());
   if(err  !=  KErrNone)
       {
5      // An  exception  occurred
       }
```

In many cases this makes the TRAPD macro somewhat more convenient to use.

The Cleanup Stack

When control returns to the trap harness due to a leave, all automatic variables on the stack are destroyed. This becomes problematic if the variables are pointers to memory allocated on the heap, in this case the pointer to the heap data is lost and we have no way of recovering the leaked/orphaned heap memory. Symbian therefore provides us with a mechanism that will free the memory correctly if a leave occurs. This is done by pushing the pointers onto a special stack called the cleanup stack while performing anything that could potentially leave. In case of a leave all pointers pushed onto the stack in-between the inclosing trap harness macro and the leave will be properly destroyed by the cleanup stack, as shown in Figure 4.14.

Objects can be pushed on the cleanup stack using CleanupStack::PushL() and popped off using CleanupStack::Pop(). Objects should be popped from the cleanup stack when there is no longer any danger that they might be orphaned by a leave. Let us have an example:

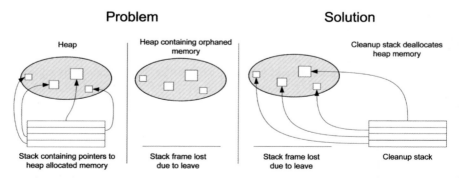

Figure 4.14. The cleanup stack ensures that no memory will be orphaned on the heap if a leave occurs.

```
1  void AllocateSomethingL ()
      {
      CObject *ptr1 = new (ELeave) CObject;
      CObject *ptr2 = new (ELeave) CObject;  // Wrong

5     //...

      delete ptr1;
      delete ptr2;
10    }
```

Consider what would happen if the second allocation fails and therefore leaves due to insufficient memory? In this case the data pointed to by `ptr1` would be lost forever. Fixing this using the cleanup stack is fortunately easy:

```
1  void AllocateSomethingL ()
      {
      CObject *ptr1 = new (ELeave) CObject;
      CleanupStack :: PushL ( ptr1 );
5     CObject *ptr2 = new (Eleave) CObject;  // Safe
      CleanupStack :: PushL ( ptr2 );

      //...
      CleanupStack :: PopAndDestroy (2);  // Pops and deletes ptr1 and
           ptr2
10    }
```

Now `ptr1` is saved on the cleanup stack, if a leave occurs while creating the second object, `ptr1` will be destroyed by the cleanup stack. The cleanup stack provides far more functionality than what we have shown so far, allowing more advanced handling of resources. The reader should refer to the Symbian OS SDK for further information. Finally, note that compound objects should destroy their member variables in the destructor, the member pointers should therefore never be pushed onto the cleanup stack, since this could result in a double deletion error. This is closely related to a paradigm called two-phase construction extensively used in Symbian.

Two-Phase Construction

Typically C-classes are compound and own, e.g., pointers to other C-classes, and are therefore responsible for safely allocating and freeing the memory they own. We might consider placing these allocations into the constructor of our compound object.

```
1  CSomeObject :: CSomeObject ()
      {
      iOtherPtr = new (ELeave) COtherObject;
      }
```

This is however not safe, and the problem is subtle, but consider if we constructed `CSomeObject` in the following way:

```
1  CSomeObject *ptrSomeObject = new (ELeave) CSomeObject;
```

The following happens in the above code:

1. Memory is allocated for CSomeObject
2. The constructor of CSomeObject is called
3. Memory is allocated for COtherObject and its constructor is called
4. A pointer to COtherObject is stored in the member variable iOtherPtr
5. Construction is complete

If step 3 fails, the flow of execution immediately returns to the trap harness and we have no way to destroy the partially created object, since the `ptrSomeObject` pointer is lost. This example motivates the need for a better object construction paradigm. In Symbian the solution is two-phase construction. The idea is that we separate out potentially leaving code from the compound object constructor and places it into separate function by convention called `ConstructL()`.

In our example above we would now have the following:

```
1  CSomeObject :: CSomeObject ()
   {
     // No leaving code here
   }
5
   void  CSomeObject :: ConstructL ()
   {
     iOtherPtr = new (ELeave) COtherObject ;
   }
```

Allowing us to construct the **CSomeObject** safely in the following way:

```
1  CSomeObject *ptrSomeObject = new (ELeave) CSomeObject ;
   CleanupStack :: PushL ( ptrSomeObject ) ;
   ptrSomeObject ->ConstructL () ;
   CleanupStack :: Pop ( ptrSomeObject ) ;
```

For convenience this is often encapsulated in two static factory functions `NewL()` and `NewLC()`. Implementing our example using `NewL()` and `NewLC()` could be done as follows:

```
1  CSomeObject*  CSomeObject :: NewLC ()
   {
     CSomeObject *self = new (ELeave) CSomeObject ;
     CleanupStack :: Push ( self ) ;
5    self ->ConstructL () ;
     return self ;
   }

   CSomeObject*  CSomeObject :: NewL ()
10 {
     CSomeObject *self = CSomeObject :: NewLC () ;
     CleanupStack :: Pop ( self ) ;
     return self ;
   }
```

This would allow us to safely use two-phase construction and simplify our construction code to the following:

```
1  CSomeObject *ptrSomeObject = CSomeObject :: NewL () ;
```

Typically a `CBase` derived class will supply both the `NewL()` and `NewLC()` functions. The suffix C in the function name of `NewLC()` signifies that a pointer will be left on the cleanup stack, which is quite useful if the returned pointer is assigned to an automatic variable.

4.3.3 Descriptors

The descriptor classes are often referred to as Symbian OS strings, and this is also one of their main purposes. However, they are also used to encapsulate binary data, and for this purpose all descriptors come in two variants, namely 8bit and 16bit (the latter allowing support for the Unicode character set). Normally we will implicitly use the 16-bit variant, e.g., `TDesC` (also called the neutral types), but we can also explicitly choose which version to use by specifying one of the following `TDesC16` or `TDesC8`. When dealing with binary data we should therefore always choose the explicit 8-bit version. Three different categories of descriptors exist, namely the pointer descriptors, buffer descriptors, and finally the heap descriptors. Additional to these are the literals or descriptor literals. These are not considered actual descriptors but are closely related and often used together with the descriptor classes. Literals are used to represent constant data built into the application binary, an example:

```
1   _LIT(KHello, "Hello_World");
```

This would create a literal in the application binary with the name `KHello` and the value *Hello World*. We will take a closer look at the literal classes at the end of this section. It is important to stress that understanding how to use descriptors is crucial, as they are used extensively through out all the Symbian OS APIs. Even though descriptors are considered pretty confusing by many new Symbian C++ programmers, once understood, they provide a very safe and efficient API compared to, e.g., C-style strings.

The Descriptor Hierarchy

The class hierarchy of the neutral descriptor types is shown in Figure 4.15 the inheritance diagram allows us to determine the various characteristics of each descriptor, which we will go through in the following. The figure shows ten descriptor classes of which only six are instantiable, these are presented at the bottom of the Figure[3].

All descriptors derive from the `TDesC` base class which provides an API allowing all operations that we would expect to be able to perform on a constant string such as `Length()`, `Compare()`, and `Find()` (see Table 4.2). The capital C suffix in the class name reflects that, we are dealing with a nonmodifiable type, with constant content. In order for us to perform manipulation

---

[3] RBuf is only available from Symbian OS v8.0 and up.

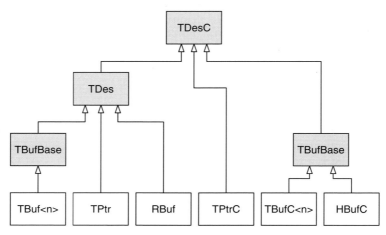

Figure 4.15. Knowing the classes in the descriptor hierarchy allows us to decide which particular descriptor should be used and in which scenario.

Table 4.2. Commonly used functions in the nonmodifiable descriptor API.

NonModifiable API	
Function	Description
Length()	Returns the current length of the descriptor
Size()	Returns the size in bytes occupied by the descriptor
AllocL()	Retuns a pointer to a new HBufC descriptor allocated on the heap
Find()	Searches for a specific data sequence in the descriptor and returns the offset to the data
Compare()	Compares two descriptors, and retuns a TInt indication the result.
Match()	Searches the descriptors data for a specified pattern.
operator[]()	Returns a reference to an individual data element within the descriptors data

Table 4.3. Commonly used functions in the modifiable descriptor API.

Modifiable API	
Function	Description
MaxLength()	Retuns the maximum number of data items the descriptor can contain
Append()	Appends data to the end of the descriptor, the new length cannot be larger than the maximum descriptor length
Delete()	Deletes a certain amount of data from a given position in the descriptor
Insert()	Inserts data into the descriptor at a given position, the new length must not exceed the maximum length.
Copy()	Copies the supplied descriptor data into the calling descriptor, replacing any existing data
SetLength()	Set the length of the descriptor to the specified value.

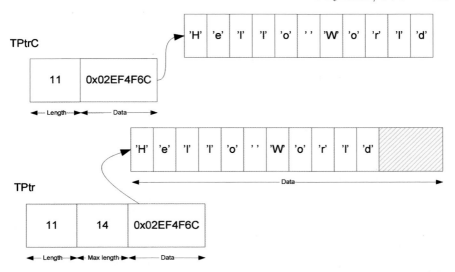

Figure 4.16. The memory layout of the two pointer descriptor classes, none of the pointer descriptors own the data to which they point.

of the descriptor data we need to use the descriptor classes deriving from the `TDes` class, which itself derives for `TDesC`. The `TDes` class provides the API that allows us to modify the content of the descriptor, e.g., `Append()`, `Capitalize()`, and `Insert()` (see Table 4.3). The following tables shows some of the additionally commonly used functions from the nonmodifiable API and the modifiable API.

Pointer Descriptors

Pointer descriptors should be instantiated either on the stack or as a member variable on the heap. They differ from other descriptors by not owning the data they point to. They can be used to point to anywhere addressable by our code in the stack, ROM, or heap. The two pointer descriptor types are the modifiable `TPtr` and nonmodifiable `TPtrC`. Their respective memory layout can be seen in Figure 4.16.

They are commonly used to represent fragments of other descriptors or to allow some nondescriptor data to be accessed through the safe descriptor API. The following code example shows various ways of constructing a nonmodifiable `TPtrC`:

```
1  // TPtrC example
   _LIT ( KHelloWorld ,  " Hello_World" ) ;

   // Construct a TPtrC from the descriptor litteral
5  TPtrC helloWorldPtr ( KHelloWorld () ) ;
```

```
    // Using the copy constructor, to get a TPtrC from another
        descriptor
    TPtrC copyPtr(helloWorldPtr);

10  TPtrC helloPtr(helloWorldPtr.Left(5)); // Extract the "Hello"

    // Change helloPtr to point to the original litteral descriptor
    helloPtr.Set(KHelloWorld());

15  TUint8 memPtr[19]; // Construct a pointer to a specific memory
        location
    TInt lenght = 19;

    // The data can now be "descriptorized" and used in APIs
    // expecting a descriptor
20  TPtrC8 ptr(memPtr, lenght);
```

The `TPtr` returned by the `TBufC::Des()` and `HBufC::Des()` functions is commonly used to manipulate the data of the otherwise nonmodifiable descriptors. However, since Symbian OS v8.0 the `RBuf` descriptor should be used manipulating heap based data. The following example code shows some typical ways of constructing the modifiable descriptor pointer:

```
1   // TBufC buffer descriptor is described shortly
    TBufC<11> helloBuf;

    // Construct TPtr from buffer descriptor, length = 0, max length
        = 11
5   TPtr ptr1(helloBuf.Des());

    ptr1.Capitalize(); // Modify the data though the pointer

    _LIT(KText1, "Mobile");
10  _LIT(KText2, "Rocks");

    // Changing the underlying data in helloBuf, either using the
    // Copy() function or copy assignment operator

15  ptr1.Copy(KText1()); // helloBuf = "Mobile"

    TInt length = ptr1.Length(); // length = 6
    TInt maxlength = ptr1.MaxLength(); // maxlength = 11

20  ptr1 = KText2(); // helloBuf = "Rock"
```

Buffer Descriptors

As mentioned also two versions of the buffer descriptors exist, namely a modifiable and nonmodifiable version. Buffer descriptors encapsulate the data as part of the descriptor object, which should be either allocated on the stack or as a member variable of some compound object on the heap. The two buffer descriptors are `TBuf<n>` and `TBufC<n>`. Both are templated, so their size is determined at compile time. In Figure 4.17 the memory layout of each descriptor is shown.

It is worth noting that we do not need the terminating `NULL` character to signal the end of the string as the descriptor objects nicely encapsulates the

TBufC<11>

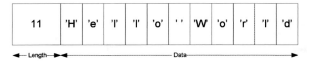

TBuf<14>

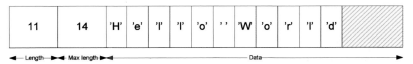

Figure 4.17. The memory layout of the two buffer descriptor classes, this descriptor type encapsulates the data in the actual descriptor object.

length information. The buffer descriptors are commonly used to hold small fix sized strings. The reason why we argue small strings are that we commonly allocate buffer descriptors on the stack, which is a limited resource in Symbian (8–20 KB depending on the Symbian OS version). A common example of the issue is the `TFileName` which is a typedef for `TBufC<256>` meaning that each `TFileName` is taking up 512 bytes[4] on the stack. Placing too many `TFileName` objects on the stack can quickly use up the, e.g., 8 KB resulting in a program Panic[5]. One thing to note about the `TBufC<n>` class is that even though it is categorized nonmodifiable is contains a function `Des()` which returns a modifiable pointer descriptor to the data, allowing us to indirectly modify and manipulate the contained data. The following code listing shows a number of ways to construct a nonmodifiable buffer descriptor:

```
 1  // TBufC, a non-modifiable buffer descriptor
    _LIT(KHelloWorld, "Hello_World!");
    TBufC<12> buf1(KHelloWorld());

 5  // Construct TBufC from another descriptor, length = 12, no max
    //    length
    TBufC<14> buf2(buf1);

    _LIT(KDevices, "Mobile_Devices");

10  // Replacing the content using copy assignment
    buf2 = KDevices(); // length = 14

    // Remember we can use the TBufC::Des() function to get a
    //    modifiable TPtr
    TPtr ptr(buf2.Des());
```

---

[4] Remember that the neutral descriptor types are 16 bit per element to support the Unicode character set.

[5] A panic will terminate the thread in which it occurs, it cannot be handled in any way. A panic often indicates a programming error.

Next let us take a look at some example code showing the construction methods for the modifiable version of the buffer descriptors in the following listing.

```
1  _LIT ( KHelloWorld ,  " Hello_World !" ) ;
   _LIT ( KDevices ,  " Mobile_Devices" ) ;

   // TBuf, a modifiable descriptor
5  TBuf<12> buf1 ( KHelloWorld ( ) ) ; // length = 12, max length = 12

   TBuf<40> buf2 ; // length = 0, max length = 40

   // Using copy assignment operator to copy buf1 into buf2
10 buf2 = buf1 ; // length = 12, max length = 40

   // Copying the literal descriptor KDevices data into buf2
   buf2 . Copy ( KDevices ( ) ) ;
```

Heap Descriptors

As the name clearly states these descriptors data are allocated on the heap. They are typically used in the case of large strings and data, or when the length of the data cannot be determined until runtime. The naming of the two heap descriptors `RBuf` and `HBufC` is a bit confusing. The difference in naming however nicely underlines the fundamental difference between the two descriptors. `RBuf` can be constructed on the stack, only the data it owns will always be allocated on the heap. Recall the meaning of the R-classes from Section 4.3.1, namely that R-classes owns a resource, in this case the resource is some data on the heap. The `HBufC` descriptor is always allocated entirely on the heap (the "H" stands for heap, and is an exception from the naming convention rules). Seeing the memory layout of both descriptors in Figure 4.18 will help us understand the difference.

As you can see the `HBufC` is stored entirely on the heap and we can access it through a pointer variable, otherwise the memory layout is identical to the `TBufC<n>`. `HBufC` is designed to be allocated on the heap and implement the factory functions `NewL()` and `NewLC()` to achieve this. The following code examples will show various ways of creating `HBufC` objects. Just as the `TBufC<n>` the `HBufC` descriptors data can also be manipulated by creating a modifiable pointer using the `Des()` method.

```
1  _LIT ( KPlus ,  " One_plus_one_is_two" ) ;
   _LIT ( KTest ,  " Life_is_too_short_for_manual_testing" ) ;
   // Creating a heap based descriptor from the literal
   HBufC *buf1 = KPlus ( ) . AllocLC ( ) ;
5
   // Allocating space for 36 elements on the heap
   HBufC* buf2 = HBufC :: NewLC ( 36 ) ;

   // Replace buf2s content with the KTest literal , notice
10 // we just dereference the buf2 pointer to get a non
   // modifiable reference to the HBufC.
   *buf2 = KTest ( ) ;

   CleanupStack :: PopAndDestroy ( 2 ) ; // buf1 , buf2
```

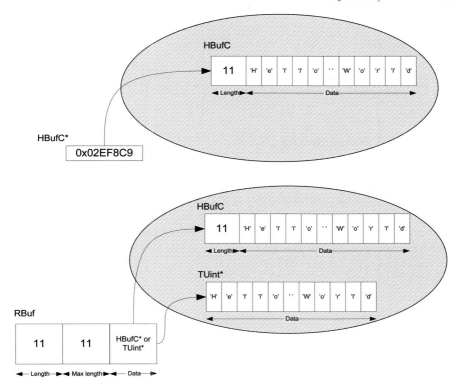

Figure 4.18. The heap descriptor types manage data allocated on the heap, they are especially useful when we do not know the size of the data until runtime.

RBuf is a relatively new addition to the descriptor family introduced in Symbian OS v9 and back ported to v8.0 and v8.1. As seen in Figure 4.18 it maintains data on the heap, which can either be an HBufC or a raw data area. It allows a more convenient way of manipulating data on the heap, as it does not require the creation of an intermediate modifiable pointer descriptor as is the case if we try to manipulate the HBufC, and should be used instead of using the Des() function in HBufC. The following example code shows how a RBuf can be constructed:

```
 1   _LIT(KHey, "Hey_you!");

     RBuf buf1; // Zero length and owns no resources

 5   // Now owns the heap data created from the HBufC
     buf1.Assign(HBufC::NewL(20));

     // Since this an R-class, we must close it to release the
         resources
     buf1.Close();
10
     // Allocates and copies the literals data on the heap
     buf1.CreateL(KHey());
```

```
      buf1.Close();

15    // Can also be used to "descritorize" raw heap data
      TUint8 *ptr = new (ELeave) TUint8[10];

      RBuf8 buf2;
      buf2.Assign(ptr, 10);// Takes ownership of the heap data
20    buf2.Close();
```

Descriptors as Arguments

When creating functions that use descriptor arguments, you should use the base class types. If the function expects a modifiable descriptor a **TDes** as argument should be used and for passing nonmodifiable descriptors **TDesC**. This will allow the greatest flexibility in your API, since you will not rely on a specific descriptor type. However remember to pass either pointers or references to avoid object slicing, as shown in the following listing.

```
1    void modifyThis(TDes &aDes);
     void doNotModifyThis(const TDesC &aDes);
```

Literals

Several times during this section you have seen the use of the macro _LIT(). The purpose of this macro is to allow us to create strings that are embedded in the application binary. The _LIT() macro expands to a TLitC class object, which defines a number of operator overloads, e.g., the operator()() which returns a TDesC reference, allowing us to use the literal in every API otherwise requiring a descriptor type. This is extremely useful in many scenarios where an API, e.g., expects some string value. For the sake of completeness a deprecated macro _L() can also be used to produce literals, this is however done in a more inefficient way and the _LIT() macro should be used instead. Creating either a 16-bit or 8-bit descriptor literal can be done in the following way:

```
1    // Descriptor literal neutral, 16 bit
     _LIT(KHey, "Hey_you!");

     // Descriptor literal neutral, explicitly 16 bit
5    _LIT16(KHey16, "Hey_you!");

     // Explicit 8 bit wide
     _LIT8(KHello, "Hello");
```

4.3.4 Active Objects

Active objects address the problem of achieving multitasking in a highly event-driven and resource-constrained environment such as the Symbian OS. Basically most applications are event driven. Events can be generated at multiple sources, e.g., user input, requests to system servers[6] completing, and so forth. When our application requests a service (e.g., sending a SMS) from one of the system servers, that service can typically either be completed synchronously or asynchronously. A synchronous service request is similar to a normal function call, the request completes once the result is ready. Making a request to an asynchronous service provider is just the opposite, the request completes immediately, and the service provider will continue to process in the background. Meanwhile, our application is free to do other tasks (e.g., respond to user input). When the asynchronous service provider has processed our request and is ready with a result, our application is notified and we can handle the completion in a suitable way. On most desktop systems, this can be implemented using preemptive multitasking, where a new thread is spawned to handle each asynchronous request, and the kernel will then schedule the different threads running. This is however not an optimal solution on a resource constrained device, where expensive thread context switches and memory overhead can become an issue. Instead, to efficiently allow multitasking Symbian provides the active object framework which enables us to achieve nonpreemptive multitasking within one thread.

How It Works

To achieve the nonpreemptive multitasking within a thread, the Symbian OS provides us with an active scheduler. The active scheduler is roughly a message loop within a thread, where all system server completion events are delivered and forwarded to the appropriate active object, which requested the service as shown in Figure 4.19. Internally the active scheduler maintains a prioritized list of active objects, and when a server signals our application thread the active scheduler will iterate through the list and find the highest priority active object eligible to run and call its handler function.

To avoid wasting valuable CPU cycles by polling for new results the active scheduler waits on a semaphore. Each server will increment the semaphore when results are ready, making sure that all events are processed before the active scheduler goes back to sleep. As mentioned active objects are used to encapsulate asynchronous service requests. We create an active object by deriving from the **CActive** class. The following code listing shows a part of the **CActive** class definition (the complete definition can be found in the header file **e32base.h**), note that some members have been omitted to enhance readability:

```
1  class CActive : public CBase
       {
       public:
           enum TPriority
5              {
```

---

[6] Symbian OS uses a micro kernel architecture, which means that most system functionality is accessed through servers, e.g., the file server and socket server.

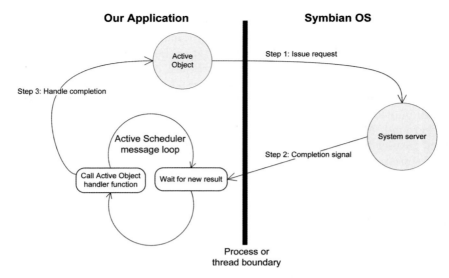

Figure 4.19. Requesting a service from one of the Symbian OS system servers, typically involves creating an active object. Using that active object to issue a request, when that request completes the service provider signals the active scheduler in our application, which will find the active object and pass on the completion result.

```
            EPriorityIdle=-100,
            EPriorityLow=-20,
            EPriorityStandard=0,
            EPriorityUserInput=10,
10          EPriorityHigh=20,
            };
    public:
        IMPORT_C void Cancel();
        IMPORT_C void SetPriority(TInt aPriority);
15      inline TInt Priority() const;
    protected:
        IMPORT_C CActive(TInt aPriority);
        IMPORT_C void SetActive();
        // Pure virtual, must be implemented by the deriving class
20      virtual void DoCancel() =0;
        virtual void RunL() =0;
        // Can be overridden to handle errors
        IMPORT_C virtual TInt RunError(TInt aError);
    public:
25      TRequestStatus iStatus;
    private:
        TBool iActive;
    };
```

The code listing above shows some of the most commonly used functions in the **CActive** class, their purposes are:

- Construction: To set the priority of the active object the **CActive** constructor is called using a chosen priority. In general this should be **EPriority-Standard** unless a good reason to do otherwise exist. The active object also needs to be added to the active scheduler, this is described later.

- Issuing requests: When issuing an asynchronous request, we pass the iStatus parameter to the server, which upon completion can be checked to see whether the request was successful. Additionally the SetActive() function sets the iActive flag, indicating that the active object has an outstanding request. The active scheduler uses the iActive flag when iterating through its list of active objects. If a flag has be set using SetActive(), the active scheduler will check the iStatus flag to see whether the request has completed.

- Receiving results: When the active scheduler finds an active object ready to run, it will call the handler function RunL(). RunL() is a purely virtual function and should be implemented by the deriving class. Note, that the RunL() function cannot be preempted by other active objects, therefore it should complete as quickly as possible.

- Canceling outstanding requests: All asynchronous requests must allow an outstanding request to be canceled. This is done through the Cancel() function which first checks to see whether the active object is active, before it calls the purely virtual function DoCancel(). DoCancel() should be implemented in the derived class.

- Handling errors: If the handler function RunL() leaves while processing an incoming result the RunError() function will be called. This gives the active object a chance to perform cleanup in case of an error. If no implementation of the RunError() function exist in the derived class the default implementation will call CActiveScheduler::Error()[7].

The CActiveScheduler class encapsulates the message-loop in the active scheduler. In GUI applications the active scheduler is already installed, and we can start immediately issuing requests. However if we are developing console applications, servers, or threaded applications we need to know how to create and install our own active scheduler. The following code listing shows the relevant class definition of the CActiveScheduler (again the full definition can be found in e32base.h):

```
 1  class CActiveScheduler : public CBase
        {
    public:
        IMPORT_C CActiveScheduler();
 5      IMPORT_C ~CActiveScheduler();
        IMPORT_C static void Install(CActiveScheduler* aScheduler);
        IMPORT_C static void Add(CActive* anActive);
        IMPORT_C static void Start();
        IMPORT_C static void Stop();
10      IMPORT_C virtual void WaitForAnyRequest();
        };
```

- Installing an active scheduler: An active scheduler can be installed if needed, by calling the static function Install().

---

[7] The default of the CActiveScheduler::Error() implementation will result in an E32USER-CBase 47 panic.

- Adding active objects: Before we can start using an active object to issue asynchronous requests, it must be added to the active scheduler using the static `Add()` function. Note, this is normally done in the active objects constructor.
- Starting the scheduler: The message loop is started by calling the `Start()` function. This function does not return until a corresponding `Stop()` has been issued. Note, that at least one active object must have an outstanding request otherwise the loop will lock[8].
- Stopping the scheduler: The message loop is stopped by calling the static `Stop()` function.

In the following section we will go through a classical active object implementation that is commonly used to demonstrate the usage of the active scheduler and active objects.

An Active Object Example

In this example we will use an active object to create a countdown timer, the entire project source code can be found on the DVD under \example\countdown. The following code listing shows the active object class definition:

```
1  // When the timer expires we would like to get at callback
   typedef void (*fPtr)(TInt);

   class CCountdown : public CActive
5      {
       public: // Public interface, two phase construction principle
               used
           static CCountdown* NewL(fPtr aCallback);
           static CCountdown* NewLC(fPtr aCallback);
           void After(TInt aSeconds);
10         ~CCountdown();
       private:
           CCountdown(fPtr aCallback);
           void ConstructL();
       protected: // From CActive
15         void RunL();
           void DoCancel();
           TInt RunError(TInt aError);
       private:
           fPtr iCallback;
20         RTimer iTimer;
       };
```

As stated previously on active objects, we derive from **CActive** and provide the appropriate implementation of the `RunL()`, `DoCancel()`, and `RunError()` functions. As a member variable `iTimer` we have a handle `RTimer` to an asynchronous service provider, providing timer, and timing services. In the following we show actual implementation of each function.

---

[8] The active scheduler waits on a semaphore by calling the WaitForAnyRequest() function. If no requests are issued, the semaphore will never be signaled and the scheduler will hang infinitely.

```
_LIT(KPanicInUse, "Request_already_pending!");

CCountdown* CCountdown::NewL(fPtr aCallback)
    {
    CCountdown *self = CCountdown::NewLC(aCallback);
    CleanupStack::Pop(self);
    return self;
    }

CCountdown* CCountdown::NewLC(fPtr aCallback)
    {
    CCountdown *self = new (ELeave) CCountdown(aCallback);
    CleanupStack::PushL(self);
    self->ConstructL();
    return self;
    }

void CCountdown::After(TInt aSeconds)
    {
    // We first check that no requests are outstanding
    _ASSERT_ALWAYS(!IsActive(), User::Panic(KPanicInUse,
        KErrInUse));
    iTimer.After(iStatus, aSeconds*1000000);
    SetActive();
    }

CCountdown::~CCountdown()
    {
    Cancel();
    iTimer.Close();
    }

CCountdown::CCountdown(fPtr aCallback)
: CActive(EPriorityStandard), iCallback(aCallback)
    {
    CActiveScheduler::Add(this); // add to active scheduler
    }

void CCountdown::ConstructL()
    {
    iTimer.CreateLocal(); // Create thread-relative timer
        kernel-side
    }

void CCountdown::RunL()
    {
    // If there occured any errors, we leave and handle
    // it in RunError
    User::LeaveIfError(iStatus.Int());
    iCallback(KErrNone);
    }

void CCountdown::DoCancel()
    {
    iTimer.Cancel(); // Cancel the outstanding request
    }

TInt CCountdown::RunError(TInt aError)
    {
    // We should do some error handling, but for the example
        we
    // just pass on the error code
    iCallback(aError);
    return KErrNone;
    }
```

The final part of the implementation is related to instantiating the active object and creating the active scheduler. As mentioned this would not be necessary in a GUI application as they utilize a framework already providing a running active scheduler. In a GUI application we would be able to use the above code as is. However, we are using a console application so we need to create our own active scheduler, which is exemplified in the following code listing:

```
 1  void  Callback(TInt aResult)
        {
        console->Printf(KCountdown, aResult);
        CActiveScheduler::Stop();
 5      }

    void  MainL()
        {
        CCountdown *countDown = CCountdown::NewLC(&Callback);
10      countDown->After(2); // seconds
        CActiveScheduler::Start();
        CleanupStack::PopAndDestroy(countDown);
        }

15  void  DoStartL()
        {
        // Create active scheduler (to run active objects)
        CActiveScheduler* scheduler = new (ELeave) CActiveScheduler()
            ;
        CleanupStack::PushL(scheduler);
20      CActiveScheduler::Install(scheduler);

        MainL();

        // Delete active scheduler
25      CleanupStack::PopAndDestroy(scheduler);
        }
```

The functions have the following purpose:

- DoStartL() installs the active scheduler in our thread.
- MainL() creates our CCountDown active object, issues an asynchronous request and starts the active scheduler. Remember that CActiveScheduler::Start() does not return until CActiveScheduler::Stop() has been called.
- Callback() is called by the CCountdown active object when the countdown is complete, it makes sure that the active scheduler is stopped.

This is a very simplified example showing how to use the active objects framework, for more advanced example refer to the SDK code examples found in the examples\Base\IPC\Async folder of the SDK.

4.3.5 Summary

This completes our introduction to the Symbian development environment and walk-through of the "Symbian Fundamentals". To find further information about these topics several online resources can be recommended:

- Tutorials, code examples and forum at `www.forum.nokia.com`
- White papers, wiki, forums, and Symbian OS documentation at `www.symbian.com/developer`
- Tutorials, code examples, and forum at `www.newlc.com`

The final sections of this chapter will explore the wireless communication technologies available on the Symbian OS platform.

4.4 Wireless Communication Technologies

Communication is the key feature of any mobile phone, whether it is done by making calls, sending SMS messages, or playing multiplayer games over Bluetooth. The Symbian OS allows us to take advantage of the wide range of communication technologies available on most new mobiles. In this section we will take a look at some of the communication APIs and see how to utilize these in our code. We will assume that the reader is already familiar with communication concepts such as sockets, client/server communication, and the common communication protocols. This section will describe the basics in how to establish a TCP connection over, e.g., GPRS or WLAN, how to use OBEX over Bluetooth, and finally how we can use messaging such as SMS or Email in our applications. This is done through a number of code examples. Each code example is written to involve only the minimum required code, focusing on using the actual communication APIs. All code examples are written as console applications and can be found on the accompanying DVD. For additional information about areas not covered here such as Infrared, telephony, and serial communication, the reader is referred to the SDK documentation and web pages presented in the previous section.

4.4.1 Using TCP/IP

To communicate over TCP/IP we will use the Symbian OS sockets API. The socket API allows us to communicate with the Symbian OS socket server named "ESock", which provides shared and generic access to the supported communication protocols and network interfaces on the device. Note, the steps involved in configuring and using a socket, differ only slightly depending on whether we are using UDP or TCP as transport protocol, you will find a UDP code example on the DVD. The following shows the basic steps involved:

1. Establishing a session with the socket server.
2. Resolving the hostname to an IP – optional if the IP is already known.
3. Opening a socket specifying the chosen protocol.
4. Connecting to a remote device.
5. Transferring data.
6. Closing the socket and socket server session.

The following code listing implements step one and two, to enhance readability we have created a **LOGLEAVE()** macro which will write any errors to the console and leave with an appropriate error code.

```
1   TInt res;
    RSocketServ socketServ;
    res = socketServ.Connect();  // Conect to socket server
    LOGLEAVE(res, "socketServ.Connect()");
5   CleanupClosePushL(socketServ);

    RHostResolver resolver;  // Open resolver to perform DNS query
    res = resolver.Open(socketServ, KAfInet, KProtocolInetUdp);
    LOGLEAVE(res, "resolver.Open()");
10  CleanupClosePushL(resolver);

    // The hostname
    _LIT(KWebsite, "mobiledevices.kom.aau.dk");

15  // Will contain the DNS response, including IP
    TNameEntry nameEntry;

    // Use the resolver to get the IP
    res = resolver.GetByName(KWebsite, nameEntry);
20  LOGLEAVE(res, "resolver.GetByName");

    // Construct an object containing IP and port
    // usable by the socket
    TInetAddr addr(nameEntry().iAddr);
25  addr.SetPort(80);

    // Print the IP address returned by resolver
    TBuf<16> ipaddress;
    addr.Output(ipaddress);
30  _LIT(KTInetAddr, "TInetAddr_addr_%S_port_%d");
    console->Printf(KTInetAddr,&ipaddress, addr.Port());
```

Before any communication can take place we must complete step one and open a session to the Symbian OS socket server, this is achieved through the **RSocketServ::Connect()** function. Besides opening a session to the socket server, the **RSocketServ** class also allows us to query the socket server about the currently supported protocols. This can be useful when determining the parameters used to open the actual socket object, which you will see in the next code listing. In step two we use the **RHostResolver** class, which can be used if we do not already know the IP of the remote device but only the hostname, e.g., "mobiledevices.kom.aau.dk". The **RHostResolver** can then be used to perform a DNS query and resolve the IP address. If you examine the API you will also see that one of the **RHostResolver::Open()** functions takes an open **RConnection** session parameter, the **RConnection** class allow the network interface that should be used for the communication to be selected, e.g., WLAN or GPRS. However, this is not mandatory as it will happen implicitly when requesting connectivity. On a device with several IAPs (Internet Access Point) defined, this will result in a user query, as shown in Figure 4.20.

You can find more information about additional functionality in the **RSocketServ** and **RHostResolver** classes by looking up their names in the SDK documentation. With step one and two completed, we are now ready to open and connect the socket as shown in the following code listing:

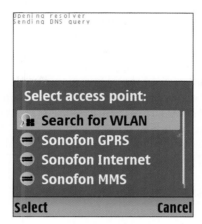

Figure 4.20. Internet access point selection dialog, here the user can select the network interface that should be use for the connection.

```
 1  RSocket  socket ;

    // Open  the  socket  using  the  protocol  parameters
    res  =  socket . Open ( socketServ ,  KAfInet ,  KSockStream ,
        KProtocolInetTcp ) ;
 5  LOGLEAVE( res ,  " socket . Open ( ) " ) ;
    CleanupClosePushL ( socket ) ;

    // Allow  us  to  synchronously  wait  for  completion
    TRequestStatus  status ;
10
    // Connect  to  the  remote  device
    socket . Connect ( addr ,  status ) ;

    // Wait  for  connection  to  complete
15  User :: WaitForRequest ( status ) ;
    LOGLEAVE( status . Int ( ) ,  " socket . Connect ( ) " ) ;
```

The `RSocket` class provides us with the actual socket object. If you are already familiar with sockets programming you will find that most of the `RSocket` functionality is similar to the BSD-style socket API found on most desktop systems. In step three we need to open the socket. There exist a number of different overloads of the `RSocket::Open()` function, which allow us to specify the desired transport protocol to be used. As mentioned you can use the `RSocketServ` class to find the supported protocols and their parameters. However, in this example we will just use hard coded values. To open an UPD socket we would use the following parameters:

```
 1  res  =  socket . Open ( socketServ ,  KAfInet ,  KSockDatagram ,
        KProtocolInetUdp ) ;
```

Note, that we are not making any connections to other devices here, we only initialize the socket object. After opening and initializing the socket object we move to step four, connecting a socket through `RSocket::Connect()` – The two overloads of this function both allow us to establish the actual connection. For TCP and UDP connections this is done by creating a `TInetAddr` object specifying the address and port of the remote device, this information is returned by the `RHostResolver` DNS query. Notice that the `RSocket::Connect()` function takes a `TRequestStatus` parameter which means that it will complete asynchronously. Normally we would wrap such function calls in an active object, however for the sake of simplicity we will call `User::WaitForRequest()`, which will block until the socket server completes the request. The next code snippet shows step five and six, transferring data, and closing the connection.

```
 1  // Construct a HTTP GET request
    _LIT8(KGetRequest, "GET_/_HTTP/1.0_\r\n\r\n");

    // Write the data
 5  socket.Write(KGetRequest, status);
    User::WaitForRequest(status);
    LOGLEAVE(status.Int(), "socket.Write()");

    TBuf8<1024> buf;
10  TSockXfrLength length;

    // Receive response
    socket.RecvOneOrMore(buf, 0, status, length);
    User::WaitForRequest(status);
15  LOGLEAVE(status.Int(), "socket.RecvOneOrMore");

    RBuf unicode;

    //Create buffer of correct size.
20  TRAP(res, unicode.CreateL(buf.Length()));
    LOGLEAVE(res, "unicode.CreateL()");
    CleanupClosePushL(unicode);

    //Note that Copy() panics if the target descriptor is too small.
25  unicode.Copy(buf);
    console->Printf(unicode);

    // RSocketServ, RHostResolver, RSocket, RBuf
    CleanupStack::PopAndDestroy(4);
```

There are several functions allowing data to be written over a socket, in this example we will write data using the `RSocket::Write()` function, which writes an 8-bit descriptor to the socket. Equivalently there are several functions allowing data to be read from the socket, in this case we use `RSocket::RecvOneOrMore()`. The `RSocket::RecvOneOrMore()` function will return as soon as data become available, reading a minimum of one byte – which is quite useful as we do not know how much data will be returned. Closing the socket is normally done in two steps, first by calling `RSocket::CancelAll()` to cancel all outstanding operations on the socket, this is however not necessary here as we do not have any outstanding read-/write requests. Step two, calling `Close()` on all open handles. We can conveniently do this by calling `CleanupStack::PopAndDestroy()` as all our open handles have been pushed to the cleanup stack. This concludes our small example of how TCP/IP connectivity can be used through the Symbian OS sockets API. In the following section we will use a new API added in Symbian

OS v9 allowing us to easily use Bluetooth OBEX, SMS, and Email from our applications.

4.4.2 Bluetooth OBEX, SMS, and Email

Symbian OS version 9 adds an additional API making it easy for us to use various forms of messaging in our applications. This section comprises the following examples:

1. Sending files over Bluetooth using OBEX.
2. Sending SMS messages.
3. Sending Email.

These functionalities are all implemented through a new easy-to-use API accessing the Symbian OS SendAs server. The SendAs server allows a client application to create and send messages using a variety of messaging types, with minimal effort. The client side API consists of two classes:

- **RSendAs**: This class allows our applications to open a session with the SendAs server.
- **RSendAsMessage** represents the message that we are constructing.

The following examples will show the usage of this API for OBEX Bluetooth transfer. The code should be explanatory with the comments included:

```
1   RSendAs sendas;

    // Connecting to the SendAs server
    TInt res = sendas.Connect();
5   LOGLEAVE(res, "sendas.Connect()");
    CleanupClosePushL(sendas);

    RSendAsMessage message;

10  // Selecting the appropriate Bluetooth MTM UID
    TRAP(res, message.CreateL(sendas, KSenduiMtmBtUid));
    LOGLEAVE(res, "message.CreateL()");
    CleanupClosePushL(message);

15  // The file we are going to push, you need to make sure that the
    //    file
    // is present on the device
    _LIT(KFile, "C:\\pushfile.txt");

    // Constructing the message
20  TRequestStatus status;

    message.AddAttachment(KFile(), status);
    User::WaitForRequest(status);
    LOGLEAVE(status.Int(), "message.AddAttachment()");

25
    CleanupStack::Pop(1); // RMessage

    // Launch the dialog allowing us to choose a receiver
    // device and send the file
30  TRAP(res, message.LaunchEditorAndCloseL());
    LOGLEAVE(res, "message.LaunchEditorAndCloseL()");

    CleanupStack::PopAndDestroy(1); // RSendAs
```

Figure 4.21. Selection dialog opened automatically by the SendAs server querying the user which device he or she wants to transfer the file to.

The above code example will open a dialog asking the user which Bluetooth devices should receive the message. This is done through the `RSendAsMessage` `::LaunchEditorAndCloseL()` command, as seen in Figure 4.21.

This example has shown one way to use the Bluetooth connectivity available on most devices. In addition to using the SendAs server, several other options for Bluetooth communication exist. A programmer can access the RF-COMM and L2CAP Bluetooth protocol layers creating both SCO and ACL links through the Bluetooth v2 socket API. The interested reader should look up the `CBluetoothSocket` and `CBluetoothSynchronousLink` classes in the SDK and see [3], which also describes more advanced usage of the OBEX protocol through the `CObexServer` and `CObexClient` classes. In the following example we will use the SendAs API to send a SMS message. The actual code does not vary a great deal from the previous example, which emphasizes nicely how easy to use the new API is.

```
1   RSendAs sendas;

    // Connecting to the SendAs server
    TInt res = sendas.Connect();
5   LOGLEAVE(res, "sendas.Connect()");
    CleanupClosePushL(sendas);

    RSendAsMessage message;

10  // Selecting the appropriate SMS MTM UID
    TRAP(res, message.CreateL(sendas, KSenduiMtmSmsUid));
    LOGLEAVE(res, "message.CreateL()");
    CleanupClosePushL(message);

15  // The body text and receiver phone number
    _LIT(KText, "Test_test_:D");
    _LIT(KNumber, "99998928");
    TRequestStatus status;

20  // Constructing the message
    message.AddRecipientL(KNumber, RSendAsMessage::ESendAsRecipientTo);
    message.SetBodyTextL(KText);
```

```
   // Send the message
25 CleanupStack::Pop(1); // RMessage
   TRAP(res , message.SendMessageAndCloseL());
   LOGLEAVE(res , "message.AddAttachment()");

   CleanupStack::PopAndDestroy(1); // RSendAs
```

The main difference from the Bluetooth example is, that we need to set a receiver, and that we do not prompt the user for any input[9], but silently send the message. Also notice that we need to select the correct MTM id (Message Type Module), a MTM is a plug-in into the Symbian messaging architecture that provides support for certain types of messages. The MTM ids are defined in the `senduiconsts.h` header file. As with Bluetooth many more advanced ways of sending and receiving SMS messages exist using the Symbian OS messaging architecture. The interested reader should refer to the SDK documentation and [4]. The final example uses the SendAs API to create and send an Email shown in the following code listing:

```
 1 RSendAs sendas;

   // Connecting to the SendAs server
   TInt res = sendas.Connect();
 5 LOGLEAVE(res , "sendas.Connect()");
   CleanupClosePushL(sendas);

   RSendAsMessage message;

10 // Selecting the appropriate SMTP MTM UID
   TRAP(res , message.CreateL(sendas , KSenduiMtmSmtpUid));
   LOGLEAVE(res , "message.CreateL()");
   CleanupClosePushL(message);

15 // The subject , body text and receiver email
   _LIT(KSubject , "Testing_email");
   _LIT(KText , "Test_test_:D");
   _LIT(KNumber , "mvpe@kom.aau.dk");
   TRequestStatus status;

20
   // Constructing the message
   message.SetSubjectL(KSubject);
   message.AddRecipientL(KNumber, RSendAsMessage::ESendAsRecipientTo);
   message.SetBodyTextL(KText);

25
   // Send the message
   CleanupStack::Pop(1); // RMessage

   TRAP(res , message.SendMessageAndCloseL());
30 LOGLEAVE(res , "message.AddAttachment()");

   CleanupStack::PopAndDestroy(1); // RSendAs
```

This example requires that an Email account with accessible SMTP server is already configured on the phone. If this is not the case, the Email delivery will fail. As with SMS messaging more advanced features of Email sending and receiving can be used through the Symbian Messaging Architecture.

---

[9] Platform security may query the user to allow the SMS, but this will not be discussed here.

4.5 Summary

The Symbian platform is today one of the most attractive both in regard to market share and development opportunities. Throughout this chapter we have gone through the initial steps of Symbian OS development, allowing new developers to rapidly begin application development. The Symbian C++ approach should be the choice of any developer requiring advanced functionality and the maximum flexibility.

References

1. Leigh Edwards, Richard Barker, and the Staff of EMCC Software Ltd. *Developing Series 60 Applications*. ISBN 0-321-22722-0. Addison-Wesley, March 2004.
2. Craig Heath. *Symbian OS Platform Security*. ISBN 0-470-01882-8. Symbian Press, March 2006.
3. Nokia. S60 Platform: Bluetooth API Developer's Guide. December 2006.
4. Nokia. S60 Platform: SMS Example v2.0. May 2006.
5. Jo Stichbury. *Symbian OS Explained*. ISBN 0-470-02130-6. Symbian Press, October 2004.

5

Open C

Eero Penttinen and Antti Saukko

Nokia {`eero.penttinen`|`antti.saukko`}`@nokia.com`

5.1 Introduction

Open C is a solution that enables efficient cross-platform development, makes porting easier, and lowers barriers to start writing native C code to S60 without needing to learn Symbian C++. Open C is a set of standard POSIX and middleware C libraries for the S60 smartphone platform, which is based upon Symbian OS. The same libraries can be found in Linux and other UNIX-based operating systems, enabling cross-platform development and efficient reuse of code written for those operating systems. Open C brings to S60 well-known C libraries, including subsets of POSIX, OpenSSL, zlib, and GLib. In its first release, Open C includes over 70 percent of the functions contained in the full versions of the libraries. Bringing this functionality to S60 provides the following advantages:

1. Developers can quickly become productive in mobile programming using their existing skills and knowledge without needing to learn Symbian C++ programming paradigms.
2. They can develop parts of their applications faster using familiar standard interfaces and programming models in an environment of their preference.
3. They can more easily develop common cross-platform components in their software.
4. They can reduce porting effort for legacy or open source code written in C using POSIX and/or GLib libraries.

Note, it is important to know that Open C is supported as an SDK extension for the first time in S60 3rd Edition; it cannot be used in earlier platform editions. For S60 3rd Edition and S60 3rd Edition Feature Pack 1, Open C will be available as an SDK plug-in. In S60 3rd Edition Feature Pack 2, Open C will be part of the SDK. More details about the Open C development environment are provided later in this chapter. One of the key criteria in the development of Open C was not to interfere with the standard APIs that have been around for ages. However, there are conceptual differences between

F.H.P. Fitzek and F. Reichert (eds.), Mobile Phone Programming and its Application to Wireless Networking, 139–158.
© 2007 *Springer.*

desktop and embedded devices that have necessitated some alterations in the existing libraries. Basically, the way that memory, CPU, and network are used depends on the surrounding environment, so the intention is to optimize for the environment in which the binary is being run. In order for Open C to benefit as many as possible and to make it feasible for third parties to develop serious applications on top of it, some modifications had to be introduced, but with minimal disruption. There are thousands of examples of how to use standard POSIX APIs and the support libraries Open C provides, so this chapter concentrates on examining the differences between these two worlds. It describes what libraries are part of the first release of Open C, gives details of how to set up the development environment, shows the steps that need to be followed to write a simple "Hello World" type of application using Open C, describes how software based on Open C can interact with Symbian C++/S60, and finally, offers some important porting tricks to keep in mind when porting software on top of Open C. Open C offers developers a great opportunity to get back to the basics and increase their productivity.

5.2 What is Open C?

Open C is a collection of libraries based on POSIX and other popular open source projects. Open C includes more than 70 percent of the functions contained in the following familiar libraries:

Libc Standard C libraries, including standard input/output routines, database routines, bit operators, string operators, character tests and character operators, DES encryption routines, storage allocation, time functions, sockets, interprocess communication, and process spawning.

Libm Provides arithmetical and mathematical functions.

Libpthread Pthreads API provides an IEEE Std1003.1c (POSIX) standard interface implementing multiple threads of execution within a traditional user process. This implementation is currently user-space only. Pthreads provides functions for thread creation and destruction, an interface to the thread scheduler to establish thread scheduling parameters, and mutex and condition variables to provide mechanisms for the programmer to synchronize access to shared process resources.

Libz "Zlib" compression library provides in-memory compression and decompression functions, including integrity checks of the uncompressed data.

Libdl Provides the user with the functionality to dynamically load the DLLs.

Libcrypto OpenSSL crypto library implements a wide range of cryptographic algorithms used in various Internet standards. The services provided by this library are used by the OpenSSL implementations of Secure Sockets Layer (SSL), Transport Layer Security (TLS), and S/MIME, and they have also been used to implement SSH, OpenPGP, and other cryptographic standards.

Libssl OpenSSL ssl library implements the SSL (SSL v2/v3) and TLS (TLS v1) protocols.

Libcrypt Cryptography libraries containing functions for crypting/encrypting datablocks and messages and password hashing.

Libglib A general-purpose utility library that provides many useful data types, macros, type conversions, string utilities, file utilities, a main loop abstraction, and so on. It works on many UNIX-like platforms, Windows, OS/2, and BeOS.

Symbian has also released a set of POSIX libraries, calling them P.I.P.S. ("PIPS Is POSIX on Symbian"). The libraries in P.I.P.S. (libc, libm, libdl, and libpthread) have been jointly developed by Nokia and Symbian to ensure that code written on top of Open C delivers the performance expected of a native environment. How do the four common libraries libc, libm, libdl, and libpthread relate to Symbian Stdlib (estdlib.dll)? P.I.P.S. provides a much more complete, up-to-date, standards-compliant solution than Stdlib. The amount of functionality has been increased in P.I.P.S., not to mention support of libdl and libpthread. Also, the performance has been improved, making it more suitable for mobile devices. Note, it is strongly recommended not to use Stdlib anymore. Keep in mind that Stdlib and Open C environments are not compatible. Open C extends the P.I.P.S. offering with libraries like OpenSSL, libz, and GLib. Common libraries in Open C and P.I.P.S. are exactly the same, exposing the same functionality to ensure full compatibility. A typical S60 application structure using Open C is shown in Figure 5.1.

As illustrated in Figure 5.1, an S60 application can use portable code, which is written on top of Open C. The portable code can be either open source software or legacy code written in standard C. In order to write full-blown S60 applications, S60 and Symbian C++ native APIs must also be used.

To use Open C efficiently, it is important to know some of the limitations:

- Open C is not a Linux/UNIX application environment where you can execute binaries built for those operating systems. However, Open C enables a lot of software written for those platforms to be easily ported to S60 devices, enabling cross-platform components.

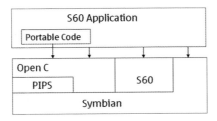

Figure 5.1. A typical application structure using Open C.

- Open C is not fully POSIX compliant with ISO/IEC 9945. The implementation is however as compliant as possible.
- There is no support for signal, fork, and exec. If the software that developers intend to port contains these system calls, careful analysis must be done about the usage to see how much modification must be done to the ported software. Different kinds of coding patterns have been documented by Symbian how to port software relying on these system calls.
- dlsym – address lookup through symbol names is not supported, and ordinal numbers rather than symbol names must be used. Ordinal numbers of the exported functions can be obtained from the module definition file (.def) of the associated DLL as shown in the listing below. This same limitation applies for the g_module_symbol function.

```
/* The below will not work in Open~C SDK plug-in */
ret = g_module_symbol(module, "openConnection",&ptr);
/* And it has to be like below. 1 is the ordinal of the exported
    function */
ret = g_module_symbol(module, "1", &ptr);
```

An example module definition file where ordinals are defined

```
EXPORTS
openConnection @ 1 NONAME
closeConnection @ 2 NONAME
```

It has to be noted that his is a limitation with the current Open C SDK plug-in only. The current plug-in allows developers to build Open C modules and deploy them to existing mobile devices built on S60 3rd Edition and S60 3rd Edition Feature Pack 1. Symbol lookup by name will be supported beginning in S60 3rd Edition Feature Pack 2 when using the new target types.

- Open C does not currently provide any APIs for UI development (such as GTK+). The UI can be written using existing S60 APIs, and such a UI can be developed using the UI designer in Carbide.c++ Developer or Professional edition. The UI can also be built using S60 Python with Python bindings for the exported functionality.
- Open C does not offer native C-bindings to Symbian/S60 platform services like telephony, messaging, Bluetooth, location services, etc. In those cases, developers have to use the provided C++ interfaces, but from a portability point of view, it is recommended to isolate platform-specific source code to different source files.

5.3 Setting up the Environment

To start using Open C, developers need to set up the development environment. The setup depends on the targeted S60 platforms. In order to write software using Open C for S60 3rd Edition and S60 3rd Edition Feature Pack

1-based devices, developers must download the Open C SDK plug-in from Forum Nokia. Before installing the Open C SDK plug-in, they must have the S60 3rd Edition SDK installed (this is the recommended SDK). They can also install the plug-in on top of S60 3rd Edition Feature Pack 1; however, this will limit the amount of devices where their software can run. The Open C SDK plug-in is installed on top of this SDK.

The Open C SDK plug-in contains the following artifacts:

- Headers and libraries enabling compilation for both the emulator and the device.
- Emulator runtime components enabling testing of Open C-based software in the emulator.
- SIS packages containing needed Open C runtime components that can be shipped with the developer's application.
- Open C documentation containing both developer guides and API reference documentation.
- Open C example applications illustrating various usage aspects of Open C.

Open C will be part of the S60 3rd Edition Feature Pack 2 SDK, and developers just need to download and install the needed SDK to have the development environment ready. However, they cannot use this SDK to build software that will work on earlier platform versions. S60 3rd Edition Feature Pack 2 offers improved support for Open C by introduction of new target types STDEXE, STDDLL, and STDLIB, intended to be used when writing Open C executables. Using these new target types provides the following advantages:

1. Symbol lookup by name will work as specified by the standard when using new target types:

```
/* This will work in S60 3rd FP 2 using new target types */
ret = g_module_symbol(module,"openConnection",&ptr);
```

   With old target types (EXE, DLL) the same limitation will apply as with the Open C SDK plug-in.
2. There is no need to annotate source with EXPORT_C or IMPORT_C for modules that used one of the the new target types. All symbols with external linkage are:
   - Exported if they are defined in the module.
   - Imported if they are referred but not defined.

From a porting point of view, the second point is very important. Often when porting a library, developers need to annotate all exported functions with IMPORT_C and EXPORT_C, which can be laborious if the library exports hundreds of APIs. When using the new target types, this is unnecessary It also makes it easier to take a new version of a ported library into use since IMPORT_C and EXPORT_C statements can be omitted in the ported software.

How do the new target types differ from the old ones?

STDLIB versus native LIB

1. Source code annotation is not required for C/C++ files under this target.

STDEXE versus native EXE

1. STDEXE has main as the entry point while native EXE has E32Main.
2. STDEXE targets have a (default) UID2, 0x20004C45.
3. Symbol lookup via names is enabled for STDEXE but not with native EXEs.

STDDLL versus native DLL

1. STDDLL targets have a (default) UID2, 0x20004C45.
2. Symbol lookup via names is enabled for STDDLLs but not with native DLLs.

Additional improvements are done to the tool chain, making it easier to configure MMP files for Open C-enabled projects, as shown in the next section.

5.4 Getting Started with Open C: Hello World

In order to become quickly productive with Open C, let us start with a simple "Hello World" application. Differences between the development environments will be highlighted.

Step 1 – Create the project bld.inf and MMP file:

```
// BDL.INF
PRJ_PLATFORMS
3   DEFAULT

PRJ_MMPFILES
HelloOpenC.mmp
```

This MMP file is required when using the Open C SDK plug-in:

```
// HelloOpenC.mmp
TARGET          HelloOpenC.exe
TARGETTYPE      EXE
4   UID          0x100039CE 0x13572468
CAPABILITY      NONE

SOURCEPATH      ..\src
SOURCE          HelloOpenC.c
9   SYSTEMINCLUDE \epoc32\include
SYSTEMINCLUDE \epoc32\include\stdapis

// This must be specified as the first library when using main()
    entry point
// In S60 3rd Ed PF 2 this will not be needed when using the new
    target type
14  // STDEXE
STATICLIBRARY libcrto.lib
LIBRARY         libc.lib
LIBRARY         euser.lib
```

S60 3rd Edition Feature Pack 2:

```
// HelloOpenC.mmp
TARGET          HelloOpenC.exe
TARGETTYPE      STDEXE // This is new in S60 3rd Edition FP 2
UID             0x20004C45 0x13572468
CAPABILITY      NONE

SOURCEPATH      ..\src
SOURCE          HelloOpenC.c
SYSTEMINCLUDE   \epoc32\include

LIBRARY         libc.lib
```

When using S60 3rd Edition Feature Pack 2, the differences are:

- Instead of using the targettype EXE, STDEXE is used. This is used when developers want to use the main() entry point with Open C.
- No need to specify the SYSTEMINCLUDE path for Open C headers.
- No need to link against libcrt0.lib and euser.lib.
- Different UID2 is used.

Step 2 – Create needed source files

Create C or CPP files as usual. An Open C application uses the main() entry point, and not E32Main() as native Symbian C++ applications do.

```
/* HelloOpenC.c */
#include <stdio.h>
int main()
{
    printf("Hello_Open~C!\n");
    return 0;
}
```

Step 3 – Building and running

The application is compiled and executed as a normal Symbian application. The application can be executed in the emulator and/or in the device.

Step 4 – Signing and deployment

Before the application can be executed in the device, it has to be signed. If additional capabilities are required, they have to be mentioned in the MMP file. Please consult the corresponding SDK documentation for an explanation of how this is done. Open C example applications come with necessary package files that can be used for creating the SIS files.

Congratulations! You have completed your first application using Open C. As you have noticed, from a building point of view, it does not differ from a normal Symbian C++ application; from a coding point of view, it does not differ from a normal C application. In fact, from day one, it is very easy to get started and be productive using Open C if you are an experienced C programmer.

5.5 Open C and Symbian/S60 Interaction

In order to write software using Open C, developers must understand how their software can interact with native Symbian/S60 code. The current Open C library offering does not allow developers to write full-blown applications with a GUI and use native platform services like Bluetooth or messaging. There are different approaches that can be used depending on the developer's skill set, preferences, and requirements for the software:

- Use native Symbian/S60 UI framework and native Symbian/S60 APIs. This is the preferred solution in many cases if the developer wants to write professional and efficient applications.
- Use S60 Python and write necessary Python bindings to the exported functions. This solution enables rapid development of the UI.

Developers can also use Open C code directly in their Symbian/S60 application. In such cases, it is a good design practice to isolate Open C code to separate modules.

5.5.1 Open C DLLs

Because Open C does not offer any APIs for building the GUI, Open C is typically used when writing application engines or middleware components in a platform-independent manner. These components are typically called from the UI part of the application. Separating the UI from the engine is a recommended design practice when writing applications – with Open C it enables experienced Symbian/S60 developers to focus on the UI and platform-dependent components, while experienced C programmers can write the application engine and/or needed middleware components.

Below is an MMP, a header, and a C file for a typical DLL when using the Open C SDK plug-in:

```
    // sampledll.mmp
    TARGET            sampledll.dll
  3 TARGETTYPE        DLL
    UID               0x1000008d 0x12213443
    ...

    // Open~C libc is required
  8 LIBRARY libc.lib

    /* sampledll.h */
    #ifdef SAMPLEDLL_H
 13 #define SAMPLEDLL_H

    /* Make the header C++ friendly using extern "C" */
    #ifdef __cplusplus
    extern ''C'' {
 18 #endif

    IMPORT_C int Func1();
    ...
```

```
23  #ifdef __cplusplus
    }
    #endif

    #endif /* SAMPLEDLL_H */
28

    /* sampledll.c */
    #include ''sampledll.h''

33  EXPORT_C int Func1() {
    /* Some code */
    }
```

And the same with S60 3rd Edition Feature Pack 2:

```
    TARGET            sampledll.dll
    TARGETTYPE        STDDLL
    UID               0x20004C45  0x12213443
    ...
5
    // Open~C libc is required
    LIBRARY           libc.lib

    /* sampledll.h */
10  ...
    /* No need for IMPORT\_C with STDDLL target type */
    int Func1();
    ...

15  /* sampledll.c */
    \#include ''sampledll.h''
    ...
    /* No need for EXPORT\_C with STDDLL target type */
    int Func1() {
20  /* Some code */
    }
    ...
```

The differences between Open C and S60 3rd Edition Feature Pack 2 when writing DLLs are:

- Target type is different – DLL versus STDDLL.
- No need to annotate source files with EXPORT_C and IMPORT_C in S60 3rd Edition Feature Pack 2.
- UID2 is different.

Note in the previous example the usage of the extern "C" linkage specifier. If the DLL is compiled with a C compiler and you want to use those C functions from C++ code, the user of those APIs only needs to include the header file in order to use those APIs. It is a recommended practice to write headers in a C++ friendly way. One typical problem that developers run into is how to export global data from a DLL to be accessed by either Open C or Symbian applications. Many OSS projects typically use a lot of global data, and in order to minimize code changes, it has to be carefully analyzed whether global data will be used or the code will be redesigned.

It is strongly recommended to avoid having global data in DLLs for the following reasons:

- EKA2 Emulator only allows a DLL with WSD to load into a single process.
- RAM usage for WSD data chunk is at least one 4K RAM page (the smallest possible RAM allocation), irrespective of how much static data is required.
- Chunks are a finite resource on ARMv5. Every process loading WSD-enabled DLLs uses a chunk to hold the data.
- ARM architecture 4 and 5 specific costs and limitations that apply only to DLLs that link against "fixed processes".
- Limit on the number of DLLs in a process with WSD.

If developers understand the caveats around using global data in DLLs, the following pattern can be used:

1. Do not export global variables.
2. Export one method that returns a pointer to that variable.
3. Define a macro for the user of the DLL.

In the MMP file of the DLL, the developer must write EPOCALLOWDLL-DATA to specify that global data will be used in the DLL; if this is not done, building for the target will fail.

1. Do not export global variables:
Within a DLL you have one global variable, for example:

```
int globalVal;
```

2. Export one method that returns a pointer to that specific variable:

```
EXPORT_C int* GlbData ()
{
    return &globalVal;}
}
```

3. Define a macro for the user of the DLL.
Within the DLL header (e.g., sampledll.h) define the following:

```
#ifdef __cplusplus
extern "C" {
#endif
IMPORT_C int* GlbData();
#define globalVal (*GlbData())
...
#ifdef __cplusplus
}
#endif
```

The usage is as shown in the code sample below:

```
#include "sampledll.h"  // DLL header
int main()
{
    globalVal = 10;
    globalVal++;
    ...
}
```

It is bad API design practice to access global variables directly; this should be avoided if new code is written by using getter and setter functions instead. However, legacy or open source software might expose global variables and in those cases, this design pattern is useful if a developer wants to minimize code changes.

5.5.2 String/Descriptor Conversions

Standard C strings and Symbian descriptors are by nature very different. The main difference is that descriptors know how many characters are in a data array. A C string does not know its length, so when length is needed, the NULL character that indicates the end of the string has to be scanned. Another difference arises with buffers. When C code reserves a buffer from the heap or stack, it has to keep the maximum length somewhere. Many C methods that alter the buffer contents do not respect the maximum size of the buffer and can override the reserved memory, causing unknown behavior. Some methods take the maximum length as a parameter but it is difficult to use those types in functions, since pointer to array and maximum length have to be passed separately. Buffer descriptors can tell the maximum length, and all the methods they provide respect the buffer limits.

When using neutral descriptor types there is no need to worry about character widths. In a C program, the programmer has to explicitly specify which method to use, for example strcat or wcscat.

Table 5.1 contains a comparison of standard C string functions and Symbian descriptors. There are cases when descriptors are passed to software written in Open C and vice versa, that C strings are passed to the Symbian C++ application. Before further processing the received strings, they should be converted to the appropriate type most suitable for the programming environment. A good practice is that the developer of a library provides convenience wrapper interfaces using descriptors when strings need to be passed to and from the library. This makes the usage of the functions easier from a Symbian C++ application.

Example:

```
     /* wrapper.h */
     ...
 3   IMPORT_C void SendData(TDes& aArg);
     ...

     /* wrapper.cpp */
     #include "comms.h"
 8   ...
     EXPORT_C void SendData(TDes& aArg)
     {
          /* sendData is written in Open~C */}
          sendData((wchar_t *) aArg.PtrZ());
13   }

     /* comms.h */
     ...
     IMPORT_C sendData(wchar_t * buf);
```

One Open C example application, OpenCStringUtilitiesEx, demonstrates how to convert Symbian descriptors to char and wchar strings and vice versa, how to convert char and wchar strings to different Symbian descriptors.

The example below shows how to convert a TBuf8 descriptor to a char buffer.

```
EXPORT_C char* tbuf8tochar(TDes8& aArg)
{
    return (char*)aArg.PtrZ();
}
```

Table 5.1. Comparison of standard C string functions and Symbian descriptors.

C Function	Symbian	Description
sprintf, swprintf	TDes::Format	Write formatted data to a string.
strcat, wcscat, strncat, wcsncat	TDes::Append	Append string to another.
strcmp, strncmp, wcsncmp	TDesC::Compare	Compare strings lexicographically.
strcpy, wcscpy, strncpy, wcsncpy	TDes::Copy	Copy string to another.
strchr, wcschr	TDesC::Locate	Find a character in a string.
strrchr, wcsrchr	TDesC::LocateReverse	Scan a string for the last occurrence of a character.
strspn, wcsspn	None	Scan index of the first character from string that does not exist in alphabet array.
strcspn, wcscspn	None	Scan the index of the first occurrence of a character in string that belongs to the set of characters.
strstr, wcsstr	TDesC::Find	Find a substring.
strtok, wcstok	TLex::	Find the next token in a string.
strlen, wcslen	TDesC::Length	Get the length of a string.
strcoll, wcscoll	TDesC::CompareC	Compare strings using locale-specific information.
strftime, wcsftime	Using TDes::Format and TTime	Format a time string.

5.5.3 GLib

Because there is a large application base that uses GLib, it became obvious that in order to provide maximum support for developers to write new code as well as port existing code, GLib would have to be part of the application middleware base. Quoting the GTK Website: "GLib is the low-level core library that forms the basis of GTK+ and GNOME. It provides data structure handling for C, portability wrappers, and interfaces for such runtime functionality as event loops, threads, dynamic loading, and an object system". In a sense, it replaces some functionality from the standard C library but it also brings in some nice, often-used features such as linked lists and timers.

Memory

Generally the concept of software design is quite different in the desktop world compared to how it is for small memory embedded devices. In the desktop world, the ability to perform in a very strict, memory-conservative environment is not usually one of the key design criteria. Developers rely on always having enough memory to run. This is the concept behind GLib, as well as operating systems such as Linux, where by default memory overcommit is the default allocation strategy. Hence, all memory allocation will always succeed, regardless of the current memory condition. This is an acceptable design criterion for Linux to perform quick memory allocations. Due to the conceptual differences between desktop and embedded devices, memory allocation needs to work differently, too. Since Symbian OS does not permit memory overcommitting, nor support virtual memory with swapping, the implementation Open C has on GLib will work a bit differently as well. The default behavior upon memory allocation failure in desktop computers on GLib is to abort the application; this is not the case with mobile devices. Open C has extended GLib a bit. If developers allocate memory on the Open C version of GLib, they will have to start checking on whether the request for memory has succeeded.

For low memory handling, it is advisable to use these new mechanisms to ensure that the application functions properly.

```
1    -- example --

     #include <glowmem.h> /* Contains the macro definition introduced
         later */

     gchar * function1 (gchar * x)
6    {
             gchar *temp;

             SET_LOW_MEMORY_TRAP (NULL) ;
             temp = g_strdup (x) ;
11           REMOVE_LOW_MEMORY_TRAP () ;

             return temp;

     }

16   -- example   --
```

The following macros

```
SET_LOW_MEMORY_TRAP( failure_value )
SET_LOW_MEMORY_TRAP_VOID ( )
REMOVE_LOW_MEMORY_TRAP ( )
```

are defined in a new header file, called glowmem.h. It is recommended that developers use these macros around all operations that do memory allocations. In the example code, upon failure to allocate memory the example function1() will return with the error "value" passed to SET_LOW_MEMORY_TRAP (value). In this example, the value is NULL, so if g_strdup() fails, the function1() will return with NULL. The REMOVE_LOW_MEMORY_TRAP() is needed to remove the trap handler earlier set by the SET_LOW_MEMORY_TRAP() in this scope. The REMOVE_LOW_MEMORY_TRAP() needs to be called just before returning. Low memory conditions should be taken seriously if developers plan to port some of their GLib source code to S60 smartphone devices.

GLib and Process Spawning

As discussed in Section 5.5.1, Open C has certain limitations, such as lack of support for fork and exec. Effectively this will have an impact in spawning processes in GLib. All the processes spawned via GLib's APIs will ignore parameters passed for the environment as well as the working directory.

GModule

GModule provides facilities corresponding to libdl. It is an enabler for dynamically loading modules, like plug-ins. The basic principle is to load libraries (by a name) and requesting function (symbol) locations. As mentioned earlier, the implementation that we have on libdl is not the full-blown one in the Open C SDK plug-in. However, although developers lack the ability to do function lookup by name, they can use ordinal numbers. For the sake of API compliancy, the ordinal number should be passed to the function as text, instead of as an integer.

```
--- example ---}
2
#include <gmodule.h>
#include <glib.h>

typedef int (*SimpleFunc) (void);
7
int main()
{
    GModule *module = NULL;
    gpointer func;   // an untyped pointer for a function
12    SimpleFunc f_a;
    int retVal;

    if (!g_module_supported()) // checking support for dynamic
        loading
```

```
17        {
                  g_print ("Dynamic_Opening_of_modules_is_not_supported");
                  return 1;
          }

          /* A module by filename called 'pluginmodule.dll' is opened
22         * and handle is returned.
           * G_MODULE_BIND_LAZY is overridden and the module is opened
             with
           * flag G_MODULE_BIND_LOCAL
           */

27        module = g_module_open("pluginmodule.dll",G_MODULE_BIND_LAZY);

          // 1 is the ordinal number for pluginmodules first function()

          if(module && g_module_symbol(module, "1" ,&func))
32        {
              f_a = (SimpleFunc)func;
              retVal = f_a();
              g_print("Function_at_ordinal_number_1_of_module__returns_%d",
                  retVal);
          }
37        else
          {
              g_print("Error_quering_symbol_at_ordinal_number_1");
              return 1;
          }
42
          return 0;
      }

      -- example --
```

GLib Main Event Loop

The GLib main event loop manages all events from GLib applications. These GLib events are emitted from a series of events, from file descriptors and sockets to timeouts. This main event loop creation is a bit different when mixed up with S60 Avkon applications. Basically, all Symbian applications that do asyncronous operations have been designed to do cooperative multitasking by using active objects, which means that these active objects are instantiated from a class that is derived from CActive. All of these active objects register to an ActiveScheduler. Once an active object makes a request, it is put into the queue of the active scheduler. Upon completion, the active scheduler calls the active object's virtual RunL() -method to signal the status. In a principle GLib's main event loop is quite corresponding to active objects in Symbian OS. In order to use GLib's main event loop, mix these two together, by creating an active object that runs one iteration of the GLib event loop every time it gets scheduled.

```
      -- example --

      class CGlibEventHandler: public CActive}
4     {
      public:
              static CGlibEventHandler* NewL(GMainContext* aMainContext);
```

```
                           ~CGlibEventHandler();
                   void RunL();
 9                 void DoCancel();
                   void Start();
                   void Stop();
          private:
                   CGlibEventHandler(GMainContext* aMainContext);
14                 void ConstructL();
                   RTimer iTimer;
                   GMainContext* iMainContext;
          };

19  -- example --
```

```
 1  -- example --

    CGlibEventHandler::CGlibEventHandler(GMainContext* aMainContext):
        CActive(EPriorityStandard)
    {
      iMainContext = aMainContext;
 6  }

    void CGlibEventHandler::ConstructL()
    {
      User::LeaveIfError(iTimer.CreateLocal());
11    CActiveScheduler::Add(this);
    }

              .
              .
              .
16            .


    void CGlibEventHandler::Start()
    {
      iTimer.After(iStatus, TTimeIntervalMicroSeconds32(1000));
21    SetActive();
    }

    void CGlibEventHandler::RunL()
    {
26        g_main_context_iteration(iMainContext, FALSE);
          iTimer.After(iStatus, TTimeIntervalMicroSeconds32(1000));
          SetActive();
    }

              .
31            .
              .
              .


    -- example --
```

```
    -- example --

    void COpenCHelloGlibEventExUi::ConstructL()
    {
 5        .
          .
      iMainContext = g_main_context_new();
      iTimeoutSource = g_timeout_source_new(1000);

10    g_source_set_callback(iTimeoutSource, timeout_cb, iAppView, NULL);
      g_source_attach(iTimeoutSource, iMainContext);

      iGlibEventHandler->Start();
```

```
15   }
     gboolean timeout_cb(gpointer data)
     {
         // code

20   }

     -- example --}
```

Practically, this means creation of a new GMainLoop with g_main_context_new() in the ContructL() function, an instance of RTimer, and in the virtual RunL() function the need to call g_main_context_iteration(). The call to iGlibEventHandler->Start() will effectively start the RTimer ticker. In RunL(), you will run one iteration of the event loop and request another callback. The SetActive() function notifies the active scheduler that this object has issued a request and waits for completion. For further information about active objects, please refer to earlier chapters in the book.

Using Open C TCP/IP Sockets

The socket interface in POSIX is quite simple. Generally in the desktop world, computers have a single interface. In the case of more than one interface, the decision of where to route packages is based on routing tables and the default route of the computer. With mobile terminals, there can be several interfaces out of the device, with different configurations per interface. For example, there can be an access point for multimedia messaging service (MMS), Internet, WAP, or, at home, Wireless LAN. With mobile terminal applications, you either have to set up which access point to use, or let the end user select it. Another big difference between the PC and mobile worlds – most of the time, mobile terminals are not connected to a network (other than cellular) and therefore do not have an IP address, whereas PCs are most likely to be connected all the time. These conceptual differences are reflected in certain differences with the Open C implementation on sockets. In Symbian OS, RConnection and RSubConnection classes can be used together with RSocket to access network services. Open C provides a way of accessing these same APIs, but with a "C style". The following examples of ioctl(), if_nameindex(), if_freenameindex(), if_indextoname(), and if_nametoindex() –functions show how they can be used with Open C to gain access to a network. Note that using sockets requires platform security capabilities NetworkServices if you need TCP or UDP sockets, or NetworkControl if you plan to use IP or ICMP, respectively. The example below demonstrates how to populate a list of all the connections and a list of the active connections.

```
     -- example --

3    struct ifconf ifc;
     int sockfd;

     sockfd = socket(AF_INET,SOCK_DGRAM,IPPROTO_UDP);
```

```
 8  ifc.ifc_len = sizeof(ifreq) * 20;
    ifc.ifc_buf = (caddr_t)malloc( ifc.ifc_len );

    // This will fetch the access point list, ifc.ifc_len will contain
         the size ifc.ifc_buf.
    // ifc.ifc_len/sizeof(ifreq) will give the number of access points.
13  ret = ioctl(sockfd, SIOCGIFCONF, &ifc);
    // This will fetch the access point list of active connetions, ifc.
         ifc_len will contain the size ifc.ifc_buf.
    // ifc.ifc_len/sizeof(ifreq) will give the number of active
         connections.

    ret = ioctl(sockfd, SIOCGIFACTIVECONF, &ifc);
18
    close(sockfd);
    free(ifc.ifc_buf);

    -- example --}
```

Once you have populated the list of potential access points, use ioctl() with SIOCSIFNAME and SIOCIFSTART to open up a connection:

```
    -- example -- }
 3  void open_interface( char * ifname )
    {
         ifreq ifr;
         int sockfd;

 8       // Name of the interface
         strcpy(ifr.ifr_name, ifname);

         sockfd = socket(AF_INET,SOCK_DGRAM,IPPROTO_UDP);
         ioctl(sockfd,SIOCSIFNAME, \&ifr);
13       ioctl(sockfd, SIOCIFSTART , \&ifr);
                .
                .
         // socket operations}}

18              .
         ioctl(sockfd, SIOCIFSTOP, \&ifr);
         close(sockfd);
         return;
    }
23
    -- example --
```

Developers who plan to use an existing active interface (also known as a subconnection) basically only need to use the existing connection name with ioctl() using SIOCSIFNAME, followed by an ioctl() with SIOCIFACTIVE-START.

```
 1  -- example --

    sockfd1 = socket(AF_INET,SOCK_DGRAM,IPPROTO_UDP);
    ioctl(sockfd1,SIOCSIFNAME, &ifr);
    ioctl(sockfd1, SIOCIFSTART , &ifr);
 6
    // Lets create another connection using the existing Open Connection

    sockfd2 = socket(AF_INET,SOCK_DGRAM,IPPROTO_UDP);
    ioctl(sockfd2,SIOCSIFNAME, &ifr);
```

```
11 ioctl(sockfd2 , SIOCIFACTIVESTART , &ifr);
   -- example --}
```

The previous examples on socket usage can be used together with functions such as if_nameindex(), if_freenameindex(), if_indextoname(), and if_nametoindex(), which are defined in net/if.h.

if_nameindex Returns all network interface names and indexes.
if_freenameindex Frees memory allocated by if_nameindex.
if_indextoname Maps a network interface index to its corresponding name.
if_nametoindex Maps a network interface name to its corresponding index.

This functionality is standard and therefore will not be discussed further.

5.6 Guidance for Porting

5.6.1 Getting Started

A general rule of thumb for porting is that if a component has been ported to more than one operating system, that is a good sign! Linux will serve as our example here, but all UNIX variants are similar. Basically, the first step in porting an application to S60 is to analyze what they are built on. Some good tools for performing this analysis are ldd and nm. ldd provides a printout of the shared library dependencies a program or library has; nm identifies the symbols it uses. Cross-checking this information against the Open C libraries and functions should give developers a good idea of the effort required to port the code. Remember: Open C is not a full-blown desktop environment and thus has a few shortcomings.

5.6.2 Configuring and Making the Source on Linux

Now that you have an idea what the binary requires, let us drop the source code and configure and make it on your Linux box. Quite often, developers can disable unnecessary features when configuring and if this is possible, it is highly recommended. The point for configuring and making on a Linux box is to get a preconfigured config.h or similar file. After a successful make, zip everything and copy it to your S60 development environment.

5.6.3 Modifying Configuration Files in the S60 Development Environment

Once developers know how to create a bld.inf and mmp-files, everything should get easier. It is strongly suggested to follow the guidelines available for S60 development. The next step is to make a simple bld.inf file, preferrably in a group\. Here the developer should define what headers are exported to the environment and what makefiles should be used when building. For the MMP file, developers need to add all the libraries their binary requires, and if they are planning to do an exe, they should remember to use the STATICLIBRARY instruction to glue the Symbian E32Main to main. The directives SYSTEM-INCLUDE and USERINCLUDE refer to includes <> and "", respectively. List all the sources after the SOURCE directive, as described earlier. If the developer's library uses writable static data, they should remember to have EPOCALLOWDLLDATA directive in their MMP file. Note, when using S60 3rd Edition Feature Pack 2 with the new target types, specifying the SYS-TEMINCLUDE path and using the STATICLIBRARY instruction to glue the Symbian E32Main entry point is no longer necessary.

5.6.4 Modifying the Source Code

Modifying the source code should only consist of adding EXPORT_C and IMPORT_C macros in the developer's code. These macros are defined as follows:

```
  #define EXPORT_C __declspec(dllexport)
2 #define IMPORT_C __declspec(dllexport)
```

When using S60 3rd Edition Feature Pack 2 with the new target types, this is no longer necessary.

6

Qtopia Greenphone

Thomas Arildsen and Morten L. Jørgensen

Aalborg University {tha|mljo}@es.aau.dk

Summary. This chapter provides a short introduction to Trolltech's recently released Greenphone and its Qtopia Phone Edition application framework. We present a short guide to getting started with the Greenphone SDK along with an example. We give an example of how to use the network features exemplified by an application accessing an FTP server.

6.1 The Greenphone and Qtopia

Linux-based mobile devices represent a convenient platform for researchers and product developers with an interest in less restrictive access to resources of the mobile device. With an open-source operating system, developers have the potential to access the more intricate details of the mobile device. In this sense open-source software, Linux-based mobile devices facilitate development of novel capabilities.

Norway-based Trolltech ASA is known for their cross-platform application development platform Qt. A variant of this is Qtopia, an application development framework for Linux-based devices. Qtopia facilitates development of software for mobile devices with Qtopia Phone Edition, providing a wide selection of phone-specific functionality easily accessible to the developer.

At the time of writing, Trolltech has very recently launched their so-called Greenphone. This is a proof-of-concept product for developers for demonstrating the capabilities of Qtopia Phone Edition. It constitutes a completely open platform including applications providing basic phone functionality which can be altered or replaced.

The Qtopia Phone Edition framework is structured as shown in Figure 6.1. The framework consists of a number of different functional layers. The foundation of this framework is the Qtopia Core running on top of the actual Linux operating system on the mobile device. Qtopia Core provides abstraction of the underlying operating system as well as a windowing system and basic widgets–a sort of GUI building blocks in the Qtopia terminology.

F.H.P. Fitzek and F. Reichert (eds.), Mobile Phone Programming and its Application to Wireless Networking, 159–174.

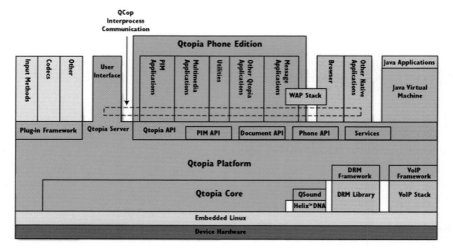

Figure 6.1. Qtopia Phone Edition framework [1].

On top of the Qtopia Core lies the Qtopia Platform. This functional layer of the framework provides for example a library containing telephony functionality as well as a library containing PIM functionality, i.e., handling of typical "organizer"-related data such as contacts, calendar appointments, etc. Part of the Qtopia Platform is also the Qtopia Server handling services that always need to be running in the phone, such as call handling. Also included in Qtopia Platform are the DRM framework and the VoIP framework for integration of third-party applications handling digital rights management of media files on the mobile device, such as music and video, and VoIP telephony along with regular cellular telephony. Other features of the Qtopia Platform include a plug-in application framework for easy inclusion of dynamic libraries that can be linked at runtime by other applications as well as an integrated Java Virtual Machine for handling Java programs.

The Qtopia API is the interface to the functionality provided by the Qtopia Platform. Part of this API, and perhaps the most interesting in the context of this book, is the Phone API. This provides access to such phone-specific components as Bluetooth, GSM/GPRS, SMS, and VoIP.

Qtopia Phone Edition is a standard set of native Qtopia applications comprising the typical applications found in mobile devices nowadays such as PIM applications, multimedia applications for audio, video, etc., and messaging applications for SMS, MMS, email, etc. These applications can be replaced or altered by the developer.

The Qtopia Greenphone is a Linux-based mobile device intended for phone software developers which comes with the above described Qtopia Phone Edition. The actual Qtopia Greenphone is a dev018pment platform that can be used for demonstrating the capabilities of Qtopia Phone Edition and, more

Figure 6.2. Qtopia Greenphone shown from all sides.

importantly, for developing software for it. Qtopia Greenphone has the following specifications:

- Touch-screen and keypad UI
- QVGA LCD color screen
- Marvell PXA270 312 MHz application processor
- 64MB RAM + 128MB Flash
- Mini-SD card slot
- Broadcom BCM2121 GSM/GPRS baseband processor
- Tri-band (GSM 900/1800/1900) support
- Bluetooth-equipped
- Mini-USB port
- 1.3 megapixel camera

The Greenphone is shown in Figure 6.2. The following section will focus on getting started developing applications for the Greenphone.

6.2 Setting up the Environment

The Qtopia SDK used for developing applications for the Greenphone is embedded in a VMware virtual machine environment which ensures easy installation on both Windows and Linux. Inside this environment everything needed for development for the Greenphone is contained. It is based on Linux with the KDE user interface. In this chapter the installation of the environment is described, followed by a quick introduction to development using the tools available in the SDK.

To run the SDK you need a machine meeting the following requirements:

- Windows 2000, Windows XP, or GNU/Linux operating system
- 512 MB RAM

- 2.2 GB hard drive space
- 1 GHz CPU

You also need the VMware Player which can be downloaded from vmware.com.

On the enclosed DVD the community edition of the SDK is available, this simply comprises an ISO CD-image which is to be burned to a CD or mounted. By following the instructions on the CD, the SDK is easily installed.

Having started the Virtual Machine you will enter the KDE environment. First you will probably want to make some changes to the settings. Click the big K-button in the lower left corner, go to "Settings" and then "Control Panel". Probably the most urgent issue is to change the keyboard layout if you are using a non-US keyboard; this can be found under "Regional & Accessibility" and the "Keyboard Layout". The SDK comes installed with two users, 'user' and 'root', with paswords 'user' and 'root' respectively. When the KDE environment starts, you are logged in as user by default, so in order to change certain settings, you may be asked for the root password, i.e., simply 'root'.

The virtual machine comes preinstalled with the following applications for development for the Greenphone:

- KDevelop – IDE, for writing code and tying the other tools together
- Assistant – Documentation
- Qt Designer – A graphical tool for creating graphical user interfaces (GUI)
- Phone Emulator (runqpe) – For emulating the target device

Let us get started straight away. Open the K-menu, under "Development" and then "KDevelop" select "KDevelop: C/C++". When the program has started, go to "Project" and select "New Project". Under "C++" select "Embedded" and then "Qtopia 4 Application", see Figure 6.3. In this example we will create a program which calculates the Body Mass Index (BMI) of a person. This measure is calculated as the mass in kg divided by the square of the height in meters; for this reason we will call the application "bmi". Also select a place for storing the project. When done press next a few times until the window disappears and we see the autogenerated code.

The first thing we want to get in place is the user interface. Click file selector on the left and open `bmibase.ui`, this opens Qt Designer. This program is used for easy construction of user interfaces. Actually the `.ui`-file is an XML-format file which is compiled by a program called `uic` to a C++-file. Obviously we could write the C++ file ourselves which some designers prefer, but for the sake of this example we will keep it simple.

Inside Qt Designer we will see a lot of toolboxes and a window in the middle symbolizing the final interface. The layout is created by laying out widgets in the user interface, this is done by simple drag-and-drop. If you prefer a docked style you can, in "Edit" go to "User Interface Mode" and select "Docked Window".

Figure 6.3. Create a new Project by selecting "Qtopia 4 Application".

Initially the window contains a label. Above this label, add a "Horizontal Layout" box by dragging it from the "Widget Box" toolbox. The horizontal layout arranges the widgets placed inside it horizontally next to each other. Now drag inside this box, a "Spin Box" and a "Horizontal Slider". Initially these elements have a range of 0–99, as can be seen in the property editor. Change the range to 1–300 for both elements, this is a range that we can count on the height (in cm) being in. This is done by altering the `minimum` and `maximum` properties in the "Property Editor". Also it is nice to have an easy to remember name for the property, change "objectName" properties to `heightSpinbox` and `heightSlider`, respectively.

Now to allow the user to change the height property by interacting with either of the two widgets we need to tie their values together. This is done by signals. Everytime a widget receives input by the user (or other parts of the program) a "signal" is emitted. Such a signal can connect to a "slot" which is a function in another widget or a function coded by us. We want any change in the slider to be reflected in the spin box and vice versa. We do this by pressing the plus-sign in the "Signal/Slot Editor" toolbox. We then select `heightSlider` as sender, `valueChanged(int)` as signal, `heightSpinbox` as reciever and `setValue(int)`. This sends the new value (an `int`) of the slider to the spin box every time it is changed. To make the relationship bilateral you must also add the corresponding opposite rule. When you have done this save the design and return to KDevelop.

Now we need to compile and run the program to see if everything is OK. Press the "Konsole" tab at the bottom of the KDevelop window and type:

```
$ source /opt/Qtopia/SDK/scripts/devel-x86.sh
$ qtopiamake
```

The first line sets up the environment variables for x86 development, that is the emulator, and the second creates a `Makefile`. Now press F8 to compile, go to the desktop, and start runqpe which opens up the emulator. Now go back to KDevelop and press shift+F9, when you go back to the emulator window you will see our application running, it should look something like Figure 6.4. Try to change the value through both the spin box and the slider to verify the functionality.

Now the next step is now to add a second set of widgets to allow the user to enter his weight, I have called mine `weightSpinbox` and `weightSlider` and set their range to 1–300 kg. I have also added some labels to help the user, I changed the name of the existing label to `outputLabel` and I used the "Vertical Spacer" to make the interface align nicely. In Qt Designer you can also right-click on an item, select "Lay out" and then "Break Layout" to leave the gridded layout. Also some more layouts should be added to make the program appear a little nicer. When you are happy save the form and go back to KDevelop.

We now need to add the function which calculates the actual BMI and displays it to the user. Open the header file `bmi.h`. Since our function needs

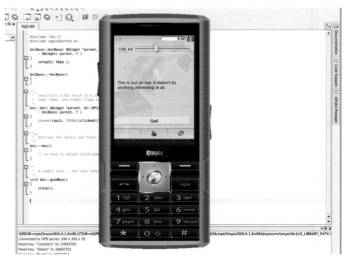

Figure 6.4. The first version of the bmi-calculator.

to be called by a signal, we need to make it a slot. We see in the .h-file that one slot already exists, goodBye(). This function is triggered when the user presses the quit-button and causes the program to halt. We will insert our slot below this function. Since this function needs to be called by both the height and the weight widgets we cannot use a value passed from these since we do not know from which widget we have been called. For this reason we make the slot void calcBMI() and return to bmi.cpp.

Now add the following code to bmi.cpp:

```
void bmi::calcBMI()
{
    double bmiValue(weightSpinbox->value() /
        pow(double(heightSpinbox->value())/100, 2));
    QString bmiText = QString("Your BMI is: ")
        + QString::number(bmiValue, 'f', 2);
    outputLabel->setText(bmiText);
}
```

The first line calculates the bmi by taking out the value of the two spin boxes. I found the value() function in the documentation browser "Qt Assistant" which can be found on the desktop. The easiest way around in the browser is to select "index" in the sidebar and search for the object of interrest. In this case it was the QSpinBox, this name is the classname of the widget and can be found in the "Object Inspector" in the designer. Since we used the pow() function we will need to include the standard C++ header math.h.

The second line concatenates the text string "Your BMI is" with the string version of the previously calculated BMI with two decimals. Finally line 3 sets the text of our outputLabel to this string.

The only thing left is to connect our slot to the signal emitted when one of the values changes. Since Qt Designer knows nothing of our slot, we need to add the connection manually. We will do this in bmi::bmi() where we see a connection was made already, the connection between the quit-button and its slot goodbye(). Add the following line:

```
connect(weightSpinbox, SIGNAL(valueChanged(int)), this,
    SLOT(calcBMI()));
```

This connects the signal valueChanged(int) in the class weightSpinbox to the slot calcBMI() in bmi (which is accessed by this). You should of course also add a signal/slot combination for the heightSpinbox.

Now fire up the program, it should look something in the lines of my program in Figure 6.5.

This simple example has shown how to use the most basic functionality of Qtopia. We have seen how to make an interface and how to create responses to user input. In the next section we will move on to look into the networking possibilities which Qtopia presents us with on the mobile phone.

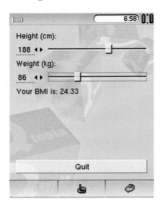

Figure 6.5. The final bmi-calculator.

6.3 FTP File Transfer Example

At the time of writing, Trolltech's Greenphone is still such a new product
that we have not yet gained a full overview of the functionality it offers.
The Qtopia application framework contains a number of libraries that can
be used for implementing various kinds of telephony- and networking-related
functionality as well as services running in the mobile phone which can be
requested by applications.

Here we provide an example to demonstrate how to make use of standard
network services provided by the Qtopia framework. As an example of using a
network protocol, this section describes the design of a shopping list application capable of storing the shopping list to an FTP server as well as loading
it from here. We build a simple user interface for managing the shopping list
which will allow us to add and delete items as well as load and save the list.

We begin by creating a new project as described in section 6.2 which we
will call "listex". Having created the KDevelop project, we turn to composing
the user interface. Follow the approach described previously to create a window for the BMI application, editing the file "listexbase.ui". Start from the
initial template window containing a `QLabel` with some example text and a
"Quit" `QPushButton`. Edit the `QLabel`'s "text" property to display a meaning-
ful headline such as "Shopping list:". Next, we add some additional widgets.
Below the `QLabel`, add a "Horizontal Layout" – `QHBoxLayout` – and below that
a "Line Edit" – `QLineEdit`. Inside the horizontal layout, add a "List Widget"
– `QListWidget` – and a "Vertical Layout" – `QVBoxLayout`. Inside the vertical
layout, add four "Push Button" – `QPushButton` – and a "Vertical Spacer"
– `Spacer`. The spacer will arrange the buttons closely spaced at the top of
the vertical layout. Edit the buttons' "text" properties to display, respectively,
"Add item", "Remove item", "Load list", and "Save list". Change the buttons'
"objectName" properties to make the names more intuitive to work with; we
will call them `pushButton_add`, `pushButton_remove`, `pushButton_load`, and

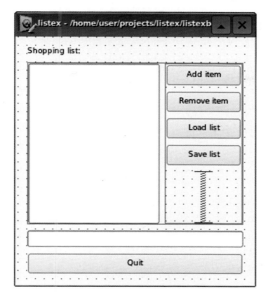

Figure 6.6. The graphical user interface of the shopping list example.

pushButton_save, respectively. Now we are done with the layout of the user interface. It should look roughly as depicted in Figure 6.6.

Next, we will add the necessary functions to handle the functionality of the shopping list application. We will need functions to add or remove items to/from the shopping list as well as functions to load and save the shopping list from/to the FTP server. Open the file "listex.h" and start by adding the following statements at the top of the file:

```
#include <QFile>
#include <QtNetwork>
```

Additionally, add the following function declarations, which we will use as slots for implementing functionality triggered by the buttons, under the "listex" class, **private slots:**

```
void addItem();
void removeItem();
void loadList();
void saveList();
```

Open "listex.cpp" to connect and implement the slot functions. Start by adding the following statements at the top of the file:

```
#include <qpushbutton.h>
#include <QString>
#include <QMessageBox>
#include <QTextStream>
```

Next, we connect the signals of the buttons on the GUI to functions that we will implement shortly; add the lower four lines to the `listex` constructor function:

```
listex :: listex ( QWidget *parent , Qt:: WFlags f )
    : listexBase ( parent , f )
{
    // Already in listex.ccp by default:
    connect(quit , SIGNAL(clicked ()) , this , SLOT(goodBye ()));

    connect(pushButton_add , SIGNAL(clicked ()) , this , SLOT(addItem
        ()));
    connect(pushButton_remove , SIGNAL(clicked ()) , this , SLOT(
        removeItem ()));
    connect(pushButton_save , SIGNAL(clicked ()) , this , SLOT(
        saveList ()));
    connect(pushButton_load , SIGNAL(clicked ()) , this , SLOT(
        loadList ()));
}
```

Next, we implement the above functions – at the end of "listex.cpp", add the function definitions:

```
void listex :: addItem ()
{
}

void listex :: removeItem ()
{
}

void listex :: saveList ()
{
}

void listex :: loadList ()
{
}
```

Adding and removing items are the easiest of the functions since they will not involve network communication but simply rely on functions provided by the list and line edit widgets. When adding an item, we want to add what the user has written in the line edit widget. We make a simple check to make sure we do not add an empty item. If this passes, we add an item to the list based on the text in the line edit widget and afterwards clear the line edit widget to prepare it for entering a new item:

```
void listex :: addItem ()
{
    if (lineEdit ->text () .compare (""))
    {
        listWidget ->addItem (lineEdit ->text ());
        lineEdit ->clear ();
    }
}
```

To remove an item, we simply use the list widget's function for taking an item out of the list:

```
void listex :: removeItem ()
{
    listWidget ->takeItem ( listWidget ->currentRow () ) ;
}
```

For communication with the FTP server, we will need a `QFtp` object. FTP communication as provided by the `QFtp` class works asynchronously, i.e., calls that interact with the FTP server return immediately and the `QFtp` object will then emit a signal whenever the asynchronous operation has completed. For this purpose, we need some variables to serve as handles to the operations we will be performing on the FTP server. We will also need to store the list contents in a file which we will upload to the FTP server. For these purposes, we add the following variables and objects (which will be explained in the following) in the `listex` class definition in "listex.h":

```
QFtp* ftp ;
QFile* file ;
QString fileName ;
int loadConnectHandle , loadLoginHandle , loadGetHandle ,
    loadCloseHandle ;
int saveConnectHandle , saveLoginHandle , savePutHandle ,
    saveCloseHandle ;
bool ftpInProgress ;
```

We initialize these in the `listex` constructor in "listex.cpp":

```
listex :: listex ( QWidget *parent , Qt :: WFlags f )
    : listexBase ( parent , f )
{
    ftp = new QFtp ( this ) ;
    ftpInProgress = false ;
    fileName = QString ( "shoplist" ) ;
    loadConnectHandle = loadLoginHandle = loadGetHandle =
        loadCloseHandle = saveConnectHandle = saveLoginHandle =
        savePutHandle = saveCloseHandle = 0;

    // Already in listex.ccp by default:
    connect ( quit , SIGNAL( clicked () ) , this , SLOT( goodBye () ) ) ;

    connect ( pushButton_add , SIGNAL( clicked () ) , this , SLOT( addItem
        () ) ) ;
    connect ( pushButton_remove , SIGNAL( clicked () ) , this , SLOT(
        removeItem () ) ) ;
    connect ( pushButton_save , SIGNAL( clicked () ) , this , SLOT(
        saveList () ) ) ;
    connect ( pushButton_load , SIGNAL( clicked () ) , this , SLOT(
        loadList () ) ) ;
}
```

We should now be ready to implement the functions for saving and loading lists. In the function `saveList()`, we only want to open one connection to the FTP server at a time so we enclose the function's contents in the following test; `ftpInProgress` will be set further down:

```
void listex :: saveList ()
{
    if  (! ftpInProgress )
        {
```

If an FTP transfer is not already in progress, create a file for saving the list and then create a **QTextStream** for writing data to it. We attempt to open the file as write-only for saving the list. A check is added to handle the error, if the file cannot be opened, by displaying an error message, deleting the file object, and returning from the function:

```
2   file = new QFile(fileName);
    QTextStream stream(file);

    if (!file->open(QIODevice::WriteOnly)) {
       QMessageBox::information(this, tr("Shopping_List"),
7                               tr("Unable_to_save_the_file_%1:_%2.")
                                .arg(fileName).arg(file->errorString
                                 ()));
       delete file;
       file = 0;
       return;
12  }
```

If the file opened without any errors, we loop through the items of the shopping list, printing the text they contain to the file and remembering to add new lines as we go:

```
10      for (int i = 0; i < listWidget->count(); i++)
        {
12         stream << listWidget->item(i)->text().trimmed();
           if (i < listWidget->count() - 1)
              stream << "\n";
        }
```

Having written the list items to the file, we now close the file and reopen it as read-only to send it to the FTP server[1]:

```
13  file->close();
    if (!file->open(QIODevice::ReadOnly)) {
       QMessageBox::information(this, tr("Shopping_List"),
                                tr("Unable_to_open_the_file_%1:_%2.")
17                              .arg(fileName).arg(file->errorString
                                 ()));
       delete file;
       file = 0;
       return;
    }
```

We now set **ftpInProgress = true** to indicate that we have an FTP operation in progress. We connect to the FTP server using **QFtp**'s **connectToHost()** function[2]. We log in using the **login()** function, providing a valid user name and password to the FTP server. Having logged in, the **put()** function is used

---

[1] Although it may seem unnecessarily complicated to first open the file write-only to write to it and subsequently reopen it read-only in order to transmit it, we have experienced difficulties making it work if the file is merely opened as read-write for use throughout the whole function.

[2] To try out the example, the reader must provide the address of an FTP server here to which the reader has read/write access.

to upload the file that we saved the list items in. Finally, the connection is closed using the **close()** function:

```
19   ftpInProgress = true;

     saveConnectHandle = ftp->connectToHost("server.address");
22   saveLoginHandle = ftp->login(QString("user"), QString("password"));
     savePutHandle = ftp->put(file, "shoplist", QFtp::Ascii);
     saveCloseHandle = ftp->close();
     }
}
```

As previously mentioned, the **QFtp** function calls are asynchronous and will return immediately. As seen above, the functions return handles to the operations which we store in order to be able to identify them later. This will be dealt with after we implement the function for loading a shopping list.

We implement the function **loadList()** starting out similarly to **saveList()**. We enclose the rest of the function in a check to ensure that an FTP transfer is not already ongoing:

```
1   void listex::loadList()
    {
      if (!ftpInProgress)
        {
```

We then create a file for saving the file that we will download from the FTP server and attempt to open a file as write-only for saving the list. A check is added to handle the error if the file cannot be opened, similarly to **saveList()**:

```
2        file = new QFile(fileName);

         if (!file->open(QIODevice::WriteOnly)) {
           QMessageBox::information(this, tr("Shopping List"),
                                    tr("Unable to open the file %1: %2.")
7                                   .arg(fileName).arg(file->errorString
                                    ()));
           delete file;
           file = 0;
           return;
         }
```

We now set **ftpInProgress = true** to indicate that we have an FTP operation in progress. Subsequently, we contact the FTP server in the same way as in **saveList()**, with the exception that we use the **get()** function in stead of the **put()** function:

```
9        ftpInProgress = true;

         loadConnectHandle = ftp->connectToHost("server.address");
12       loadLoginHandle = ftp->login(QString("user"), QString("
             password"));
         loadGetHandle = ftp->get("shoplist", file, QFtp::Ascii);
         loadCloseHandle = ftp->close();
         }
}
```

The actual task of filling the shopping list with the contents of the downloaded file is handled when we have received notification that the pending FTP operations have finished.

In order to track the status of the ongoing FTP operations and react when they complete, we must connect functions for handling the signals emitted by the `ftp` object. For this purpose, we add the functions `fileTransferHandler()` and `connectionHandler()`. In "listex.h", add the following under the "listex" class, `private slots::`

```
void fileTransferHandler(int id, bool status);
void connectionHandler(bool status);
```

In "listex.cpp", connect the functions to the signals emitted by the `ftp` object, below the connections we previously added:

```
connect(ftp, SIGNAL(commandFinished(int, bool)), this, SLOT(
    fileTransferHandler(int, bool)));
connect(ftp, SIGNAL(done(bool)), this, SLOT(connectionHandler
    (bool)));
```

When each of the FTP operations we initiated in the functions `saveList()` or `loadList()` finish, the `ftp` object emits the `commandFinished` signal which is now handled by the function `fileTransferHandler()`. When all pending operations have finished, it emits the `done` signal which is now handled by the function `connectionHandler()`.

We implement `fileTransferHandler()` as follows to handle the different operations we have initiated. `id` is the handle to the operation we initiated and `status` indicates if the operation finished with an error (`true`). For the connect operation, we simply display a message of whether the operation went well or not:

```
1  void listex::fileTransferHandler(int id, bool status)
   {
     if ((id == loadConnectHandle)||(id == saveConnectHandle))
       QMessageBox::information(this, tr("Shopping_List"), tr(status
          ? "Connect(%1):_Error" : "Connect(%1):_OK").arg(id));
```

Likewise for the login operation:

```
2    else if ((id == loadLoginHandle)||(id == saveLoginHandle))
       QMessageBox::information(this, tr("Shopping_List"), tr(status
          ? "Login(%1):_Error" : "Login(%1):_OK").arg(id));
```

For the put operation (meaning that `saveList()` initiated the FTP operations), we close and delete the file that we uploaded:

```
1    else if ((id == savePutHandle))
     {
       QMessageBox::information(this, tr("Shopping_List"), tr(status
          ? "Upload(%1):_Error" : "Upload(%1):_OK").arg(id));
       file ->close();
       file ->remove();
       delete file;
7      file = 0;
     }
```

Finally, we also display a message of whether the close connection operation went well:

```
6   else if ((id == loadCloseHandle)||(id == saveCloseHandle))
7       QMessageBox::information(this, tr("Shopping_List"), tr(status
                ? "Close_connection(%1):_Error" : "Close_connection(%1):
                _OK").arg(id)));
```

For the get operation (meaning that `loadList()` initiated the FTP operations), we first display a message of whether the file was downloaded successfully or not. If so, we re-open the file as read-only to read from it, handling the error and displaying an error message if the file cannot be opened:

```
5   else if (id == loadGetHandle)
    {
7       QMessageBox::information(this, tr("Shopping_List"), tr(status
                ? "Download(%1):_Error" : "Download(%1):_OK").arg(id));
        if (!status)
        {
            QByteArray tmpReadLine;

            file->close();
            if (!file->open(QIODevice::ReadOnly)) {
12              QMessageBox::information(this, tr("Shopping_List"),
                        tr("Unable_to_open_the_file_%1:_
                        %2.")
                        .arg(fileName).arg(file->
                            errorString()));

                delete file;
17              file = 0;
                return;
            }
```

If the file was opened successfully, we clear the contents of the list widget and read a line from the downloaded file. As long as the lines read from the file are not empty, we keep reading lines from the file and adding the text as items in the list widget:

```
18      listWidget->clear();
        tmpReadLine = file->readLine(20).trimmed();
        while (QString(tmpReadLine).compare(""))
        {
22          listWidget->addItem(QString(tmpReadLine));
            tmpReadLine = file->readLine(20);
        }
    }
```

Finally, we make sure to close and remove the file opened for downloading the shopping list from the FTP server:

```
23      file->close();
        file->remove();
        delete file;
        file = 0;
27  }
}
```

In order to detect if all ongoing FTP operations have completed, in the function `connectionHandler()` we a receive signal when the pending operations have ended. We set `ftpInProgress` to indicate that there are no longer any ongoing FTP transaction and display an error message if something during the transfer failed:

```
void  listex :: connectionHandler ( bool  status )
{
   ftpInProgress = false ;
   if ( status )
      {
         QMessageBox :: information ( this ,  tr ( "Shopping_List" ) ,
                              tr ( "Something_went_wrong_during_
                                   FTP_transfer !" ) ) ;
      }
}
```

All that is left to do now is to compile and run our shopping list application as described in Section 6.2.

We have here demonstrated how easily one can create an application that uses the FTP protocol as an example of an application communicating across a network. Most of the example consisted of setting up the functionality of the user interface. The actual FTP communication was realized through a very few lines of code thanks to the facilities offered by Qtopia's `QFtp` class.

6.4 Conclusion

The Greenphone is not a consumer product. It is meant as both a proof-of-concept product showcasing the capabilities of the Qtopia Phone Edition framework to potential customers such as phone manufacturers as well as a developer test platform for developing Qtopia applications. As such, the Greenphone is, at the time of writing, not a completely functional product. For example, when we received a Greenphone, the Bluetooth capabilities were not yet available due to software problems, but such difficulties are expected to be fixed through software updates.

Due to the relative ease of getting started with programming applications for the Greenphone and the fairly open nature of the software, this is a good platform for getting started developing applications for mobile devices in a framework providing a basic suite of ready-made phone functionality.

References

1. Trolltech ASA. Qtopia phone edition 4.2 whitepaper. Whitepaper, Trolltech ASA, November 2006.

7

Maemo Linux

Matti Sillanpää

Nokia Research Center `matti.jo.sillanpaa@nokia.com`

Summary. In this chapter, a software development environment that enables embedded systems' software to be compiled, run, and debugged on regular Linux PCs is presented. The aim of this setup is to provide a way to develop software for embedded devices running Linux in an easy and efficient manner. As an example target environment a system that is currently available is demonstrated, namely the Nokia 770 Internet Tablet and its development platform, maemo. Supporting software around this environment is also presented, to give an idea of how it is possible to perform development tasks via an integrated development environment.

7.1 Introduction

In recent years, Linux has gained a lot of popularity among mobile device manufactures. There are a number of reasons as to why embedded Linux has been found a viable platform by many companies: the source code is open, the memory footprint is small, no royalty costs are required, and it is already quite mature and stable. From the developer's perspective, it offers an embedded platform on which regular Linux programmers can be effective right away, without needing to learn the idiosyncrasies of an embedded OS.

To enable efficient development on a Linux-based platform, it has been found useful to replicate the target system's environment on a PC. This approach enables development tasks without resorting to the actual target device, and this way a more distributed development effort. It also provides more speed and control over the development process. Building software using a PC is a lot faster than using an embedded device, and on the other hand, using the actual environment instead of regular cross-compilation reduces the need for continuous regression testing. We will take a closer look at how mobile software development can be performed without necessarily having any other tools available but a regular Linux desktop PC. The development environment that we will use is the Scratchbox, which has been developed by Movial [12] with sponsoring support from Nokia [13].

F.H.P. Fitzek and F. Reichert (eds.), Mobile Phone Programming and its Application to Wireless Networking, 175–205.
© 2007 *Springer.*

7.2 Scratchbox

So Scratchbox is the tool of choice for trying out embedded Linux development over the course of this chapter. You already might have an idea of what cross-compilation means, but what exactly is it that Scratchbox adds to that process, and by what means? Let us take a closer look at what it offers us, and how it does it.

Cross-compilation means essentially the process of compiling binary files for a CPU architecture that is foreign to the system running the compiler. The reasons to perform this process when developing for embedded devices are the facts that embedded devices normally lack development tools, and on the other hand, they cannot compete with the efficiency of desktop computers in performing tasks like software compilation.

In a wider sense, we can talk about **cross-development**, the idea of not only producing software that is going to run on a different CPU than the development system has, but also developing for a different **platform** – a system including both its hardware and software. What we are going to do is development for a platform with both foreign hardware and software compared to the development host, and this is what Scratchbox is designed to facilitate.

The main promises of the Scratchbox environment are that, using only your x86 Linux PC, you are able to:

- Compile your software for the target architecture.
- Run your cross-compiled software.
- Debug your cross-compiled software.

It is designed to work as a sandbox environment that the developer can log into, and then use it essentially the same as if she had a terminal open in the actual target environment.

The abovementioned features mean that running a compiler or even executing a **cross-configuration** script in this environment produces the same results as if it were done on the target device. If you have worked with Linux development before, you have probably run into the concept of "configure" scripts. These are scripts that are generated by the developer using GNU Autoconf, and their purpose is to resolve things like the host architecture and the presence of common libraries in order to set up the build process appropriately. When run inside the Scratchbox environment, such a script is able to find these things exactly as they appear inside the target environment, and even execute small test programs for determining some attributes. This way, the task of porting software to a specific embedded environment is greatly simplified by using Scratchbox.

Since there can be several, varying systems you may be developing for, Scratchbox provides you with the possibility of managing several development **targets**, and changing between them easily.

A major entity that you can swap when you change your development target is the **toolchain**. It contains the tools necessary for compilation – most essentially, the target CPU-specific set of a cross-compiler with a C-library, an assembler, and a linker. Because of interdependencies between each CPU target, kernel major version and C-library, they have to be selected as a set. These dependencies have been detailed in the Scratchbox documentation [18].

In addition, you can add tools and libraries specific to the platform you are developing for into the target. We will have a closer look at the components of Scratchbox in Section 7.2.1.

7.2.1 Under the Hood

In this section, we will take a look at the internals of Scratchbox and give an idea of how it has been built, and how it works. It is useful to have a look at some central concepts of Scratchbox, to support our development tutorial later on. An overview of the main ingredients of Scratchbox is presented in Figure 7.1, and we will have a closer look at them in the following.

To enable the functionality presented so far, Scratchbox has to provide at least these two essential things:

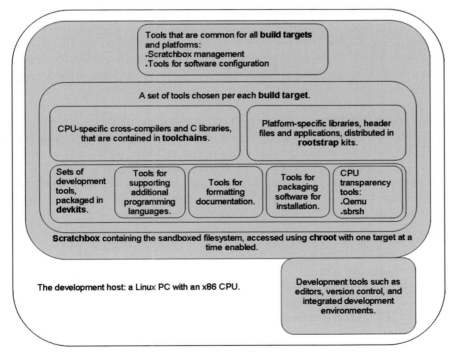

Figure 7.1. An overview of the components inside the Scratchbox environment.

- A sandboxed environment that is a "lookalike" of the target device, containing the files and tools needed for building software.
- A method to provide execution of binaries that are foreign to the build host.

To address the first problem, Scratchbox uses a program found on Unix operating systems called **chroot**. It provides an operation that changes the effective root directory of the file system – but only from the perspective of the programs that are run by the user of chroot, naturally. When the user logs in to Scratchbox, a chroot operation takes place, and the files of the active target are accessible as if they were populating the host file system directory hierarchy.

To support using a set of libraries that exists on a target device, there is a method to distribute an entire development environment in a package that can be installed under Scratchbox. The concept of a package containing the environment with all its libraries and "include" header files is known as a **rootstrap** in Scratchbox parlance. The term originates from the **bootstrap** concept. Bootstrapping means the idea of building a simple system that enables a more complex system to be built, or running a system that allows for running a more complex system. With compilation, this idea translates to having a basic collection of libraries available that allow for compiling more complex programs. Without this set, problems with circular dependencies between basic compilation units could cause a lot of extra work. Using rootstrapping you can thus store a root file system that solves the chicken-and-egg problems of cross-compilation. Using a rootstrap with Scratchbox is in no way mandatory, however. It is merely a facility that can be used to save a basic working set of files for a specific environment, to skip the groundwork of building it in the future.

Another Scratchbox concept concerning a collection of software that can be easily installed into a target environment is the **Development kit**, or **devkit** for short. A devkit basically contains a set of software tools, and several of them can be installed into a target. Such a set can contain tools which override ones from another devkit, so that the user can partially override a devkit with another one at will. This is based on the order in which they are installed. The following devkits are available at the Scratchbox site:

- The **Debian** devkit, which contains the Debian-specific tools for assembling and handling software installation packages, such as **dpkg**.
- The **doctools** contains tools for generating documentation files, for instance the **groff** and **teTeX** programs.
- The **Perl** devkit contains tools for Perl-based development.
- The **CPU transparency** devkit contains the tools for running binaries that are foreign to the development host, discussed below in more detail.

For the second major challenge – being able to run cross-compiled binaries – a method has been devised to utilize another device over a network connection, in order to use the actual target CPU for execution. Additionally, there is the possibility of using a processor emulator on the development host.

Both of these solutions are based on the fact that the Linux kernel supports registering specific binary formats to be run using separate applications. In Scratchbox, this feature has been used to register either the **Scratchbox Remote Shell (sbrsh)** [16], or the **Qemu** [3] processor emulator as the executing program. Currently the ARM, SPARC, PowerPC, and MIPS processors can be emulated.

The sbrsh is a shell, similar to rsh and ssh. It's purpose is very specific, however. Using it, a regular remote program execution happens as follows, from sbrsh's perspective:

1. Contact a daemon program (sbrshd) on the target device, piping all input and output between the executable program to be run at the remote end, and the user shell on the development host.
2. Mount the target filesystem from the development host, including the binary.
3. Set the environment variables on the remote host to correspond to the development environment.
4. Execute the binary.

Alternatively, the Qemu processor emulator can be used to execute ARM binaries under a regular x86 Linux PC. In this case, Qemu is executed in place of sbrsh.

When a program with a graphical user interface is executed, an additional need arises: the graphical output of the program needs to be displayed somewhere. If there is no X server running inside the target environment, with an external output – in practice, a screen on the actual target device – the output needs to be routed to the development host. Fortunately, a rather straightforward solution exists for programs based on the X Window System: you can direct the output of the program to your normal X session, or even invoke an embedded X server inside your X session, which can have parameters such as screen depth and resolution set to correspond to the target hardware. The program that is run in Scratchbox can then be made to use it simply by setting the DISPLAY environment variable.

Another important component also exists within Scratchbox that is needed in cross-development work. It is a program called **fakeroot**, and it has been developed as a way to run programs under virtual super user privileges. It is useful in situations where one would need to run a program that requires the user to be the "root" user, even though the actual functionality of the program would not necessitate this requirement. In other words, when running fakeroot, you cannot change anything on the system that is outside the rights of your user account, but you can make applications think that you are root. A popular use has been for creating archive files that contain files which have root ownership. Using fakeroot, one can preserve all permission settings in the archived files without actually being the super user.

The Scratchbox version of fakeroot [15] was somewhat modified from the original implementation, and was at first given the name **fakeroot-net**. The changes have since been committed back to the regular fakeroot codebase, however. As you might tell by the alternate name, use of fakeroot inside Scratchbox necessitated changes that allow running fakeroot over network connections – namely the sbrsh-sbrshd connection. When using sbrsh for CPU transparency, user sessions could jump to running something on the target device, so a uniform fakeroot session needs to be maintained on both ends.

For developers of new platforms, it is very interesting to note that in addition to the toolchains that have been compiled for Scratchbox specifically, you can bring in and adapt toolchains compiled outside of Scratchbox. These are known as **foreign toolchains** [6]. Since Scratchbox includes the tools for assembling a toolchain based on any GCC cross-compiler, it is even possible for a developer to provide a Scratchbox toolchain for a completely new processor architecture.

7.2.2 Getting Started with Scratchbox

Now that we have had a glimpse at the essential background concepts, it is time to get our hands dirty and try using Scratchbox. We will go through the necessary steps to install Scratchbox on a Linux PC, and to compile and run our first cross-developed program. We will go through the phases in a rather straightforward fashion here to get to actually using Scratchbox fast. In case you feel like you would like to peruse information on the steps of the installation and set-up procedure in more detail, please visit the Scratchbox website [17].

Installing Scratchbox

To get started, we need to set up the environment. Make sure you have a computer that includes the following:

- An Intel compatible x86 processor, 500 MHz or faster.
- 512 MB RAM or more.
- 3 GB of free hard disk space.
- The GNU/Linux operating system. A fairly recent distribution is recommended – developers of the maemo SDK recommend Ubuntu or Debian in particular, but this is not a requirement.

Also, you will need to have access to the super-user account of this system, i.e., the "root" account. If you are logging onto a shared Linux server, you most likely do not have that access. The set up for Scratchbox would have to be done by the system administrator for such a host. Instead, our installation process is one you could perform for a personal desktop computer.

The software that we are going to need is at the Scratchbox website [17], the current version being under the URL http://www.scratchbox.org/download/scratchbox-apophis/ Go to this site, and navigate to "Binary tarballs" Save the files that begin with the following names:

- `scratchbox-core`
- `scratchbox-libs`
- `scratchbox-toolchain-cs-2005q3.2-glibc-arm`
- `scratchbox-toolchain-cs2005q3.2-glibc-i386`
- `scratchbox-devkit-debian`
- `scratchbox-devkit-cputransp`

The middle two packages are specific toolchains, and the latter two are devkits – both of which concepts were discussed in Section 7.2. To unpack the files you have just downloaded, first switch to the super-user ("root") account by running `su -` in a terminal. Then, execute the command `tar xvzf <package> -C /`, with `<package>` replaced by each of the file names in turn. This will result in a new directory: `/scratchbox`.

Once you are through unpacking the packages, remain as the root user and run the command `/scratchbox/run_me_first.sh`. You will be first asked whether you would like to use the `sudo` utility for running the script with heightened privileges, but this is not necessary since you are the root user, so just press enter for the default answer `no`. Then you are queried for the name of the group that Scratchbox users will be added under. The default value of `sbox` is fine for this also, unless you have already actually created a group of this name on the computer for some other purpose.

Once you have completed the previous steps, execute the command:

```
/scratchbox/sbin/sbox_adduser <username>
```

with `username` replaced by your system account name. Confirm being added to the `sbox` user group by pressing enter. After this, you should restart your user session by logging out of your system, and then back in, to have your group information updated.

Once you have opened a new session, you do not need to attain super-user privileges again, but just use your regular user account. You can verify that the changes in Scratchbox user information have taken place by running the command `groups`, and confirming that the Scratchbox group `sbox` appears in the output. If this is the case, log in to Scratchbox by running the command `/scratchbox/login`. You should be greeted by the following output:

```
$ /scratchbox/login

You dont have active target in scratchbox chroot.
Please create one by running "sb-menu" before continuing
```

```
    Welcome  to  Scratchbox ,  the  cross−compilation  toolkit !
9   Use  'sb−menu'  to  change  your  compilation  target .
    See  /scratchbox/doc/  for  documentation .

    sb−conf : No  current  target
    [ sbox −:  ~ ] >
```

A few additional adjustments are necessary to have the man command available inside Scratchbox, and for localization to work properly. Also, we will set the DISPLAY variable to point to an X server on the local host, so it will not need to be set separately each time, as long as we are using an X server on the local host for graphical output. Open the file .bash_profile in either the nano or the vim editor, and add these lines:

```
    export  LANGUAGE=en_GB
2   export  PAGER=less
    export  DISPLAY = 1 2 7.0.0.1:2
```

Congratulations! Scratchbox is now installed on your system. Note that for future use, Scratchbox needs to be started every time you restart your system by calling /scratchbox/sbin/sbox_ctl start as the root user. If you are familiar with the Unix System V initialization method that most Linux distributions use, you can install the aforementioned file as a system initialization script to avoid this manual step.

7.2.3 Your First Cross-Developed Program

Now that we have the environment there, let us proceed to define a build target with a toolchain that provides us with the means to actually compile software for a foreign platform. We will add a target for ARM compilation. We could give the new target any name, and we will call it MY_ARM_TARGET. For now we will just create and try out using a target without installing an SDK, but we will get to installing platform-specific files using a rootstrap package in Section 7.3.2.

After you have logged in by running the /scratchbox/login command, run the sb-menu tool:

```
    [ sbox −:  ~ ] > sb−menu
```

Then, use it to create a new target by following the steps of Figures 7.2, 7.3, and 7.4.

After completing the set up procedure, we are ready to implement a program. Open a new file hello.c in nano or vim and type in the classic Hello World, as shown in Listing 7.1.

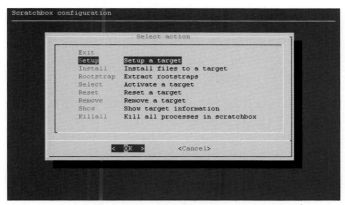

(a) Start by choosing to set up a target, and "New" in the next screen to create one from scratch.

(b) First, give the new target a name.

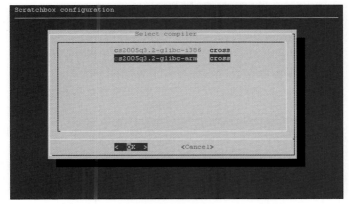

(c) Then, choose the appropriate toolchain.

Figure 7.2. Creating a cross-compilation target for the ARM processor architecture.

(a) Choose to include the CPU transparency devkit.

(b) Choose an ARM version of Qemu.

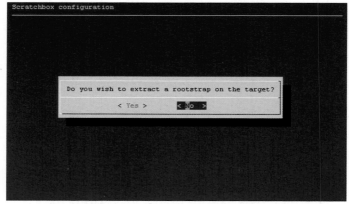

(c) We will not be installing a rootstrap at this point.

Figure 7.3. Creating a cross-compilation target for the ARM processor architecture.

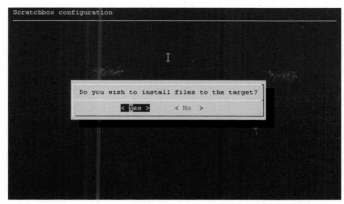

(a) Choose to install some other files into the target filesystem to get a basic set of libraries and configuration files.

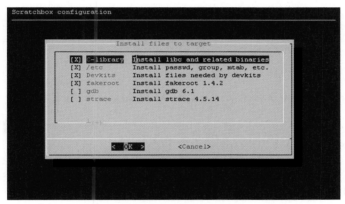

(b) The default set of files is enough for our needs.

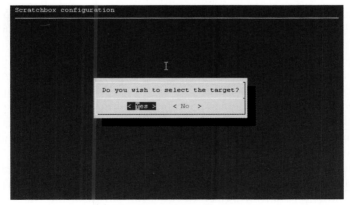

(c) Finally, activate the target.

Figure 7.4. Creating a cross-compilation target for the ARM processor architecture.

Listing 7.1. The Hello World

```
#include <stdio.h>
int main(int argc, char *argv[])
{
4    printf("Hello_World!\n");
     return 0;
}
```

Then, run the compiler:

```
[sbox-MY_ARM_TARGET: ~] > gcc hello.c
[sbox-MY_ARM_TARGET: ~] >
```

Now, check to see that the output file is an ARM executable:

```
[sbox-MY_ARM_TARGET: ~] > file a.out
a.out: ELF 32-bit LSB executable, ARM version 1 (SYSV), for GNU/
     Linux
3 2.4.17, dynamically linked (uses shared libs), not stripped
[sbox-MY_ARM_TARGET: ~] >
```

Finally, execute the program, and admire the result.

```
1 [sbox-MY_ARM_TARGET: ~] > ./a.out
Hello World!
[sbox-MY_ARM_TARGET: ~] >
```

So you were, in fact, able to run an ARM binary on your PC!

7.3 The Nokia 770 Internet Tablet

The 770 Internet Tablet [1] is an interesting mobile device from Nokia: it is both the first to run Linux, and also the first that is not a phone. First and foremost it is designed to provide a handy and mobile gateway to the everyday applications of the Internet, such as browsing the Web, communicating by instant messaging and e-mailing, and even calling voice calls over IP. It is operated using a large touch screen, with some buttons providing handy shortcuts. You can see the device in its "idle" state, that is, not running any application on the foreground, in Figure 7.5. As you can see, the application framework supports running several small applications in the idle state. This display is called the Home screen.

The concept of a touch-screen operated tablet device might bring to mind a personal digital assistant. In comparison, however, the Nokia 770 is very Internet-oriented. This is why it has a wireless LAN connection available, as well as built-in Bluetooth.

For developers the open nature of this platform is very attractive. For the first time, it is possible for a Linux programmer to implement, or even just port programs onto a mobile device from Nokia, using the same paradigms of programming as in normal Linux development. The programs can be written using the C programming language, with graphical user interfaces

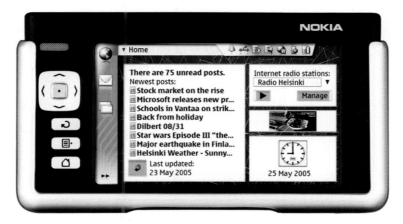

Figure 7.5. The Nokia 770 Internet Tablet.

implemented using GTK+ components. Then, the distribution and installation can be handled through the Debian Linux package management applications. The Python programming language is also supported for rapid software development.

In this section we will get to try something for a real-world device: we will write, compile, and run a program for the Nokia 770. The good thing is, you do not need to have the actual device at hand to be able to do this. As discussed in Section 7.2.1, with Scratchbox, you can do software development under a purely emulated run environment.

7.3.1 Maemo

Before we begin with an example, let us have a look at the SDK for the Nokia 770 – maemo [9]. If you recall our discussion in Section 7.2.1 on packaged software development environments known as rootstraps, you might have guessed that maemo is provided as just such a package for the Scratchbox environment.

An overview of the contents of the maemo platform is shown in Figure 7.6. They can be summarized as follows:

The Hildon Application Framework The desktop framework that provides the implementation for parts of the UI that are omnipresent, i.e., the frames and other user interface elements that are not dependent on the application that is running on the foreground. The global "look" of the desktop, including theming, is the responsibility of Hildon. It also controls the user applications that are run on the device, allowing the user to, e.g., start and switch between them (the Task Navigator), as well as to manage global settings via a control panel.

GTK+ A multiplatform toolkit for creating graphical user interfaces. The version used in maemo has been modified to accommodate the theming

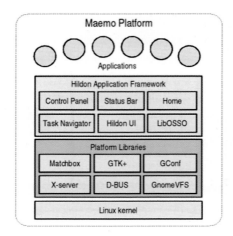

Figure 7.6. An overview of the contents of maemo, the application development platform for the Nokia 770 Internet Tablet.

system as well as some additional widgets. It is based on GTK+ 2.6 and is binary compatible with normal GTK+, but desktop applications that are ported to this environment should still undergo some adaptation to fit the accessibility and look-and-feel of maemo.

Matchbox The lightweight window manager that is used for windowing under the X Window System. It has been modified to achieve a custom style for the user interface.

X Server The X Window System server that is responsible for drawing graphics on the screen. It has also been modified to optimize it for a specific embedded platform.

D-BUS A message passing system between applications and libraries. It contains both a system level daemon and a per-user-login-session level daemon, for general interprocess communication between each user application. It has been built to work over a one-to-one message passing framework, allowing any two applications to communicate directly. It is used for system notifications, such as "battery low", as well as providing a bus between application logic and the UI in some cases. Thirdly, the Task Navigator uses D-BUS for starting programs and delivering requests to them.

GnomeVFS Gnome Virtual File System is a library that provides transparent access to different types of file systems – both networked and local ones. Mostly the API it provides conforms to the POSIX standard for accessing files.

GConf A system for storing application preferences globally, into a database of key-value pairs, that is arranged like a simple file system. It is used in the Gnome desktop environment.

7.3.2 Installing Maemo

Having had a glimpse of the contents of the maemo platform, it may come as a surprise to see how smoothly we are able to install all this under our Scratchbox environment. Let us see exactly how this happens.

First, download the SDK from the maemo website. Head for the site at http://www.maemo.org/downloads/download-sdk.html. Choose the stable release
download, and then navigate for the latest version – at the time of writing, 2.1. Download the files:

- i386/Maemo_Dev_Platform_v2.1_i386-rootstrap.tgz
- armel/Maemo_Dev_Platform_v2.1_armel-rootstrap.tgz

So it seems there is a version of maemo labeled for i386 as well as ARM. The idea behind having an Intel rootstrap available is simply that it is more reliable – and of course more efficient – to run programs on your development host without having to use an emulator. But at least when you are compiling something for an actual ARM device, you need to switch to using the ARM-based rootstrap. You might also prefer to use the ARM rootstrap in case you actually have the device available during development, and are thus able to use the sbrsh remote execution daemon discussed in Section 7.2.1 to run your program on the actual device.

Once you have downloaded the files, switch to the root account of the system, copy the rootstrap packages to a place where Scratchbox can access them, and log out of the root account:

```
  $ su
2 Password:
  # cp Maemo_Dev_Platform_v2.1_i386-rootstrap.tgz /scratchbox/
      packages/
  # cp Maemo_Dev_Platform_v2.1_armel-rootstrap.tgz /scratchbox/
      packages/
  #
  # exit
```

We will get started with maemo by creating a target for its Intel version. Log in to Scratchbox, and create a new target for the Intel rootstrap using the sb-menu tool. We will call this target SDK_PC, to have the name as a mnemonic for a target using the host CPU and having the SDK installed. Follow the steps detailed in Figures 7.7, 7.8, 7.9, and 7.10 next, to install maemo in place.

After completing the creation of the new target, you have the maemo environment ready for Intel-based development. You could go on using just this target for now, but on the other hand, why not set up a target for actual ARM compilation already too? For simplicity, we will create a target that uses the Qemu processor emulation for execution, instead of setting up networking with the real device at this point. If you have an actual device that you would like to use over sbrsh available, you can find information on setting up the

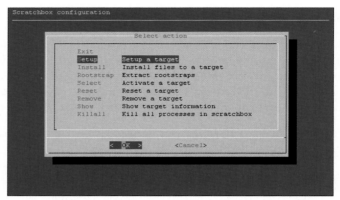

(a) Start by choosing to set up a target, and "New" in the next screen as before.

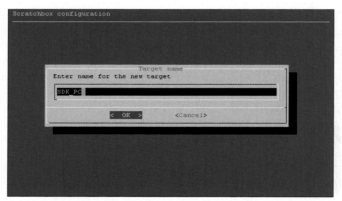

(b) Give this target the name SDK_PC to describe the fact that we are compiling for the build host CPU.

(c) Then, choose the corresponding toolchain.

Figure 7.7. Creating a new build target that uses the maemo SDK.

(a) Just add the Debian devkit.

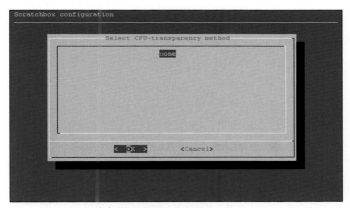

(b) No emulation is necessary.

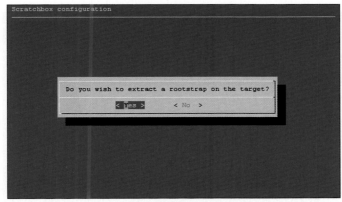

(c) At this point, we will actually add the maemo environment into our mix. Choose to install a rootstrap file.

Figure 7.8. Creating a new build target that uses the maemo SDK.

(a) We have the file available locally.

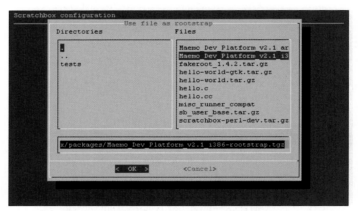

(b) Choose the i386 version of the maemo rootstrap.

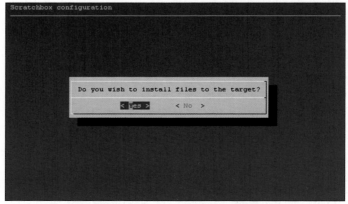

(c) We will also populate the target with some other files.

Figure 7.9. Creating a new build target that uses the maemo SDK.

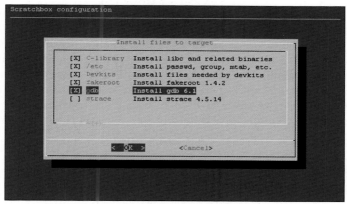

(a) Choose this set of supporting files for use in the target.

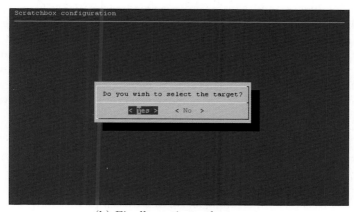

(b) Finally, activate the target.

Figure 7.10. Creating a new build target that uses the maemo SDK.

connection from the Scratchbox website [16]. For maemo-specific advice, see
the tutorials on the maemo website [11].

To create the new target, we will follow the steps of Figures 7.7, 7.8, 7.9,
and 7.10 again, with exceptions to the following phases:

7.7(b) Give the target a name that describes the new ARM-based target:
SDK_
ARMEL.

7.7(c) Choose the ARM-based toolchain.

7.8(a) Also include the `cputransp` devkit.

7.8(b) Choose an ARM processor emulator, like was shown in Figure 7.3(b)
(`qemu-arm-0.8.1-sb2` is a good choice for running the application framework of maemo 2.1 on Scratchbox 1.0).

7.9(b) Choose the ARM version of the maemo rootstrap.

After completing this process, you are ready to compile, run, and debug programs for both ARM and Intel.

Especially in the case that you are working on a Fedora or Red Hat Enterprise Linux-based system, your Linux kernel might have an option enabled for optimizing the mapping of virtual memory, that could interfere with running maemo under Qemu. If you experience problems running Qemu, try changing the way virtual memory is mapped back to an older convention. You can do this temporarily by running the following command as the root user:

```
# echo 1 > /proc/sys/vm/legacy_va_layout
```

7.3.3 Setting up the X Windowing Environment

As discussed in Section 7.2.1, in order to run a program that has a graphical user interface you should run an X server embedded inside your regular X session that matches the properties of the target device. The need for graphical output has been taken into account in the maemo environment, and you will find that an embedded X server called Xephyr [20] has been included with the Intel rootstrap. We will now write a shell script to simplify the invocation of this X server for use in maemo-based development. Open a new file in your favorite editor, calling it for instance start-xephyr.sh. In it, write the following:

```
#!/bin/sh -e
prefix=/scratchbox/users/${LOGNAME}/targets/SDK_PC/usr
export LD_LIBRARY_PATH=${prefix}/lib; export LD_LIBRARY_PATH
exec ${prefix}/bin/Xephyr :2 -host-cursor -screen 800x480x16 -dpi
    96 -ac
```

As you can see, we are invoking a program from under the target file system that we created earlier. LOGNAME is an environment variable that contains your user account name. The parameters given to the Xephyr program specify an output screen similar to that which maemo assumes of the device.

After you have saved this file and exited the editor, give the new file rights for execution:

```
chmod +x start-xephyr.sh
```

Now, make sure that this script works by running it:

```
./start-xephyr.sh&
```

Verify that a window opens, shaped 800 by 480 pixels. Log in to Scratchbox, as before by running /scratchbox/login. Then, if you have set the DISPLAY variable using your .bash_profile file according to the instructions in Section 7.3.3, you can test invoking maemo's application framework in this display:

```
[sbox-SDK_PC: ~] > af-sb-init.sh start
```

Figure 7.11. The Xephyr X Server running with maemo's application framework attached to it.

Verify that you can see the desktop appearing in the new window, as illustrated in Figure 7.11. You can now shut down maemo:

```
[sbox-SDK_PC: ~] > af-sb-init.sh stop
```

Now, feel free to log out of Scratchbox and close the Xephyr window.

7.4 Using Eclipse with Scratchbox

You might have already wondered about whether or not it is possible to use an integrated development environment with Scratchbox, instead of running each of the commands for compilation, execution, and debugging on the command line every time. It is not only possible, but a set of plug-ins [8] for the Eclipse IDE [5] have already been developed by Nokia [19] and the Tampere University of Technology [7] specifically to help Scratchbox-based development, by integrating the aforementioned basic functions inside the graphical user interface of Eclipse.

What made Eclipse a particularly good candidate for this kind of an extension is its openness. Eclipse is designed to be an open platform that does not provide an environment for any specific type of development, but instead it explicitly defines mechanisms for extending it for specific uses. It is best known for the Java Development Tools that provide a very popular Java IDE on top of Eclipse, but the platform is not limited to building programming language-specific IDEs. It is also used as a rich client platform with a component framework based on the OSGi standard [14], for projects that provide

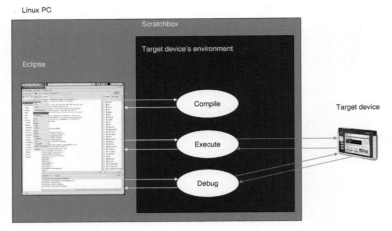

Figure 7.12. The interoperation of Scratchbox and Eclipse, as facilitated by the Laika plug-ins [8].

tools to accommodate a complete enterprise software development lifecycle, and for projects that are targeted for embedded software development.

As a prerequisite for producing these plug-ins, it seemed obvious that the IDE should have support for C language. This is why the plug-ins actually extend another Eclipse project: The C and C++ Development Tools project [4]. It already provides features such as automated parsing of GNU Make error messages, editor functions such as syntax highlighting and code completion, and even a front-end for the **GDB** GNU Debugger. Our plug-ins add an option to create a new type of C or C++ project, that has the run, compile, and debug commands linked to the corresponding commands inside the Scratchbox environment via command line input and output. The Eclipse session is run under a normal log-in session on the development host, and the project files are located inside the sandboxed filesystem so that Scratchbox can access them. Recalling the idea of sandboxing we discussed in Section 7.2.1, you cannot access any outside files from within Scratchbox, but the files inside the sandbox are accessible from outside. The operating principle of the plug-ins is illustrated in Figure 7.12.

In addition to the major functions of compiling, running, and debugging programs, the plug-ins provide shortcuts for switching between Scratchbox targets, and even maemo-specific tasks, such as starting the framework.

7.4.1 Setting up Eclipse with Scratchbox Support

To get started with using Eclipse for Scratchbox-based development, you naturally need to get the IDE to begin with. Head over to http://www.eclipse.org/downloads/ and download the latest Eclipse build for Linux. Since Eclipse is written in Java, make sure your system also has an up-to-date Java virtual machine. If this is not the case, visit `http://java.sun.com` first and download a JVM.

After downloading the Eclipse package, unpack it by running the command `tar xvzf <package>`, with `<package>` replaced with the name of the file you downloaded. This will result in a new directory called `eclipse`. Go to this directory, and start up Eclipse with the command `./eclipse`.

From the welcome screen, navigate to the "Workbench", to get a normal view of your working files. Next, we need to add the components we discussed in Section 7.4: CDT and the Laika plug-ins. The former is a prerequisite to latter, so we will begin with CDT. Eclipse has a convenient built-in software installation and updating component, and we will take advantage of it to perform the installation.

Invoke the software updater by going through the menu system in your Eclipse workbench, as follows:

1. Open the "Help" menu, and highlight the "Software Updates" item. From the pop-up menu, choose "Find and Install".
2. A new dialog will open, titled "Install/Update". Choose the option "Search for new features to install", and click the button marked "Next".
3. Now you will see a list of update sites. Add one for CDT, by clicking "New Remote Site", and entering a descriptive name – such as "CDT" – and `http://download.eclipse.org/tools/cdt/releases/callisto` as the URL. Click "OK" to close the dialog, and then "Finish".
4. You will be asked to choose a site mirror. Choose the one that appears to be closest, and click "Ok".
5. Now you will actually get to select the features of CDT you wish to install. Click on the arrow next to the entry "CDT Main", to unfold the options beneath it. Tick the selection that includes the SDK of CDT, and click "Next".
6. You will be asked to confirm that you accept the terms of the license to download the software. Read through them, choose the option "I accept the terms in the license agreement", and click "Next", if you find the terms acceptable.
7. In the final screen, confirm that the installation directory for Eclipse is correct, and click "Finish".
8. Once the download is complete, click "Install". You will be asked for permission to restart the Eclipse workbench. Accept this by choosing "Ok".

Now that CDT is installed, we will add the Laika plug-ins. Follow the same routine as above, with the following exceptions:

- In step 3, give this remote site a different name – for instance, "Laika" – and enter `http://www.cs.tut.fi/~laika/update` as the URL. Also, make sure only the new Laika site is ticked before clicking "Finish".
- In step 5, unfold the "Laika" selection, then again "Laika" underneath, and tick the "fi.tut.cs.laika.feature" option. There is also an option to add a

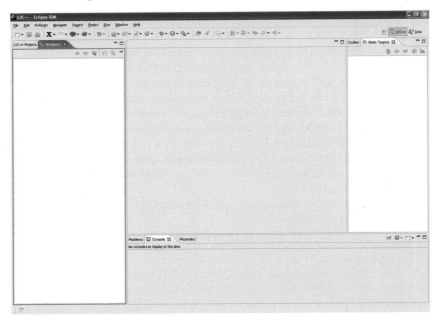

Figure 7.13. The Eclipse workbench with components of the Laika plug-ins visible on the toolbar.

plug-in for Python development under Scratchbox, but this would require another plug-in for generic Python development support, and the subject is outside the scope of our tutorial. Finish the selection again by clicking "Next".

After completing the installation, some new options should be visible on the toolbar of your workbench, as illustrated in Figure 7.13. Top-left, near the icons for "New", "Save", and "Print", there should be symbols of an "X" for starting an embedded X Server, and an "M" for starting the maemo framework, followed by icons for generic command invocation under Scratchbox, as well as one for starting Gazpacho, a user interface builder tool for GTK+ .

Open the menu workbench menu "Window", and choose "Preferences". Unfold the "Scratchbox Preferences" option. Click on "Build Target", and confirm that you see a listing of the targets we defined earlier for Scratchbox: SDK_PC and SDK_ARMEL. If this is not the case, click on the outer level "Scratchbox Preferences" title, and make sure Scratchbox version 1.0 is chosen. Also, make sure that you have executed `/scratchbox/sbin/sbox_ctl start` at some point after restarting your system, so that Scratchbox is up and running.

Make sure that you are not logged in to the Scratchbox environment on any terminal, and activate the SDK_PC target from the "Build Target" menu for now. The build target cannot be changed if you have other active sessions running under Scratchbox.

One more item of configuration is necessary for convenient access to the embedded X server that was introduced in Section 7.3.3. Under the "Scratch-box Preferences" section, choose the section labeled "X-Environment". In the first field, titled "The syntax for starting X-Server", enter the path to the script we created in Section 7.3.3, i.e., something like: /home/<username>/start-xephyr.sh. In the field titled "The syntax for exporting display's IP and number", enter: DISPLAY=127.0.0.1:2. Clear the field marked "The syntax for starting viewer". Also, make sure the option "Run X-server command inside Scratchbox" is not ticked.

After applying these settings, test them by clicking on the "X" symbol on the top left of the workbench toolbar. You can also click the arrow next to it to open a pull-down menu, and choose "Start X". When you see the Xephyr window open, click the "M" symbol to start the maemo framework using this window as the output. If you see the desktop appear, just like after the procedure in Section 7.3.3, you are all set.

7.4.2 Creating a Scratchbox Project

Now everything should be ready for us to create a new project inside Scratch-box using Eclipse. Start by opening the "File" menu, then choose "New" and "Project". Now, unfold the "C" option. Choose "Standard Make C Project inside Scratchbox", and click "Next".

Now, you will be asked for a project name. Note that the default location for the project is under the Scratchbox environment, inside your home direc-tory – a logical assumption, since creating the project anywhere outside the sandboxed file hierarchy would make it inaccessible from Scratchbox. Give the project the name "hello_maemo". Then, you are asked for C project specific settings, such as builder commands and the binary and error parsers used. Just accept the defaults here. In the final screen, you are offered a choice from ready-made templates for a GTK-based program, but just click "Finish" to complete the project creation process.

Now, create a new C source file under this project called hello_maemo.c. In it, type the following program:

Listing 7.2. Hello World For Maemo

```
#include <hildon-widgets/hildon-program.h>
#include <gtk/gtkmain.h>
#include <gtk/gtkbutton.h>

int main(int argc, char *argv[])
{
    /* Create the necessary variables */
    HildonProgram *program;
    HildonWindow *window;
    GtkWidget *button;

    /* Initialize GTK. */
    gtk_init(&argc, &argv);
```

```
     /* Create the hildon program and prepare the title */
     program = HILDON_PROGRAM( hildon_program_get_instance ());
     g_set_application_name ("Hello_maemo!");

19   /* Create a HildonWindow and add it to the HildonProgram */
     window = HILDON_WINDOW( hildon_window_new ());
     hildon_program_add_window ( program , window);

24   /* Create a button and add it to the main view */
     button = gtk_button_new_with_label ("Hello , _world_of_maemo!");
     gtk_container_add (GTK_CONTAINER( window ) , button);

     /* Connect a signal to the X in the upper corner */
     g_signal_connect (G_OBJECT( window ) , "delete_event",
29   G_CALLBACK( gtk_main_quit ) , NULL);

     /* Begin the main application */
     gtk_widget_show_all (GTK_WIDGET( window ));
     gtk_main ();
34
     /* Exit */
     return 0;
     }
```

Even if you are not familiar with using graphical components provided
by the GTK+ library in C programs, you probably have a good idea of what
kind of a program we are creating, after reading Listing 7.2. Essentially, we are
doing some initialization that is related to GTK and the Hildon application
framework that maemo uses, creating a button with the text "Hello, World
of maemo!", and putting it onscreen.

Next, we need to define how to compile the program. Create another file,
this time just a plain text file called Makefile. In the Makefile, enter the
contents of listing 7.3.

Listing 7.3. A GNU Makefile for Our Project

```
   all:
         gcc −o hello_maemo hello_maemo.c 'pkg−config −−cflags \
3        gtk+−2.0 hildon−libs ' −ansi −Wall 'pkg−config −−libs \
         gtk+−2.0 hildon−libs '
   clean:
         rm hello_maemo
```

This is a definition for a build target according to the GNU Make syntax.
Note that the empty space before both the compilation and clean command
has to be a tab symbol. The command pkg-config that is applied here is a
helper tool that can be used to insert the needed compiler options in a simple
fashion, like above, when we just know the names of the libraries we want to
use in our program.

Now, click on the project to highlight it, open the "Project" menu, and
choose "Build Project". If it is greyed out, you probably have the "Build
Automatically" option beneath ticked, so the Make program has probably
already run.

Once the build is complete, click on the "C/C++ Projects" tab on the left-
hand side of your workbench, to see a nicely grouped view of your project. If
it is not available, go through the menu "Window" and "Show view" to enable

it. Unfold the "Binaries" section to see a file titled `hello_maemo`, along with information that it is an x86 file in little endian byte order. Click on the file to select it, and open the Run menu. Make sure that your Xephyr server is still running, with Hildon inside. Choose "Run", and double-click on the "C/C++ Scratchbox Application" to create a new profile for executing the program. Tick the "Run standalone" option. Click the "Run" button. Now you should be able to see your first Maemo application appear inside the Xephyr window, as shown in Figure 7.14.

Now, let us cross-compile it. Close the program and shut down your maemo session by clicking on the arrow next to the "M" in your workbench, and choosing "Stop maemo". Now, open the Window/Preferences menu, and navigate to the "Build Target" option under "Scratchbox Preferences" again. Choose the "SDK_ARMEL" target, and click "Ok".

Clean our previous build by choosing the project, and the option "Clean" under the "Project" menu. Build the program, start maemo, and run the program exactly as above. That is all it took to cross-compile our piece of software. You can verify that the new binary is in fact compiled for the ARM processor, by checking the C/C++ Projects view just as before.

If at any point during this tutorial you find that there are errors while switching between targets, or that the application framework fails to restart, you can use the `sb-conf` tool to make sure Qemu is not running anything by executing the following:

```
[sbox-SDK_ARMEL:  ~] > sb-conf killall --signal=9
```

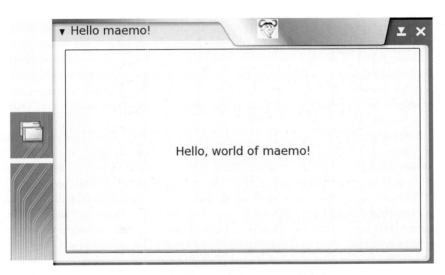

Figure 7.14. Our basic maemo application.

7.4.3 Debugging Inside Scratchbox

Let us create another Scratchbox project, using a different approach. Switch back to the SDK_PC target, and create a new project of type "Automake Hildon Project inside Scratchbox". Give the project the name "hello_automake". After you have gone through the normal Scratchbox C project creation dialogs just as in Section 7.4.2, you will be offered a new set of program templates. You can choose to use a ready-made template for your maemo project, or create a customized skeleton project using a helper tool. Choose "Maemo Hello World" from the ready-made templates. After you finish this step, you will also be asked for information to support automatically creating an application package compatible with the Debian package management system. You can have a look under the "Project" menu in your workspace to see the shortcuts that the Laika plug-ins add for packaging applications. The type of project we created automatically includes a set of files that are usable by the set of GNU tools generally known as "Autotools". These tools can be used to generate a Makefile that is appropriate for the build environment. When you created the project, they were automatically run, so now you have a Makefile ready. Now, let us take a look at the debugging facilities available through this IDE. For debugging, we will use the GDB GNU Debugger paired with the GDB server, both run inside the Scratchbox environment [2]. The versions used are specific to the particular toolchain we are using. To automate the step of invoking the GDB server, we will configure a debugging option from the "Scratchbox Preferences" section of the Window/Preferences menu. Set the field titled "Starting GDB server inside Scratchbox" to `gdbserver 127.0.0.1:1234 ${binary}`. This is the syntax used to start up a server that actually executes the binary, and which GDB can connect to. If we were using sbrsh for remote execution, we should specify the external IP address of our development host here, but since we are operating inside a single host, the local address is enough. It is also possible to utilize Qemu for processor emulation combined with GDB server functionality. With the current version of Qemu, this option would then be set to `qemu-arm -g ${binary}`. As you can see, no port is specified – at the time of writing, Qemu is hard-coded to use the port 1234.

Build the project just like in Section 7.4.2. Open the Run menu, but this time, choose the option "Debug". Again, create a new run target by double-clicking on "C/C++ Scratchbox Application". Tick the "Run stand-alone" option, and change to the "Debugger" tab. Choose "GDB Server inside Scratchbox", and un-tick the option labeled "Stop on startup at main". Inside the "Debugger Options" subsection, navigate to the "Main" tab and tick the "Start GDB-Server Automatically". Then, from the same subsection, choose the "Connection" tab, and specify that the connection takes place over TCP to the local host's port 1234. Finally, click "Debug".

The Eclipse workbench will now change in appearance to better suit debugging work, as shown in Figure 7.15. Next, the program will start, and you will see a large button appear onscreen. Open the `src` folder, and the file `main.c`

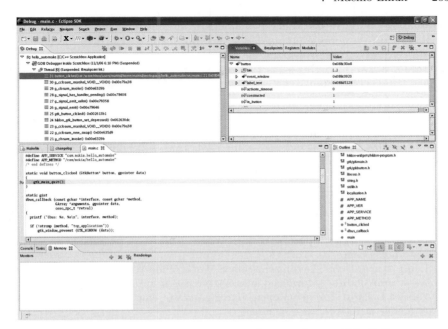

Figure 7.15. Debugging a maemo application from within Eclipse.

under it. Find the function `button_clicked` in the beginning of the file, and
place a breakpoint in it – on the line with the function call to `gtk_main_quit()`.
You can do this in Eclipse by right-clicking on the left marginal of the editor, and choosing "toggle breakpoint" from the pop-up menu. Click on the
button on the program screen. The execution will stop at the breakpoint we
defined, and you can freely investigate the state of the program at this point,
including reading the values of variables and registers, and so on. We can see
that the front-end Eclipse provides us a very convenient way to use GDB. It
is also worth noting that we can use the same interface even when debugging
a program running on an actual device with a finalized software environment.
The gdbserver just needs to be installed and started on the device.

7.5 Conclusions

Hopefully this chapter has given you an idea of what development using the
Scratchbox environment means, and maybe also some thoughts on what kind
of opportunities the maemo SDK offers you. We have seen some of the intricate details inside Scratchbox – but on the other hand, we have also seen
how, by complementing this environment with high-level tools, you can use it
to perform software development tasks while almost forgetting that you are
in fact doing cross-development. It is good to remember that even though we
used the maemo SDK for a concrete example platform in this chapter, with

Scratchbox you have the tools available for building a development environment for quite a generic embedded Linux platform. If you feel interested in starting a real software project for the maemo platform, a thorough tutorial [10] on the maemo website is a good starting point. Also, when considering just about any embedded Linux development, looking into setting up the Scratchbox environment for the system at hand may be well worth the initial effort. Whether your next project is to write the killer application for embedded Linux, or to port an existing piece of Linux software for the Nokia 770 to have it on the go, you now know some tools that will help take your project off the ground.

References

1. Nokia 770 internet tablet web site. World Wide Web. Available at: http://www.nokia.com/770.
2. Lauri Arimo. Debugging in scratchbox. World Wide Web. Available at: http://www.scratchbox.org/download/files/sbox-releases/1.0/doc/debugging.pdf.
3. Fabrice Bellard. The qemu open source processor emulator. World Wide Web. Available at: http://www.qemu.org/.
4. Eclipse c/c++ development tooling – cdt. World Wide Web. Available at: http://www.eclipse.org/cdt/.
5. Eclipse. World Wide Web. Available at: http://www.eclipse.org.
6. Foreign toolchains. World Wide Web. Available at: http://www.scratchbox.org/wiki/ForeignToolchains.
7. Juha Järvensivu, Matti Kosola, Mikko Kuusipalo, Pekka Reijula, and Tommi Mikkonen. Developing an open source integrated development environment for a mobile device. In *International Conference on Software Engineering Advances*. IEEE Computer Society Press, 2006.
8. Laika – the scratchbox eclipse plug-in for maemo. World Wide Web. Available at: http://www.cs.tut.fi/~laika/.
9. Maemo development platform: White paper. Available at: www.maemo.org/platform/docs/maemo_exec_whitepaper.pdf.
10. Developing new applications. World Wide Web. Available at: http://www.maemo.org/platform/docs/howtos/howto_new_application.html.
11. maemo.org. World Wide Web. Available at: http://www.maemo.org/.
12. Movial. Movial corporation. World Wide Web. Available at: http://www.movial.com.
13. Nokia. World Wide Web. Available at: http://www.nokia.com.
14. Osgi alliance. World Wide Web. Available at: http://www.osgi.org.
15. Timo Savola. *Fakeroot in Scratchbox*. Nokia, 2005. Available at: http://www.scratchbox.org/download/files/sbox-releases/apophis/doc/fakeroot.pdf.
16. Timo Savola. *Scratchbox Remote Shell*. Nokia, 2005. Available at: http://www.scratchbox.org/download/files/sbox-releases/1.0/doc/sbrsh.pdf.
17. Scratchbox. World Wide Web. Available at: http://www.scratchbox.org.
18. Cross-compilation and cross-configuration explained. World Wide Web. Available at: http://www.scratchbox.org/documentation/general/tutorials/explained.html.

19. Matti Sillanpää. Extending eclipse and cdt for embedded systems development with scratchbox. EclipseCon 2006, March 2006. Available at: http://www.eclipsecon.org/2006/Sub.do?id=216.

20. Xephyr. World Wide Web. Available at: http://www.freedesktop.org/wiki/Software_2fXephyr.

8

Windows Mobile Programming
.Net Programming on Mobile Phones

Rico Wind, Christian S. Jensen, and Kristian Torp

Aalborg University {rw|csj|torp}@cs.aau.dk

Summary. This chapter presents an overview of current techniques and tools for developing applications using the .Net Compact Framework targeting the Windows Mobile platform. The chapter provides an introduction to the various aspects of application development and points to more detailed sources for further reading.

8.1 Introduction

An increasing number of mobile phones and PDAs are sold with the Windows Mobile operating system. In this chapter, we use the C# programming language to demonstrate different aspects of Windows Mobile programming using the .Net Compact Framework. The chapter introduces the reader to the basic functionality of the .Net Compact Framework and provides a number of references to more detailed information. In particular, other programming languages can be used with the .Net Compact Framework – for details, see [11]. It is possible to execute Java Micro Edition (Java Me) programs on a Windows Mobile device. However, this does not expose the full potential of the device, as the strength of the Java ME platform, "write once, run everywhere", is also a weakness because, roughly, the lowest common denominator among devices defines the capabilities. The .Net Compact Framework does not have the same limitations, but the current market is also much smaller, in terms of the number of devices supported and the number of devices sold. This chapter's coverage does not assume prior knowledge about the .Net Framework and C#. Readers familiar with C# and Windows Forms programming might thus find parts of this chapter trivial. We do assume that the reader has basic knowledge about object-oriented programming. However, we try to minimize the number of object-oriented concepts used. This means that concepts such as encapsulation, inheritance, and layered architectures are not discussed in detail, although these concepts are important in real-world applications. To minimize the sizes of the code examples exception handling is omitted, although, again, proper exception handling is important in practice.

F.H.P. Fitzek and F. Reichert (eds.), Mobile Phone Programming and its Application to Wireless Networking, 207–235.

The chapter is structured as follows. Section 8.2 gives a short introduction to C# and the .Net Framework. Section 8.3 presents the .Net Compact Framework and explains how to transfer programs to a mobile device. Section 8.4 presents a range of controls that can be used when programming for a Windows Mobile environment. Section 8.5 introduces network functionality and limitations. Finally, Section 8.6 presents the Visual Studio development environment, covering specifically how to use web services and databases in applications.

8.2 .Net and C# in a Nutshell

.Net is a standardized framework [1, 2] for developing applications, web services, and websites. The framework was originally developed by Microsoft for the Windows platform, but other implementations exist for other platforms (cf. the Mono Project [9]).

Figure 8.1 presents an overview of the .Net Framework architecture. The Common Language Runtime (CLR) on top of the operating system is responsible for executing the .Net byte-code on the underlying platform. It does so using a Just-In-Time (JIT) compiler to convert the byte-code into executable code. In effect, a .Net application can run on any platform that has a CLR. On top of the CLR lies the .Net Framework Class Library that offers access to a wide range of functionality such as collections, network communication, graphical user interface (GUI) components, database access, and web services.

The .Net Framework also includes what is known as the Common Type System (CTS), a unified type system that is shared by compilers and the CLR. It specifies how the CLR declares and uses types. This enables cross-language capabilities, i.e., a component implemented in one programming language can directly be used as a component in a program written in another language. A list of languages (and compilers) that target the .Net platform can be found in reference [14].

One of the .Net languages is C#, a standardized object-oriented programming language developed by Microsoft. The language is specified in a freely

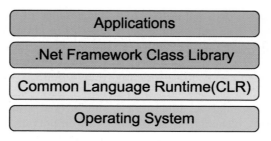

Figure 8.1. .Net Framework Architecture.

available ISO standard [1]. C# targets the .Net Common Language Infrastructure (CLI) and takes advantage of the built-in structures of the .Net Framework. The .Net Framework SDK from Microsoft is freely available (see [13]) and includes a C# compiler. In the following, we assume that this SDK has been downloaded and installed. If a computer already contains a copy of Microsoft Visual Studio the compiler is available without further installation through the Visual Studio Command Prompt. It is also possible to use the Mono framework. However, some of the code in this chapter might not work, as the Mono framework does not yet include a full implementation of version 2.0 of the .Net Framework.

Let us start by considering the C# version of the famous and very simple "Hello World" application. The source code is given in Listing 8.1, which displays the content of the file *Hello.cs*.

Listing 8.1. Hello World in C#

```
using System;

public class Hello{
    public static void Main(string[] args)
    {
        Console.WriteLine(''Hello World'');
    }
}
```

To try out the program in Listing 8.1 put the content into a file called *Hello.cs*. The file name and class name are identical. This is according to a recommendation made by Microsoft [3] and is not required, as it is in Java. Start a command prompt and change the directory to the location where the file is placed. Then type the following.

```
csc Hello.cs
```

This will invoke the C# compiler (csc) on the source file and create an executable file named *Hello.exe*. By typing

```
Hello
```

the system executes the program and writes "Hello World" on the screen. The program should be straightforward to read for people with a knowledge of other object-oriented languages such as Java or C++, although the method names are different and the import statements have a different syntax. The first line includes the *System* namespace (explained shortly) that is part of the .Net Framework class library. Line 3 defines the class that we call *Hello*. In lines 4–6, we define the static *Main* method that is equivalent to the main method of a Java or C++ program, i.e., this method automatically gets executed when the program is started. All this method does is to write "Hello World" to the screen using the *WriteLine* method of the *Console* class that is part of the *System* namespace.

Namespaces are a convenient means of uniting related classes into a clearly defined area and are equivalent to the package concept in the Java language.

While several conventions exist for the naming of new namespaces, we recommend the one suggested by the Microsoft Design Guidelines for Developing Class Libraries [3]. The guide recommends the following convention for namespaces:

<Company>.(<Product>|<Technology>)[.<Feature>][.<Subnamespace>]

In the following, we will use this convention in our examples. Note that it is always possible to leave out the *using* statements in the beginning of the file and instead use fully qualified names, e.g., removing the *using* statement above and writing *System.Console.WriteLine("Hello World")* yields the same result.

We continue with a more advanced example that illustrates the use of the *System.Windows.Forms* namespace that includes a large set of components usable for programming graphical user interfaces.

Listing 8.2. Windows Form in C#

```
using System;
using System.Drawing;
using System.Windows.Forms;

namespace AAU.CSharpBook.WindowsExamples
{
  public class WindowsForm : Form
  {
    public static void Main()
    {
      Application.Run(new WindowsForm());
    }

    public WindowsForm()
    {
      this.Text = ''My first form'';
      Button b = new Button();
      b.Text = ''Press me'';
      b.Location = new System.Drawing.Point(50, 50);
      b.Click += new EventHandler(HandleButtonclick);
      this.Controls.Add(b);
    }

    private void HandleButtonclick(object sender, EventArgs e)
    {
      Button b = (Button)sender;
      b.AutoSize = true;
      b.Text = ''Press me again please'';
    }
  }
}
```

Listing 8.2 thus shows an example that is useful for demonstrating simple user-interface concepts, in preparation for moving on to the mobile platform. As illustrated in lines 1–3, we now use three different namespaces: the *System* namespace from the first example, the *System.Windows.Forms* namespace introduced above, and the *System.Drawing*, which enables us to draw on the screen and position components from the *System.Windows.Forms* namespace. Note that we have encapsulated the class in our own namespace as defined in line 5.

Line 7 defines a new public class called *WindowsForm*. The notation *: Form* after the class name makes the new class inherit from the *Form* class. The *Form* control class is the fundamental class when working with graphical-user interfaces. On a *Form*, it is possible to draw other controls such as *Buttons*, *Labels*, and *TextBoxes*. The standard library contains a large collection of such controls, and it is possible to add new custom controls in case no existing controls offer the desired functionality; new controls can be reused across projects. Custom controls are beyond the scope of this chapter, and we refer to references [4, 21].

As in the previous example, the program is started in the *Main* method that only contains a single line. The *Application.Run* method starts a new message loop in the current thread (this is done behind the scenes by .Net). The method makes the form that is passed as an argument visible, and it ensures that everything will be cleaned up when the user closes the form. Note that the form could just as well be instantiated and started from a separate file with a *Main* method.

The next thing we see is the constructor for the class that actually sets up the form. The first line sets the title using the *Text* property inherited from the *Form* class using dot notation on *this* that refers to the current object. Properties typically encapsulate instance variables, i.e., the *Form* class probably has a private variable called *text* that can be set or retrieved by the *Text* property. In C#, properties, and methods are written with the first letter in uppercase, whereas variables are written in lowercase [3]. In Line 17, we create a *Button* that is then assigned a text and a location relative to the left upper corner of its parent using the *Point* class.

Line 20 uses a delegate of the *Button* class to specify which method should be called when the user presses the button. A delegate is a method pointer (or more correctly, a list of method pointers), i.e., a delegate points to a method that is called when the delegate is invoked. Delegates are used intensively in .Net to handle events in graphical user-interfaces. While this is not the only use of delegates, it is the only use that we consider. Note that the assignment of a delegate uses +=, not =. This means that the method is added to the method list of the delegate, i.e., it is possible to have several event handlers for a single delegate. These are called serially when the delegate is invoked.

Our program specifies that the *HandleButtonClick* method should be called when the user presses the button. The delegate is of the standard *Event-Handler* type, meaning that the signature of the method, i.e., the parameters it takes and what it returns, is fixed. A method used as an *EventHandler* must be void and must take two arguments: an *object* and an *EventArgs*. The *object* is a reference to the calling object, in this case the button. The *EventArgs* can be used to pass string arguments to the calling method. The last line of the constructor adds the button to the list of controls on the form.

The last part of the file is the method that handles the button click. As explained above the first argument is the sender of the event, i.e., the button. The general type *object* is downcasted to a *Button* object. As it is the case in Java, every class in C# implicitly inherits from the class *object*. We enable a property of the button called *AutoSize*. With this property enabled, the button automatically resizes itself to fit the text content. To illustrate that the method is indeed called when the button is pressed, we use a new *Text* property. To test the program, it should be compiled and run.

8.3 .Net Compact Framework

The .Net Compact Framework (.Net CF) is a scaled-down version of the .Net Framework. Components from the .Net Framework Class Library that are not useful on a mobile platform have been removed. In addition, the number of overloaded methods in the class library has been decreased. The size of the .Net Compact Framework is therefore only around 8% of the size of the .Net Framework. This reduction enables the framework to fit into memory-constrained devices. The .Net CF also includes extra functionality directly targeting embedded and mobile platforms that is not available in the .Net Framework. A more detailed description of the differences between the two frameworks is given in reference [5].

We proceed to consider several examples of programs that target the .Net CF. First, however, we must set up an environment capable of compiling to the .Net CF. By default the compiler will target the full .Net Framework. This creates an executable that cannot be executed on the .Net CF. One solution is to use the Visual Studio IDE (to be introduced in Section 8.6.) Visual Studio automatically targets the .Net CF assemblies and includes a graphical drag-and-drop environment for creating the graphical user-interface. In addition, it includes an emulator that can be used to execute the program without having to transfer it to a mobile device. The main problem is that the Mobile Development Environment is only available in the Standard, Professional, and larger Editions—not in the free Express Edition.

Use of the full versions of the Visual Studio is by far the easiest way to get started with Windows Mobile development. However, it is possible to develop mobile applications without Visual Studio, as explained below. Readers with access to Visual Studio can skip the next steps and simply run the examples from within this IDE.

The solution without Visual Studio is to tell the compiler to use the .Net CF assemblies. A batch file for handling the compilation is available from reference [6]. Note that the variable *NETCF_PATH* must be changed to point at the right directory on the computer used. In the following, it is assumed that the batch file has been saved as *com.bat* and that the necessary adjustments have been made to the *NETCF_PATH* variable.

Let us begin by reconsidering the program from Listing 8.2 that we run on the .Net Framework. First, we compile the file with:

```
com WindowsForm.cs
```

The compiler complains that the *System.Windows.Forms.Button* does not contain a definition for the *AutoSize* property. As explained above the .Net CF omits some parts of the full .Net Framework, the *AutoSize* property of a button being one. When commenting out or deleting line 27 in Listing 8.2, the program now compiles without problems.

8.3.1 Using the Windows Mobile Emulator

To test the program we could simply copy it to a Windows Mobile device and run it (assuming that the device has version 2.0 of the .Net CF installed). However, when creating and testing programs, it is often much easier to use an emulator. A Windows Mobile emulator is freely available [19] (the stand-alone version). Figure 8.2 shows a screen-shot of the emulator. Notice the two soft buttons just below the *Notification* and *Contacts* menus in the bottom of the screen. Section 8.4 explains how to use these in programs.

In addition, the freely available [19] Microsoft *ActiveSync* should also be installed. Assuming that Visual Studio is not installed, going through the following steps enables the testing of the newly created application. Install *ActiveSync* and the stand-alone *Device Emulator*—follow the on-screen instructions. Start up both programs—*ActiveSync* will start by default when

Figure 8.2. Emulator screen.

Windows starts. In the *Device Emulator*, an entry called "Windows Mobile 5.0 Pocket PC Emulator" should appear. Right-click on it and choose "Connect". The device emulator will now start a virtual device in a window. In this device, it is possible to use the standard programs delivered with the Windows Mobile operating system.

To install programs onto the emulator we use the *ActiveSync* program. First we will install the .Net CF 2.0—the operating system as the default only comes equipped with the .Net CF. The .Net CF 2.0 End User Redistributable is downloadable [10]. Before running the installer, the *ActiveSync* program must be started, as it is used to transfer the framework to the emulator. To make the emulator visible for the *ActiveSync* program it is necessary to simulate that the device is in a cradle. To do so, go back to the device manager, right-click the running image, and press the cradle option. This emulates that an actual Windows Mobile device is put into a cradle attached to the computer's USB port. The device should now pop up in the *ActiveSync* program. In addition, a synchronization wizard might pop up. The wizard can simply be closed.

The next step is to start up the framework installer and to follow the on-screen instructions. This will install the framework onto the emulator. Note that user input is required in both the installer and on the emulator (including a restart of the emulator). It is now possible to copy the newly created program to the emulator. In *ActiveSync*, the *Explore* button in the toolbar is pressed. This brings up an explorer window that gives access to the file system. The directory "Program Files" in "My Windows Mobile-Based Device" is accessed and a directory is created in which the program executable is placed. The program can now be started on the device emulator by choosing "Programs" in the start menu and then starting the "File Explorer". The directory in which the program was placed is found, and the program is click-it. The program should start and display the same form as the Windows application did. Note that the button does not automatically resize when clicked because of the line with the resizing was removed.

8.3.2 Installing Programs on Mobile Devices

Installation on a real device is very similar to installation on the emulator. The only difference is that the device emulator does not need to be installed. Make sure that the *ActiveSync* is running and that the device is in its cradle. If the .Net CF 2.0 End User Redistributable has already been installed on the emulator, one simply goes to "Add/Remove Programs" in "Tools" via the *ActiveSync* and chooses the framework. If it has not been installed, the procedure from the previous section should be followed. The actual program installation is performed as presented in the previous section.

8.4 Using the Windows Mobile Controls

We proceed to cover a number of features available on the Windows Mobile platform. Thus, not all of the example programs in this section will run as a desktop Windows application. The most important mobile functionality is probably the two soft keys that allow users to quickly access functionality (In Figure 8.2, the soft keys are named *Notification* and *Contacts*.) As a precursor to covering these, we introduce a number of the controls available through the *System.Windows.Forms* namespace.

8.4.1 Controls in the *System.Windows.Forms* Namespace

We consider a subset of the controls in the *System.Windows.Forms* namespace. These controls are available on both the mobile platform and the Windows platform. A complete overview of the class libraries that includes simple examples can be found in the .Net Class Library Reference [12]. This reference is an important resource when programming for the .Net platform. For each of the classes, methods, and properties included in the reference, it is stated whether or not the given functionality is available in the .Net Compact Framework.

Button The *Button* control supports the creation of custom buttons. It has *Size* and *Location* properties that control the size and placement on a button and the other controls discussed here. Other often used properties include *Name*, *Text*, and *BackColor*. The most useful event is the *Click* event, but the *GotFocus/LostFocus*, and *KeyDown* (only occurs when the control has focus) events might also come in handy.

TextBox The *TextBox* control enables the user to enter a string into a text field. Its most useful properties are *Name* and *Text*. If a password from the user is needed, the *PasswordChar* property is set to a suitable character, e.g., *, that is then used when displaying the password characters as they are entered. In many cases it is not necessary to subscribe to any events on a *TextBox*. Rather, it may be more convenient to use the *Text* property when the user presses a button. In some cases, however, the *TextChanged* event is useful, e.g., when filtering data in another control based on the content of the *TextBox*.

Label A *Label* control is used to display a headline for another control or to display textual information to the user. The *Text* property enables the setting and retrieval of the textual content of the control. Event handlers on a *Label* are needed only rarely, but in some cases the *TextChanged* event is useful.

RadioButton A number of *RadioButtons* can be used together to allow the user to enter an exclusive choice among options. When several *RadioButtons* are used in a form they are by default exclusive, i.e., the user can only choose one of them. If more than one group of *RadioButtons* is needed on a single form, the different groups should be added to distinct *Panels* (not covered here).

The properties *Name*, *Text*, and *Checked* are the most important ones. The *Checked* property is used to set and retrieve the state of the *RadioButton*. In most cases, the *RadioButton* is used to capture a user's choice in conjunction with other controls, e.g., whether the user is male or female. If other controls are to change appearance or functionality immediately when the currently *Checked RadioButton* changes, the *CheckedChanged* event is useful. This event is raised both when the *RadioButton* is *Checked* and when another *RadioButton* is checked while the current *RadioButton* was the checked one.

ComboBox The *ComboBox* offers a compact means of presenting the user with an exclusive choice among a set of alternatives. In addition, it is possible to have several *ComboBoxes* on the same form without the use of *Panels*. Another advantage is that it is easy to use data binding with a *ComboBox*. Data binding is a feature that allows the binding of a control to an object or a collection of objects. Data binding will be illustrated in Section 8.6.

The *ComboBox* has an object-collection property called *Items* to which content can be added. There are two different ways of specifying the textual representation of the objects in the *ComboBox*, i.e., specifying what text the user will actually see for a given object in the box. As the default, the *ComboBox* uses the *ToString* method of the objects and displays this text. The alternative is to set the *DisplayMember* property of the *ComboBox*; this enables the specification of the name of a public property on the objects in the *Items* collection. This is illustrated in the ensuing example where we assume that a person has a public property called *Name*:

```
Person  p1 = new Person(''Heidi '');
Person  p2 = new Persen(''John '');
ComboBox  cb = new ComboBox();
cb.DisplayMember = ''Name'';
cb.Items.Add(p1);
cb.Items.Add(p2);
```

Here, the *ComboBox* will use the *Name* property of the *Person* class to retrieve the string that represents the object (assuming that the *ComboBox* is used somewhere on a form).

The most used event handler on the *ComboBox* is the *SelectedValueChanged* that is invoked whenever the user changes the currently selected item. The *SelectedItem* property gives access to the currently selected object. Assuming that an event handler has been created for the *SelectedValueChanged* and that *cb* is an instance variable, we can retrieve the currently selected *Person* as follows.

```
Person  p = (Person)cb.SelectedItem;
```

Note that we have to downcast from the general *object* class to the *Person* class. Generic versions of the controls, i.e., versions that are instantiated to the type of objects that the control is to contain, do not yet exist. Care should be taken when using typeless collections. In our example, it is perfectly legal to add an object of the class *Animal* to the *ComboBox*. If the animal had a *Name* property, this would even be displayed in the list. A runtime exception would, however, be thrown when the user selects the *Animal* in the *ComboBox* because of the downcast to the *Person* class.

ListBox The *ListBox* is a particularly useful control on the mobile platform because it provides easy selection from a range of entries. The *ListBox* includes much of the functionality of the *ComboBox* and can be used in the same manner, i.e., the code example above can be used directly if *ComboBox* is exchanged with *ListBox* (it would then be appropriate to also change the variable names to better reflect the type).

Because all elements of a *ListBox* are directly visible to the user, this control is useful when displaying incoming content. Assume that we are creating a program that displays headlines and content from an RSS feed. Assume also that we have a number of RSS objects with a headline and a body text in HTML and that new RSS objects will be made available to us continually. We could simply use a *WebBrowser* control (to be introduced shortly) and show the most recently arrived content. But because we do not have the mobile user's unconditional attention some of the entries will go by unseen. We could of course draw attention to incoming headlines by playing a tune on the device speakers, but repeated interruptions of the user is not the way to go in most cases. If we instead use a *ListBox* in conjunction with a *WebBrowser* control, a list of the all *HeadLines* of all RSS entries received can be shown in the *ListBox*; and when the user clicks one of the entries, the *Body* can be shown in the *WebBrowser* control.

MessageBox The *MessageBox* is the well-known pop-up message box. There are no event handlers or properties on this control. Its functionality is controlled using three overloaded versions of the *Show* method. The simplest version takes a string as input; this creates a dialog displaying the string and requires the user to press an OK button. The second version takes two strings, namely the message, as before, and a caption for the dialog. The user must again press the OK button to proceed. The last version allows the specification of which buttons to show on the form and which icon to use in the message box. The button and icon are set using values from two enumerations: *MessageBoxButtons* and *MessageBoxIcon*. When the user clicks one of the buttons, the *MessageBox* will close and return to the calling form. The *Show* method returns a *DialogResult*, which is again a value from an enumerator. The *DialogResult* enumerator includes values for all the buttons usable on the *MessageBox*, i.e., when using a given set of buttons, it is possible to determine which button was pressed by comparing the result with the values from the enumerator. As an example consider the following code fragment with an "OK" and a "Cancel" button:

```
DialogResult  dialogResult   =  MessageBox.Show("Press  OK  or  Cancel",
    "My caption",  MessageBoxButtons.OKCancel,
    MessageBoxIcon.Question ,  MessageBoxDefaultButton.Button1);
if(dialogResult == DialogResult.OK)
    MessageBox.Show("The user pressed OK");
else if(dialogResult == DialogResult.Cancel)
    MessageBox.Show("The user pressed Cancel");
```

In the example, we first query the user to confirm or cancel. We then show the result of the choice in another *MessageBox*. The second *MessageBox* only allows the user to press "OK" as no *MessageBoxButtons* are specified. Care should be taken when using *MessageBoxes* on a mobile device as it requires that the user responds using the device's pointing mechanism.

WebBrowser The *WebBrowser* control makes it possible to easily add support for displaying HTML content, including the fetching and display of online websites. To directly supply the HTML to be displayed, the *DocumentText* property is used. If the browser is to display an online website, the *Url* property is used. When this property is set, the *WebBrowser* control will automatically fetch and display the web page specified. Assuming that we have a *WebBrowser* control *wb*, the following will fetch and display the mobile version of Google:

```
wb.Url = new Uri(''http://mobile.google.com'');
```

Not all of the controls presented above will be used in the subsequent examples in this section, but they serve as a good foundation for programming on the mobile platform. We proceed to consider a few Windows Mobile specific controls in more detail.

8.4.2 Controls Specific to Windows Mobile

Listing 8.3 shows a program that uses the soft keys of the Windows Mobile device. This program randomly chooses between three URLs and shows the corresponding web page. In the program, we have three instance variables: one that represents a WebBrowser element, a string array that holds three different URLs, and a random-number generator. As in the previous examples, we have a *Main* method that creates a new instance of the class and starts a message loop.

Then follows the constructor where the instantiations of the controls are accomplished. The constructor creates a new *MainMenu* object. We want to include two *MenuItems* in this object. In the following lines, we first create a *leftMenu* that is used to close the application. The *Text* property is set and an *EventHandler* is attached. Next, we create a *MenuItem* called *rightMenu* to which we attach another *EventHandler* and set the *Text* property to "Internet". Both *MenuItems* are added the *mainMenu* object and the actual menu on this form is set to the *mainMenu* object.

In lines 30–31, we instantiate the WebBrowser control and include it into our control collection. Finally, we set the WebBrowser to occupy the entire screen (or more correctly, to occupy the form on which we work).

Listing 8.3. Windows Mobile in C#

```csharp
using System;
using System.Drawing;
using System.Windows.Forms;

public class WMForm : Form
{
  private WebBrowser wb;
  private string[] urls = new string[] { ''http://mobile.google.com'',
                                         ''http://mobile.news.com'',
                                         ''http://www.yahoo.com''};
  private Random rnd = new Random(DateTime.Now.Millisecond);

  public static void Main()
  {
    Application.Run(new WMForm());
  }

  public WMForm()
  {
    MainMenu mainMenu = new System.Windows.Forms.MainMenu();
    MenuItem leftMenu = new System.Windows.Forms.MenuItem();
    leftMenu.Text = ''Close'';
    leftMenu.Click += new EventHandler(HandleCloseMenuClick);
    MenuItem rightMenu = new System.Windows.Forms.MenuItem();
    rightMenu.Click += new EventHandler(HandleStartBrowsingClick);
    rightMenu.Text = ''Internet'';
    mainMenu.MenuItems.Add(leftMenu);
    mainMenu.MenuItems.Add(rightMenu);
    this.Menu = mainMenu;
    wb = new WebBrowser();
    this.Controls.Add(wb);
    wb.Size = this.Size;
  }

  private void HandleCloseMenuClick(object sender, EventArgs e)
  {
    this.Close();
  }

  private void HandleStartBrowsingClick(object sender, EventArgs e)
  {
    Uri uri = new Uri(urls[rnd.Next(0, urls.Length)]);
    wb.Url = uri;
  }

}
```

The next method is the *EventHandler* attached to the first menu item. The method closes the program by calling the close method inherited from the *Form* class.

It is important to note that while the cross button in the upper right corner enables the user to close the application (or current window) in Windows applications, the cross by default minimizes the application in Windows Mobile applications (on the mobile platform, only one instance of a program can be run at a time). This is the default behavior of all Windows Mobile programs, so this behavior of the button should not be overridden without a very good reason.

Consider an application that constantly pulls a network resource to update the display. For a user on a pay-as-you-go GPRS subscription, this might be expensive, and so this constitutes a good reason to close the application. To perform additional functionality when the user presses the cross, a function is to be attached to the *Deactivate EventHandler* of the form. This could be done as follows.

```
this.Deactivate +=new EventHandler(Form1_Deactivate);
```

This then assumes that a method with the following signature has been implemented:

```
void Form1_Deactivate(object sender, EventArgs e)
```

In order to touch slightly upon another feature in .Net, we will show an alternative way of accomplishing the above. Assume that we want to mimic the desktop Windows behavior, i.e., to close the application when the cross is pressed. This may be done as follows:

```
this.Deactivate += delegate(object o, EventArgs e){ this.Close();};
```

In this so-called anonymous method , we may include any number of statements in between the curly brackets. The delegate takes the same arguments as the *EventHandler* method. It is advisable to only use anonymous methods for very short and trivial methods as the code quickly becomes difficult to read.

The last method in the code sets the *Url* property of the web browser to a randomly chosen URL from the array of strings. Three screen shots from a Windows Mobile device that illustrate the program can be seen in Figure 8.3.

Let us return to the soft keys. Above, we added two *MenuItems* to the *mainMenu*. What happens if we add only one or if we add more than two? If we add only one *MenuItem*, this becomes the left soft key. If we add more

Figure 8.3. Windows mobile screen for listing 8.3.

than two, the soft keys are disabled and a menu organized as a desktop Windows application menu is used. It is possible to experiment with this behavior by simply adding additional *MenuItems* to the *MenuItems* property of the *mainMenu*.

In more complex applications, more than two *MenuItems* may be needed, but it may still be attractive to use the device's soft keys as these offer convenience. As is the case for desktop Windows application menus, it is possible to nest the *MenuItems*. This is accomplished by the code in Listing 8.4 that contains an alternative definition of the *rightMenu* object.

Listing 8.4. Nested MenuItems

```
  . . .
  MenuItem rightMenu = new System.Windows.Forms.MenuItem();
  rightMenu.Text = ''Go to'';
4 mainMenu.MenuItems.Add(leftMenu);
  mainMenu.MenuItems.Add(rightMenu);
  MenuItem google = new System.Windows.Forms.MenuItem();
  google.Text = ''Google'';
  google.Click += delegate(object sender, EventArgs e)
9     {wb.Url = new Uri(''http://mobile.google.com'');};
  MenuItem yahoo = new System.Windows.Forms.MenuItem();
  yahoo.Text = ''Yahoo'';
  yahoo.Click += delegate(object sender, EventArgs e)
      {wb.Url = new Uri(''http://mobile.google.com'');};
14 MenuItem news = new System.Windows.Forms.MenuItem();
  news.Text = ''IT news'';
  news.Click += delegate(object sender, EventArgs e)
      {wb.Url = new Uri(''http://mobile.news.com'');};
  rightMenu.MenuItems.Add(google);
19 rightMenu.MenuItems.Add(yahoo);
  rightMenu.MenuItems.Add(news);
  . . .
```

Instead of using a random URL, we add three submenu items to the *MenuItems* of *rightMenu*. For each of the *MenuItems*, an anonymous method is used to specify what should happen when the item is clicked. Menus can be nested to any level.

Let us proceed to consider another widely used construct, the *Notification* class. In many use situations, our application may not be the user's primary focus of attention. The situation may well be that the user has some other application in the foreground, and that we want to display some content to the user in response to an external event, e.g., a new posting on a blog or an RSS feed. Simply displaying the new content on a form will not get the user's attention as the form is hidden behind other applications or is minimized. The *Notification* control class offers solutions. This control's usage is somewhat odd when compared to other controls. This is illustrated in the following coding examples.

Consider Listing 8.5. It uses the *Microsoft.WindowsCE.Forms* namespace that defines functionality specific for the Windows Mobile platform. In line 10, we define and instantiate an instance variable *notific* of the *Notification* type. The *Main* method remains the same as in the previous examples. The first

6 lines of the constructor are also reused – these create a button for closing the application.

In line 25 a new class called *Timer* is introduced. A timer is used to raise an event repeatedly with a given interval (when it is enabled). In the example, we use it to display a notification every seven seconds by setting the *Interval* property to 7,000 milliseconds in line 26. We attach the *TimerTick* method as the event handler of the *timer* in line 27 and enables it in line 28.

Listing 8.5. Notification

```csharp
using System;
using System.Drawing;
using System.Windows.Forms;
using Microsoft.WindowsCE.Forms;

namespace AAU.CSharpBook.WindowsMobileExamples
{
  public class WMForm : Form
  {
    Notification notific = new Notification();

    public static void Main()
    {
      Application.Run(new WMForm());
    }

    public WMForm()
    {
      MainMenu mainMenu = new System.Windows.Forms.MainMenu();
      MenuItem leftMenu = new System.Windows.Forms.MenuItem();
      leftMenu.Text = ''Close'';
      leftMenu.Click += delegate(object o, EventArgs e){this.Close
          ();};
      this.Menu = mainMenu;
      mainMenu.MenuItems.Add(leftMenu);
      Timer timer = new Timer();
      timer.Interval = 7000;
      timer.Tick += new EventHandler(TimerClick);
      timer.Enabled = true;
    }

    private void TimerClick(object o, EventArgs e)
    {
      notific.Caption = ''New message'';
      notific.Text =
      ''<html><body><a href="">Click here to view</a></html></body
          >'';
      notific.ResponseSubmitted +=
    new ResponseSubmittedEventHandler(Response);
      notific.InitialDuration = 10;
      notific.Visible = true;
    }

    private void Response(object o, ResponseSubmittedEventArgs e)
    {
      this.BringToFront();
      notific.Visible = false;
    }

  }
}
```

The *TimerTick* method that is called every seven seconds is shown in line 31. We start by setting the *Caption* property for the notification balloon. Next, the *Text* property is set. This is done with a string containing an HTML document and not using desktop Windows form constructs. It is possible to use plain text in the notification balloon, but this has significant drawbacks.

Very often, we want the user to be able to bring the program window to the front when the user clicks on the message, or we may want the user to answer a simple question, which is then treated further in the background. The key problem with using plain text is that there is then no way to generate events, i.e., invoke functionality in the program. Furthermore, we will have to specify functionality for dispatching user input if the user has several options, e.g., the user can click three different buttons.

In the code, we show how the user can click a link in the notification balloon to see a new message that our program would then display. Having initialized the HTML body, we attach a specialized event handler to the *Notification* balloon. The *ResponseSubmitted* event is raised whenever the user clicks an HTML button or link. The code includes just one link, but it is possible to have several links or buttons, as we shall see shortly. The initial duration is set to ten seconds in line 39. This is actually done in seconds whereas most other Windows controls measure time in milliseconds. Finally, we set the *Notification* to be visible.

The final part of the program is the *Response* method that is invoked when the user clicks the link in the *Notification* balloon. This method brings the actual program window to the front and sets the visibility of the *Notification* to false. This is needed because a *Notification* can exist by itself, i.e., if we close the program without removing the *Notification*, it will remain visible. The user has no way of removing the notification except by rebooting the device or using a process viewer from a PC to shut it down.

Let us continue to consider a slightly more complicated version of the previous code listing. Assume that we change only line 34 and the *Response* method to reflect the code in Listing 8.6.

Listing 8.6. Notification With Two Buttons

```
1   ...
    notific.Text =''<html><body>'' +
    ''<input type='button' name='discard' value='Discard'>'' +
    ''<input type='button' name='accept' value='Accept'>'' +
    ''</html></body>'';
6   ...
    private void Response(object sender, ResponseSubmittedEventArgs
        e)
    {
      if(e.Response == ''accept'')
        this.BringToFront();
11    notific.Visible = false;
    }
    ...
```

The HTML of the *Notification* balloon now has two buttons. Each button is given a name corresponding to the action to be taken when it is clicked.

The *Response* method takes action based on the arguments that are passed by the *Notification* class. The *ResponseSubmittedEventArgs* includes a *Response* property that holds the name of the HTML control that was pressed. In the example, it is either "accept" or "discard".

It is possible to use different controls in one *Notification* balloon, e.g., both a link and a button. It is also possible to include more advanced controls such as radio buttons and check boxes. In Listing 8.7, we again change the HTML of the *Notification* control and the *Response* method. The HTML now contains a link and a drop-down box. The drop-down box has been encapsulated in an HTML form tag to allow a submit button. To understand what happens in the *Response* method, the following three lines, which describe the possible values of the *Response* property on the *ResponseSubmittedEventArgs*, should help:

```
     notify?answer=discard
2    notify?answer=accept
     setup
```

The *Response* method assigns appropriate actions to the possible values. There are numerous possibilities for errors using dispatching based on textual values. To avoid runtime exceptions, the developer must therefore be particularly careful that the checking on string values is done correctly.

Listing 8.7. Revised Notification with Two Buttons

```
     ...
2    notific.Text = ''<html><body>'' +
                    ''<a href='setup'>Setup</a>'' +
                    ''<form method='get' action=notify>'' +
                    ''<select name='answer'>'' +
                    ''<option value='accept'>Accept</option>'' +
7                   ''<option value='discard'>Discard</option>'' +
                    ''</select><input type='submit'></form>'' +
                    ''</html></body>'';
     ...
     private void Response(object sender, ResponseSubmittedEventArgs
          e)
12   {
        lbl.Text = e.Response;
        if(e.Response == ''setup'')
           lbl.Text = ''mega'';//Call some setup form
        else if(e.Response.Substring(0,6) == ''notify'')
17      {
           if(e.Response.Substring(14,6)==''accept'')
              this.BringToFront();
        }
        notific.Visible = false;
22   }
     ...
```

8.5 Network Functionality

Many mobile business applications need network functionality to communicate with, for example, a central server. The .Net CF offers a range of libraries for networking. In the following section HTTP and socket communication is presented.

8.5.1 Network Basics

With the increasing focus on Service-Oriented Architectures (SOA), the use of web services is becoming increasingly important. A web service is typically specified in a Web Service Description Language (WSDL) [17] file using XML. To make the functionality of a web service transparently usable from a programming environment, a proxy class must be generated. This way the functionality of the web service is encapsulated in a class, and all encoding and decoding of XML are hidden. The programmer can then use the web service via the proxy class as any other local class. The literature on distributed systems offers additional detail on proxy classes [20].

Creating a proxy class by hand is tedious and error prone, and this can be automated. On the .Net platform, the tool for proxy class generation is the *WSDL tool* [16]. However, proxy classes generated by the *WSDL tool* cannot be used by the .Net Compact Framework. An article by Microsoft [7] proposes solutions that aim to remedy this situation, but these do not work for all scenarios, and there currently seems to be no foolproof rules for when the tool works and when it does not.

As good news, it is very simple to use web services from Visual Studio that is also capable of generating proxy classes that always work. This will be illustrated in Section 8.6.

HTTP

It is simple to use HTTP requests in .Net. An example is shown in Listing 8.8. Only lines 39–44 have to do with the actual HTTP connection. The remainder of the code is used to set up the controls in which the response to a HTTP request are shown.

In line 39, we set up a HTTP connection to the URL `http://www.google.com`. The *WebRequest* class supports the creation of different network requests. The directly supported protocols include *http*, *https*, and *file*. It is possible to add additional request types by using the *RegisterPrefix* method. Returning to the example, we retrieve a response from the request in line 40. The actual HTML body of the response is provided as a stream. Streams are used intensively in the .Net Framework when working with network or file I/O. The basic idea is to read from or write to a stream of bytes. On top of these simple streams, we build more abstract representations. As an example, a *StreamReader* allows us to read strings from a stream using the *ReadLine* and *ReadToEnd* methods. For a more elaborate review of stream functionality, we refer to the .Net Class Library Reference [12].

226   R. Wind et al.

Listing 8.8. HTTP Connection

```
using System;
using System.Drawing;
using System.Windows.Forms;
using System.Net;
using System.IO;

namespace AAU.CSharpBook.WindowsMobileExamples
{
    public class WMForm : Form
    {
        Label lblHTML = new Label();
        ListBox lbKeys = new ListBox();
        ListBox lbValues = new ListBox();
        WebHeaderCollection headerCollection;

        public static void Main()
        {
            Application.Run(new WMForm());
        }

        public WMForm()
        {
            this.AutoScroll = true;
            this.Deactivate += delegate(object o, EventArgs e){this.
                Close();};

            lblHTML.Location = new Point(5,5);
            lblHTML.Size = new Size(240, 80);
            this.Controls.Add(lblHTML);

            lbKeys.Size = new Size(100, 180);
            lbKeys.Location = new Point(5,90 );
            lbKeys.SelectedValueChanged += new EventHandler(
                lb_ValueChanged);
            this.Controls.Add(lbKeys);

            lbValues.Size = new Size(300, 180);
            lbValues.Location = new Point(110, 90);
            this.Controls.Add(lbValues);

            WebRequest req = WebRequest.Create(''http://www.google.com
                '');
            WebResponse res = req.GetResponse ();
            StreamReader sr = new StreamReader(res.GetResponseStream())
                ;
            lblHTML.Text = sr.ReadToEnd();
            lbKeys.DataSource = res.Headers.AllKeys;
            headerCollection = res.Headers;
        }

        private void lb_ValueChanged(object sender, EventArgs e)
        {
            lbValues.DataSource =
                headerCollection.GetValues(lbKeys.SelectedValue.
                    ToString());
        }
    }
}
```

Line 42 uses the *ReadToEnd* method to read all the content of a stream into the *Text* property of a *Label*. In addition to the actual content, the *WebResponse* also contain the headers of the HTTP response. The *Headers.AllKeys* property contains a collection of strings that constitute the descriptive name of a given header, e.g., *Date* or *Host*. The actual content of the header item is retrieved using the *GetValues* property of the *HeaderCollection*, which again retrieves a string array. This is done in the event handler for the *ListBox* later on. Note that we simply bind the collection of strings to the *DataSource* of the *ListBox* instead of looping through the collection and add the items individually.

The binding of an object or a collection of objects to a control is called data binding. What we illustrate here is very simple data binding. Additional examples of data binding are given in Section 8.6. For a more detailed description of data binding, we refer to the literature [18].

Sockets

Sockets are useful in many scenarios where two-way communication is needed, such as between a client application and a server application. In .Net, a socket provides two streams: an input stream and an output stream. If a mobile device is connected through GPRS, it is probably not possible to run a sever application on the device. The reason for this is that the mobile-service provider enforces firewall restrictions. It is, however, possible to run a client application, i.e., it is possible to connect to a server.

The example in Listing 8.9 concerns a simple ping server that responds when a string is sent to it. We have left out the *using* statements and the constructor. When using sockets the namespaces *System.Net*, *System.Net.Sockets*, and *System.IO* must be included.

Listing 8.9. Socket Server

```
 . . .
2   Console.WriteLine(''Starting Server'');
    IPHostEntry ipHostEntry = Dns.GetHostEntry(Dns.GetHostName());
    If(ipHostEntry.AddressList.Length > 0)
    {
        Console.WriteLine(ipHostEntry.AddressList[0]);
7       TcpListener tcpListner = new TcpListener(
                            ipHostEntry.AddressList[0], 1320)
                            ;
        tcpListner.Start();
        TcpClient client = tcpListner.AcceptTcpClient();
        NetworkStream nStream = client.GetStream();
12      BinaryReader binReader = new BinaryReader(nStream);
        BinaryWriter binWriter = new BinaryWriter(nStream);
        string s = binReader.ReadString();
        Console.WriteLine(''Recieved: '' + s);
        binWriter.Write(s + '' - back to you'');
17      client.Close();
        tcpListner.Stop();
    }
     . . .
```

The *TcpListener* class is capable of listening for incoming request on a given port and network adapter using the TCP protocol. Line 3 retrieves the IP address of the local network adapter. We check to see whether one is available, in which case we create a new *TcpListener* using the IP address of the network adapter and port 1320. If the computer being used contains more than one network adapter, we use the first one. We start the listener

and call the blocking *AcceptTcpClient* method. The program now waits for
a client to connect on port 1320. When a client connects, we retrieve the
underlying network stream that we use in the construction of a *BinaryReader*
and a *BinaryWriter* in lines 12 and 13. These two classes are used to read
and write types such as integers, doubles, and strings.

In line 14, we read a string from the *BinaryReader* . We then write this to
the screen in line 15; and in line 16, we send back the string to the client with
the addition of "-back to you". Finally, we close the client and the listener.

Listing 8.10 shows a client capable of connecting to our server. We have
again omitted the *using* statements and the constructor. We start by setting
up a *Label* for displaying the result returned to the client. In addition, we
override the minimize button to instead close our form. In line 7, we create a
new instance of the *TcpClient* class with the server name and port number as
arguments. We need to exchange the *servername* to reflect the actual name of
the local computer. As on the server, we retrieve the *NetworkStream* and use
it in the creation of a *BinaryReader* stream and a *BinaryWriter* stream. We
use the *BinaryWriter* to send a string to the server. The response is read in
line 12 using the *ReadString* method of the *BinaryReader* object. The response
is then displayed in the *Label* using the *Text* property, also in line 12.

Listing 8.10. Socket Client

```
   ...
   Label  l = new  Label();
   l.Size = new  Size(170,  60);
   l.Location = new  Point(5,  5);
5  this.Controls.Add(l);
   this.Deactivate += delegate(object o,  EventArgs s){this.Close()
        ;};
   TcpClient tcpClient = new  TcpClient(''servername'',  1320);
   NetworkStream nStream = tcpClient.GetStream();
   BinaryReader binReader = new  BinaryReader(nStream);
10 BinaryWriter binWriter = new  BinaryWriter(nStream);
   binWriter.Write(''Hello  Server'');
   l.Text = binReader.ReadString();
   tcpClient.Close();
   ...
```

8.6 Visual Studio IDE

So far, we have focused on using command-line utilities for developing and
building Windows Mobile applications. In this section, we introduce the
Visual Studio Integrated Development Environment (IDE), which simplifies
the development and building processes. In particularly, we show how to use
web services and database access from within Visual Studio.

8.6.1 An Overview of Visual Studio

The Visual Studio IDE is a development environment created by Microsoft.
The tool is available in several editions. The free edition does not have support

for developing applications for the .Net CF and will not be considered here. Visual Studio offers a wide range of functionality including code completion, refactoring, short-cut keys, solution browsing, integrated help, class library reference, database management, debug functionality, and drag-and-drop GUI creation. Visual Studio supports the creation of new projects that target the .Net CF. This gives access to a drag-and-drop interface that looks like a Windows Mobile device. Figure 8.4 contains a screen shot from Visual Studio that illustrates the development of a Windows Mobile application.

When creating a new Windows Mobile project, Visual Studio will automatically target the .Net CF assemblies without the use of custom-made batch files as discussed in Section 8.3. Dividing a solution into several projects, e.g., creating a layered architecture, is very easy, and referencing a project from another project is done by pointing and clicking.

Visual Studio uses the partial class file concept of C# to partition the code into two logical parts that are stored in separate files. In the first file, the developer writes the non-GUI code; in the second file, Visual Studio hides the instantiation and placement of the GUI controls. It is possible to automatically create event handlers using the property sheet of a control. This functionality, except the actual method, is also stored in the second file. Care should be taken when changing the second file by hand, as the file is used for displaying

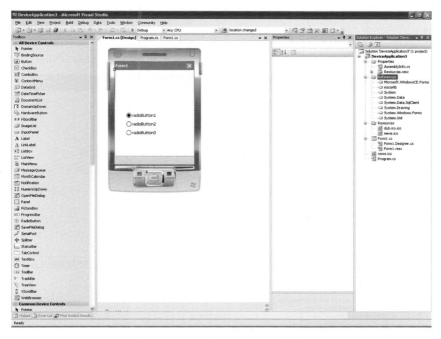

Figure 8.4. Visual Studio IDE.

the GUI in Visual Studio. The file is also automatically updated by Visual Studio when the user adds or removes GUI controls.

Although Visual Studio is a very nice development environment, it is useful for developers to familiarize themselves with the nuts and bolts of Windows Forms programming in a simpler environment. The capabilities of Visual Studio cover most programming needs. In the rare cases where special functionality is needed, it is very useful to be able to provide this by hand and use the .Net Framework to its fullest.

8.6.2 Using Web Services

It is easy to use web services from a Windows Mobile application in Visual Studio. This is done by right clicking the appropriate project in the solution manager (in Figure 8.4, this is the rightmost toolbar) and adding a web reference. A pop-up box appears, in which the URL of the service and a name are written. Visual Studio automatically compiles a proxy class that can be used as any other class in the solution. This offers very high transparency. Figure 8.5 illustrates this. We connect to the EC2 (Elastic Compute Cloud) web service and give it the name *AmazonEC2*.

8.6.3 Using Databases with ADO.Net

Most business applications use a Database Management System (DBMS) as the back-end for data storage. This might be done indirectly, by calling a web

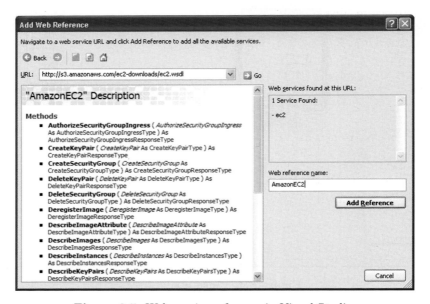

Figure 8.5. Web-service reference in Visual Studio.

service, or by using HTTP or socket communication. For efficiency reasons, it is often desirable to directly access a database. It may be argued that this yields a less secure system because the application developer has direct access to the data. However, the security issues can be solved with user access rights and data views, and we disregard data security in this chapter. Instead we refer the reader to the literature for the specific DBMS being used. A general overview of database concepts can be found in, e.g., reference [22].

In .Net a database is accessed using ADO.Net. We assume that the DBMS is Microsoft SQL Server. A free version of the SQL Server, SQL Server 2005 Express Edition, is available for download [8] and is also part of the Visual Studio installer. ADO.Net is not limited to SQL Server, but offers access to a wide range of DBMSs, including Oracle, IBM DB2, MySQL, and PostgreSQL. The Oracle driver is included in the default .Net SDK installer. ADO.Net drivers for other databases can be downloaded freely. It is possible to use ADO.Net on the .Net Compact Framework without using Visual Studio but this is difficult, and we will not demonstrate it here.

To illustrate the use of ADO.Net, we assume that tables have been created that correspond to the database shown in Tables 8.1 and 8.2. The database includes an employee table and a department table. Each employee belongs to exactly one department. This is specified by the foreign key from the *department* column of the *Employee* table to the *id* column of the *Department* table. Each department also has an employee as a leader, so there is also a specified by the foreign key from the *leader* column in the *Department* table to the *id* column in the *Employee* table.

To start using the database, we need to add an additional namespace to our project in Visual Studio. In the Solution Explorer, we right-click and press *Add Reference*. On the *.Net* tab, we locate the *System.Data.SqlClient* as illustrated in Figure 8.6. The functionality that we load here is not part of the standard .Net Compact Framework, but comes with Visual Studio.

We start with an example that shows the content of the *Employees* table. Listing 8.11 shows a code fragment that excludes the class definition and

Table 8.1. Employee table.

id	firstname	lastname	address	zip	department
1	John	Peterson	Oneway Road 47	90060	1
2	Tim	Smith	Twoway Road 88	90070	2
3	Sophie	Henderson	Oneway Road 19	90060	1
4	Jim	Alexander	Threeway Road 21	90070	1

Table 8.2. Department table.

id	name	leader
1	Sales	1
2	Support	2
3	IT	1

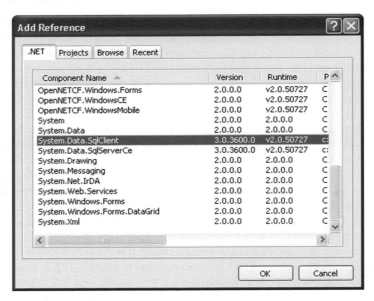

Figure 8.6. Adding a reference to *System.Data.SqlClient*.

instantiation. The namespaces *System, System.Data, System.Data.SqlClient,* and *System.Windows.Forms* are needed. We use a Windows Forms control (not introduced earlier) called *DataGrid*. This control can be bound directly to a *DataTable* object, which is what ADO.Net will return. In line 3, we specify that the *DataGrid* should occupy the whole screen, and we add this to the form in line 4.

In line 5 the connection string is created. The capitalized words must be exchanged with the actual values from the environment. Lines 8–12 are surrounded by a *using* statement. In the *using* statement we instantiate a new *SqlConnection*. When encapsulating the database functionality in a *using* block, the connection will be opened and closed automatically. This effectively removes the presence of non-used, open database connections, which can cause serious performance problems.

Listing 8.11. Display Customers Example

```
1    ...
     DataGrid  dg = new  DataGrid();
     dg.Dock = DockStyle.Fill;
     this.Controls.Add(dg);
     string  cs = ''Server=DBSERVER; User  ID=USER; Password=PASS; Database=
         DB'';
6    using (SqlConnection  con = new  SqlConnection(cs))
     {
       SqlCommand  com = new  SqlCommand(''SELECT * FROM employees '', con
           );
       SqlDataAdapter  da = new  SqlDataAdapter(com);
       DataSet  ds = new  DataSet();
11     da.Fill(ds);
```

```
    dg.DataSource = ds.Tables[0];
}
...
```

In line 8 we create a new *SqlCommand* with the query to be executed and our connection. There are several different ways of executing SQL statements against the database. In our case, we want to retrieve the complete set of data. For this purpose we use a *DataAdapter* and a *DataSet*. The *DataAdapter* uses the *SqlCommand* object to fill the *DataSet*. This is exactly what happens in line 11. While a *DataSet* can consists of several *DataTables*, our example only contains one (more than one query can be specified in the string given as an argument to the *SqlCommand*, in which case several *DataTables* will be added to the *DataSet*). We set the *DataSource* of the *DataGrid* to the first (and in our case only) *DataTable* in the *DataSet*. Running the program will verify that the data is actually being displayed.

Another way of executing SQL statements against a database is to invoke commands by using the *SqlCommand* directly, or we can use additional classes that use the *SqlCommand* object. If the query returns a single value, say an integer, the following code works.

```
int returnValue = (int)com.ExecuteScalar();
```

Sometimes a statement will not return any result. This happens in the case of an insert or an update statement. In this case, we can use the following code.

```
com.ExecuteNonQuery();
```

Listing 8.12 contains a slightly more advanced example of data binding. Instead of using a *DataGrid*, we now use a *ListBox* and a *Label*. The query has also been changed to retrieve all employee names and their leader, i.e., for each employee, we return the leader of the department to which the employee is attached. In lines 9–10, we set the *DisplayMember* of the *ListBox* to "*empname*" and bind it to the retrieved *DataTable*. This will make the *ListBox* display the *empname* values of all employees.

The *Label* is also bound to the *DataTable*, although we use a different method. We do this by adding a new binding to the *DataBindings* property of the *Label*. The first argument specifies which property of the *Label* that the value should be bound to, in our case the *Text* property. The second argument specifies the actual data source. The final argument specifies the name of the attribute in the data source to use, i.e., it is the equivalent of the *DisplayMember* used above. A very nice feature of this example is the implicit connection between the data-bound controls. Whenever the user clicks on an employee in the *ListBox*, the value of the *Label* will automatically change to reflect the right leader. Note that this done without us having had to explicitly set up any event handlers.

234   R. Wind et al.

Listing 8.12. Double Data Binding

```
    ...
    ListBox  lbEmployees = new ListBox();
    Label  lblLeader = new Label();
4   ...
    string  sql = ''SELECT e1.firstname as empname, '' +
                   ''e2.firstname as leader '' +
                   ''FROM employees e1 INNER JOIN department d ON '' +
                   ''e1.department = d.id INNER JOIN employees e2 ON '' +
9                  ''d.leader = e2.id '';
    ...
    lbEmployees.DisplayMember = ''empname'';
    lbEmployees.DataSource = ds.Tables[0];
    lblLeader.DataBindings.Add(''Text'', ds.Tables[0], ''leader '');
14  ...
```

In addition to being able to connect to remote databases, it is also possible to put an embedded SQL server on the Windows Mobile device. Installing and using an embedded database management system is beyond the scope of this chapter. The actual querying is very similar to the examples given above, except that the *System.Data.SqlServerCe* namespace is to be used instead of the *System.Data.SqlClient* namespace. Further information about the embedded SQL Server can be found in the literature [15].

References

1. *C# language specification: ISO/IEC 23270:2006.* ISO/IEC, Dec 2006.
2. *Common Language Infrastructure (CLI) Partitions I to VI (ECMA-335).* ECMA, 4th edition, June 2006.
3. Design Guidelines for Developing Class Libraries. http://msdn2.microsoft.com/en-us/library/ms229042.aspx, Jan 2007.
4. Developing Custom Windows Forms Controls with the .Net Framework. http://msdn2.microsoft.com/en-us/library/6hws6h2t.aspx, Jan 2007.
5. Differences with the .Net Framework. http://msdn2.microsoft.com/en-us/library/2weec7k5(VS.80).aspx, Jan 2007.
6. How to: Compile at the Command Prompt. http://msdn2.microsoft.com/en-us/library/ms172492.aspx, Jan 2007.
7. How to: Use a proxy generated by wsdl.exe. http://msdn2.microsoft.com/en-us/library/ms229663.aspx, Jan 2007.
8. Microsoft sql server 2005 express edition. http://www.microsoft.com/sql/editions/express/default.mspx, Jan 2007.
9. The Mono Project. http://www.mono-project.com, Jan 2007.
10. .Net Compact Framework 2.0 Redistributable. http://www.microsoft.com/downloads/details.aspx?FamilyID=9655156b-356b-4a2c-857c-e62f50ae9a55, Jan 2007.
11. .Net Compact Framework web-site. http://msdn2.microsoft.com/en-us/netframework/aa497273.aspx, Jan 2007.
12. .Net Framework Class Library. http://msdn2.microsoft.com/en-us/library/ms229335.aspx, Jan 2007.
13. .Net Framework SDK. http://msdn2.microsoft.com/en-us/netframework, Jan 2007.

14. .Net Languages. http://www.dotnetpowered.com/languages.aspx, Jan 2007.
15. Sql server 2005 compact edition.
    http://www.microsoft.com/sql/editions/compact/default.mspx, Jan 2007.
16. Web services description language tool (wsdl.exe). http://msdn2.microsoft.com/
    en-us/library/7h3ystb6(VS.80).aspx, Jan 2007.
17. Web services description language (wsdl) 1.1. http://msdn2.microsoft.com/en-
    us/library/ms229663.aspx, Jan 2007.
18. Windows forms data binding.   http://msdn2.microsoft.com/en-us/library/
    ef2xyb33.aspx, Jan 2007.
19. Windows mobile tools. http://msdn.microsoft.com/windowsmobile/downloads/
    tools/install/default.aspx, Jan 2007.
20. George Coulouris, Jean Dollimore, and Time Kindberg. *Distributed Systems:
    Concepts and Design.* Addison-Wesley, 2005.
21. Charles Petzold. *Programming Microsoft Windows Forms.* Microsoft Press,
    2005.
22. Abraham Silberschatz, Henry F. Korth, and S. Sudarshan. *Database System
    Concepts.* McGraw-Hill, 2006.

## Part III

Infrastructure-Based Communication

9

Service Discovery

Andreas Häber

Agder University College `andreas.haber@hia.no`

Summary. Service discovery deals with locating services which can be used to accomplish a task. For example discovering a printer to print documents on, or locating a shop offering the service of hairdressing. When you locate a service you usually obtain an address which you use to invoke the offered service. Here in this chapter we will start by looking at an example of service discovery from human life, with a short story of a pizza restaurant which offers the service of baking pizzas and see how customers can discover it. Next, we will have a look at how those methods from the pizza-story can be applied in computer networks. Finally, an introduction to UPnP is given as an example of a service discovery system.

9.1 Service Discovery in Real Life

We start here with a simple story of establishing a new service in a city, and look at how customers discover this service. See [20] for a longer introduction to service discovery, including more "daily life"-examples of service discovery. Let us see what happens when somebody moves to a new city and offers a new service, and how to get customers to use this service[1]. As an example we would like to start a new pizza restaurant. We rent some office space, delivery cars, and then install a kitchen there and whatnot. Great! We are now ready to take up orders for pizzas! Sipping espresso and waiting in excitement we wait and wait and... then we need to make more coffee, before we can wait even more. Suddenly the phone rings! Wahoo! What kind of pizza does this person want to order? Maybe the specialty of the house, Pizza Espresso? No, unfortunately this was not a customer, just somebody who dialed the wrong number. We can therefore continue to drink coffee and wait for a customer to request our service instead. Or, perhaps there is a reason why we do not get any customers requesting our service? How many people know about our

---

[1] This example is *not* intended as a crash course in clever marketing. Check the book title!

F.H.P. Fitzek and F. Reichert (eds.), Mobile Phone Programming and its Application to Wireless Networking, 239–255.

new restaurant anyways?[2] Currently just one person who dialed the wrong number. And that person probably forgot all about our nice pizza restaurant right after the conversation, no matter how politely we answered the phone. Simply nobody knows about our service and we are not doing anything to make our restaurant easier to be found, by just passively waiting for customers to arrive at a restaurant they do not know exists. Let us therefore try to make our restaurant easier to find: first, we put up a sign on the front of our building saying "Pizza". Now we will announce ourself to potential customers walking by our restaurant. If they are hungry then they will see that we are an option and hopefully pop in, or they may remember our restaurant the next time they get hungry. Especially if they occasionally happen to be in the area of our pizza restaurant. Another way to make our restaurant visible is to make sure that the restaurant name and contact information is present in the yellow pages. If the city has a web portal, such as visitalesund.com/ it should be arranged so our pizza restaurant is included there. Finally, we will target a specific group of customers: those who like the pizza & coffee combination. This is accomplished by advertising in a magazine which they are likely to read, or we can even start our own magazine.

9.1.1 Active and Passive Service Discovery

Advertisement is, for our concern here, all about making a service discoverable by users. Marketers probably also want users to find our service the most attractive and become addicted to it. Above we have described three methods to advertise our pizza restaurant to potential customers. These can be divided into two different strategies, namely active and passive. With *active service discovery* the customer actively searches for a service to be used. In the story above this is realized with yellow pages and web portals in. Desperate customers might also try a primitive distributed search by getting on top of a garbage bin and shout for pizza. In *passive service discovery* the user receives announcements/advertisements/notifications about the presence of services, which relates to putting up the advertisement sign outside our pizza shop in the above and the advertisement in the magazine. The differences between these methods will be described soon. But first we need to look at different routing schemes in computer networks.

9.2 Service Discovery in Computer Networks

9.2.1 Where to Send a Packet to?

In networks which use some kind of shared media, such as Ethernet or wireless networks, each node is usually assigned an address and all transmitted packets

---

[2] Assume that this is a very shy city where nobody unnecessarily talks to each other, and in particular no gossip!

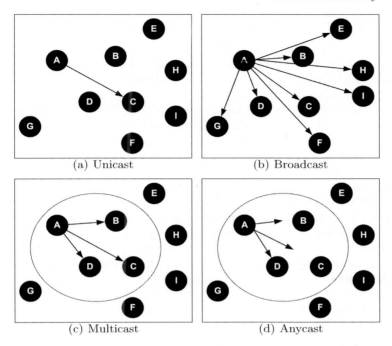

Figure 9.1. Different routing schemes.

include a source and destination packet. There are also some special addresses which open up different communication methods, as shown in Figure 9.1.

Unicast: From one node to another node, as shown in Figure 9.1A where node A sends a packet to node B. In IPv4-networks this could be from source address 10.0.0.1 to destination address 10.0.0.5.

Broadcast: From one node to all other nodes in the network, as shown in Figure 9.1B where node A sends a packet to all the other nodes. In IPv4-networks this is usually done by sending to IPv4 address 255.255.255.255.

Multicast: One-to-many traffic has many use cases, but it is not always that you want "many" to be the whole network. With multicasting a node must join a multicast-group before it retrieves packets, as shown in Figure 9.1C where nodes A, B, C, and D are all part of one multicast group. Node A is here sending a packet which all the group members receives. For IPv4 networks the address range 224.0.0.0 through 239.255.255.255 has been reserved for multicasting.

Anycast: This is multicasting with a twist: the packet is only delivered to the nearest member of the group. This routing scheme is usually used for load balancing.

9.2.2 Address Configuration

For unicast each node in a network must have a unique address for that network. There are three different ways to provide nodes with these addresses: manual configuration, dynamic configuration, and random configuration. With manual configuration, or static configuration, an administrator must manually configure each node in the network. In dynamic configuration there is a central service which hands out addresses to nodes. Here we run into a so-called "chicken & egg" problem: you need to send a packet to the service which hands out addresses, but you do not have any address yet to send from. And you do not know the destination address of the service as well. Let us see how this service discovery problem is solved in IPv4 networks, using DHCP [3] to hand out addresses. In IPv4 networks you have to specify both the sender and the destination of a packet. In special cases, like this one where you do not have any address, the anonymous address 0.0.0.0 can be used. Now we need to figure out the destination address. By looking at the four different routing schemes available it is only broadcast which is usable for this purpose (e.g., you cannot register the anonymous address to a multicast group). So we therefore broadcast a message (DHCPDISCOVER) and wait for a reply from the server. Because small networks often do not have any dynamic configuration service, and it is too hard (or boring) for many users[3] to manually configure the hosts, random configuration makes it easier. In IPv4 networks a special range, 169.254/16, has been reserved [2] for this purpose. The host should first check for any available DHCP server, and if none is available then randomly create an IP-address from that special range. Then it must test if that address is in use or not. When it has created an unused address it can claim it as its own and start to use it. Periodically it should check if a DHCP server has become available on the network and try to acquire an address from that service instead. With an understanding of these four different routing schemes, we will now go back to the pizza shop story and see how the different service discovery methods can be implemented in computer networks.

9.2.3 Basic Service Discovery in Computer Networks

In the above story about the pizza-place we covered different methods of discovering our service, a pizza restaurant. All of those methods can be implemented in computer networks as well. However, not all of them are necessarily used together and varies between service discovery systems, which is true for different pizza restaurants too. In [5] a more detailed classification for service discovery is presented than what we cover here. The article also provides a comparison of nine service discovery protocols based on that classification. In our pizza restaurant we supported *passive service discovery* by placing a sign outside the shop and inserting advertisements in a magazine.

---

[3] Consider a small home network with no computer professional.

Magazine-advertisements can be implemented in computer networks by sending multicast requests. To implement "signs" and let customers "walk around" in the network we get into the domain of crawlers or robots, which "walk" around the network and pick up interesting information, but that topic is out of scope for this chapter. Let us have a look at the *active service discovery* methods. Customers shouting can easily be implemented with multicasting. In UPnP all members of the multicast group will receive the search request and may reply with a unicast. This is also how discovery of a DHCP server works, except that broadcast is used for sending the search request instead of multicast[4].

There are many alternatives available for implementing "yellow pages" in computer networks. For example LDAP [17], UDDI [1]. Clients can search these directories and locate information about services. Notice the difference between these two methods for active service discovery. In the first method a distributed search is used, while in the latter case a single search request is sent using unicast. The first method works good on small networks with little or no infrastructure, while the second method scales much better and is preferred for larger networks, or when multiple search requests are issued in a short time frame. Of course the server hosting the directory must be capable of handling the requests. We will now have a look at a service discovery system, UPnP and see how it implements some of these service discovery methods. As mentioned above all these methods are not necessarily used together. In addition to service discovery UPnP also includes a lot of other functionality which we will also cover.

9.3 Universal Plug & Play

UPnP technology enables devices to connect together and form a pervasive peer-to-peer network where they can control each other. Such a distributed network is very powerful because it enables the devices to get more intelligent *together*. The UPnP architecture was initiated by the UPnP Forum, formed 18. October 1999. The UPnP Forum currently consists of over 800 organizations, companies and individuals, and the main focus of the organization is to standardize and promote UPnP technology.

9.3.1 UPnP Architecture

UPnP technology is based on the UDA [7]. The architecture is divided into six phases, as shown in Figure 9.2. Before describing these phases we will describe parts of the layers of Figure 9.3, which shows the protocols used by UPnP Device Architecture (UDA) and the layers built on top of UDA.

---

[4] Unicast should be used for the reply from the DHCP server, but for some clients this does not work and they indicate this with a special flag in their initial request. Therefore the reply must be broadcasted as well.

Figure 9.2. Six phases of the UPnP Device Architecture.

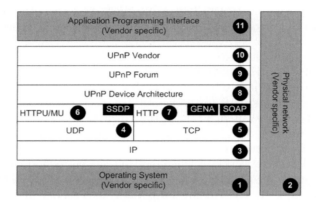

Figure 9.3. UPnP Protocol stack.

1) **Operating System:** UPnP is OS agnostic, and it can be implemented on any OS which is able to parse UTF-8 encoded text received through an IP-network.

2) **Physical network:** Any network can be used, as long as it supports TCP/IP (see below). However, notice that UPnP is designed for small networks, especially targeting small offices and residents with no more then approximately 50 nodes. We will see why when we describe service discovery in Section 9.3.2 below.

The protocols (3)IP [15], (4)UDP [14], (5)TCP [16], and (7)HTTP [6] are used for the lower level communication, and we will not describe them here. (6) HTTPMU and HTTPMU are extensions *not* to HTTP because HTTP itself only specifies TCP as the transmission protocol. HTTPU is used for UDP transmission of HTTP, while HTTPMU is used for multicast transmission of HTTP. Both of these extensions are necessary for Simple Search Discovery Protocol (SSDP), which is described in Section 9.3.2.

On the top of these protocols we have the following layers:

8) **UDA:** Described in detail below and is the core of the UPnP architecture.

9) **UPnP Forum:** This layer consists of all "Device and Service Descriptions (DCPs)" defined by the forum. The UPnP Forum set up Working Committees (WCs) for specific domains which create these DCPs. Examples are the Audio/Video WC and the Home Automation and Security WC.

10) **UPnP Vendor:** Vendors can add their own extensions on top of the standard functionality by the UPnP Forum. In the device and/or service descriptions these must be specifically marked as vendor extensions. Vendors may also create nonstandard devices and services besides those standardized by the UPnP Forum.

11) **Vendor-specific API:** To make it easier to use UPnP technology vendors can create libraries to handle the UPnP architecture. For example, instead of issuing raw protocol requests, such as HTTP requests, it is much more convenient and efficient to let a library create requests and parse the replies, and deliver the result in a format useful for applications. We mention a couple of alternatives in Section 9.3.3.

Devices and Services

Before going through the six phases of UDA, we will describe the relationship between *UPnP devices* and *UPnP services*. Let us first look at devices. UPnP allows a hierarchy of devices, and distinguishes between *root devices* and *embedded devices*. A root device can have any number of embedded devices. Next there are services, which have actions which can be executed (see Section 9.3.2) and notify about events (see Section 9.3.2). A device, both root and embedded, contains zero or more services. This relationship can be seen in Figure 9.4. In addition to devices and services you also have control points in the UPnP architecture. These are entities which use devices and services. These are all logical entities so nothing is stopping you from creating a physical node with both devices and a control point present. A TV-set is a good example for this. It provides a service for media rendering and also has a control point which allows you to interact with it.

9.3.2 UPnP Device Architecture

We will now go through the six phases of UDA, as shown in Figure 9.2. Please note that we cannot cover all the details of the UDA here. See [7] for the full details.

Phase 0: Addressing

All devices must have an IP-address. See Section 9.2.2 for different ways which can be used to acquire these. It is important that random configuration is supported to make UPnP technology easy-to-use for end users.

Phase 1: Discovery

Simple Search Discovery Protocol (SSDP) is used for service discovery in UPnP. It supports both passive and active service discovery and works peer-to-peer with no directory mechanism. As shown in Figure 9.3 SSDP uses UDP,

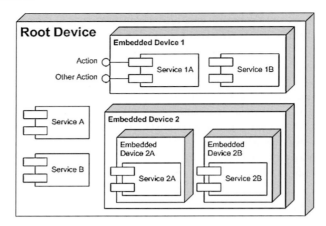

Figure 9.4. Relationship between UPnP devices, services, and actions. Note that all services must have one or more actions, while in this figure we only show actions on Service 1A.

HTTPU, and HTTPMU. The protocol is layered on top of HTTP and extends it with new request methods.

The multicast address 239.255.255.250 is reserved for SSDP together with port number 1900, for both TCP and UDP.

Active service discovery in UPnP

A control point searches for devices and/or services by multicasting a M-SEARCH request to the multicast address 239.255.255.250 on port 1900. Below we show an example M-SEARCH request.

Listing 9.1. SSDP M-SEARCH Request

```
M-SEARCH * HTTP/1.1
HOST: 239.255.255.250:1900
MAN: "ssdp:discover"
MX: (seconds to delay response)
ST: (search target)
```

On the request-line of Listing 9.1 we have the *M-SEARCH* method with the request-uri set to *, because the request is not intended for a specific resource, and *HTTP/1.1* indicates that we are using HTTP version 1.1. In Table 9.1 we describe the headers of this request.

Devices which receive this request should check the *ST* header to see if the request should be replied to. For example if ST is *ssdp:all* then a reply should be given from all devices (root and embedded) and all their services. This is the broadest search possible. The response to a M-SEARCH is a normal HTTP response, except that it should be sent as UDP to the sender of the M-SEARCH request. Below we show the content of a response and describe the most interesting headers in Table 9.2.

Table 9.1. M-SEARCH request.

Header	Description
HOST	Set to the multicast address for SSDP, as described above.
MAN	Comes from the *HTTP Extension Framework* [12], and defines the scope of this extension. It is required to be in double-quotes exactly as written above.
MX	The maximum delay time a receiver should wait before sending a reply. It is *important* that receivers have a *random* delay in the range [0, MX-header value] seconds to make it possible for the control point receive and handle all replies.
ST	Search target. Indicates which devices and/or services are searched for. See [7] for the legal values.

Table 9.2. M-SEARCH response.

Header	Description
EXT	Comes from the *HTTP Extension Framework* [12], and means that the server understood the extension. It is required to be empty.
LOCATION	From this URL the device description of the *root device* can be downloaded. This is Phase 2 of UDA and is described in Section 9.3.2 below.
ST	Depends on the ST header of the request.
USN	See [7] for the legal values.

Listing 9.2. SSDP M-SEARCH Response

```
   HTTP/1.1  200  OK
   CACHE-CONTROL:  (seconds  until  advertisement  expires)
   DATE:  (when  the  response  was  generated)
   EXT:
5  LOCATION:  (URL  of  the  UPnP  description  for  the  root  device)
   SERVER:  OS/version  UPnP/1.0  product/version
   ST:  (search  target)
   USN:  (advertisement  UUID)
```

So with such a response (see Listing 9.2) the control point can download the device description of the root device and start with Phase 2, which is described in Section 9.3.2. But UPnP also supports passive service discovery, which we will have a look at in the next section.

Passive service discovery in UPnP

Passive service discovery works very similar to active service discovery. Periodically a device multicasts NOTIFY requests to the SSDP address. A NOTIFY can be thought of as a heartbeat for UPnP devices. Actually, it is not only root devices which should send NOTIFY requests, but also all embedded devices and services too. The formula to calculate the number of requests to send is $3 + 2d + k$ for a root device with d embedded devices and s embedded services, but only k distinct service types. In addition to notify that an entity is available it is also used to notify when an entity becomes unavailable. Below we show a general NOTIFY request.

Listing 9.3. SSDP NOTIFY Request

```
   NOTIFY * HTTP/1.1
2  HOST: 239.255.255.250:1900
   CACHE-CONTROL: (seconds until advertisement expires)
   LOCATION: (URL of the UPnP description for the root device)
   NT: (notification type)
   NTS: (notification sub type \URI)
7  SERVER: OS/version UPnP/1.0 product/version
   USN: (advertisement UUID)
```

As can be seen NOTIFY-requests are almost identical to the M-SEARCH response described above. There are two big differences: the NT header is used instead of ST and the NTS-header not found in the M-SEARCH response. We describe those headers in Table 9.3. A SSDP NOTIFY-request must *not* be replied to. Control points interested in this entity should rather download the device description provided in the Location-header (see Section 9.3.2).

For a NOTIFY-request with the NTS-header set to *ssdp:byebye* (see Listing 9.4) the request can be shortened to the following, because that information is irrelevant since the entity will be unavailable anyways.

Listing 9.4. SSDP NOTIFY Request – Unavailable Device

```
   NOTIFY * HTTP/1.1
2  HOST: 239.255.255.250:1900
   NT: (notification type)
   NTS: ssdp:byebye
   USN: (advertisement UUID)
```

This should be enough information for a control point to figure out which entity it is, and if necessary remove it from its internal list of available entities. As can be noticed there is no directory-functionality in SSDP. This limits the scalability of UPnP, and makes it unsuitable for large networks. SLP [8] is a service discovery protocol designed for both small and large networks, and includes a directory function.

Table 9.3. NOTIFY Request Headers.

Header	Description	
NT	Notification type. Is analog to the ST header. See [7] for the legal values.	
NTS	Notification subtype. Two URIs has been defined:	
	NTS URI	Description
	ssdp:alive	The entity is available.
	ssdp:byebye	The entitiy is unavailable.

Phase 2: Description

Notice that the information provided in the discovery phase is very limited. It is just enough for a control point to know that the entity exists, and use the Location-header to get more information about it. And that is what the Description-phase covers. The Location-URL should always point to the device description of the root device, and must be downloaded using HTTP over TCP. In addition to device description there is also service description documents, which is used to describe services. Device descriptions are specified in the UPnP Device Template, which is encoded in XML, and service descriptions are specified in UPnP Service Template, also XML encoded.

Device description

The device description includes some information intended for the user, such as a friendly name, icons to display, and manufacturer information, and information for control points to use. We will focus here on the information for control points. Important information for control points is the device type of the device, its Unique Device Identifier (UDN),[5] a list of services, a list of embedded devices, and finally a URL for the presentation phase of the device, which we cover in Section 9.3.2. The device list contains all embedded devices, and includes the same information as above. In the service list each service of the device is listed with the following information:

Service type: URI specifying the type of this service.
Service id: Service identifier, which must be unique for the device description it is included in.
SCPD URL: URL where the service description can be downloaded from.
Control URL: URL used in the Control phase to post requests to.
Event subscription URL: URL used in the Eventing phase to subscribe to events of the service.

Service description

The service description is only intended for the control point, and it is not supposed to be presented to humans. Notice that services do not have a friendly name, for example. Control points should rather provide some nice user interface to use the services together. If you are familiar with WSDL [4], then UPnP service descriptions can be thought of as WSDL's little brother. A UPnP service description gives information about the state variables and actions of the service. Actions can be invoked, as described in the Control phase below, and may have input and/or output arguments. An argument must have a relation with a state variable. State variables describe the state of the service, and it is of a particular type specified in [13]. State variables

---

[5] UDN can for example be used later in a M-SEARCH request to easily find it again if it is present on the network.

can be evented, which we cover in the Eventing-phase below. It is possible to restrict the available value of a state variable using either a *value range* or a *value list*. Value list works similar to enumerations in the C programming language, where the service description lists all legal values as text strings. Value ranges set the minimum and maximum ranges for the variable, and optionally can specify the increments (step) between the legal values.

Notice how these description documents allow control points to dynamically use services with no a priori knowledge that they even exist. UPnP test tools, such as the "Generic Control Point" by Siemens [18] and Intel's UPnP Tools [9], can work with any UPnP service. However, notice that the user experience is rather poor because they do not know the semantics of the services. For example, there is no way for them to know that the XML result returned by a UPnP Media Server can be parsed to display a nice list of, say, songs. Instead, they will just return the XML returned from the service. On the other hand, you can create applications which target a specific subset of UPnP devices, such as A/V Media Servers and Media Renderers, and provide a user experience optimized for these. Still, you may come across, for example, a media renderer with vendor-specified methods. Because no semantics are included in the description documents you should ignore all extensions of the description documents which you do not understand, execpt, of course, if you are the vendor of those extensions in the first place and know the semantics of them. In [19] the difference between service contract and service semantics is more throughly described, along with a discussion about static versus dynamic binding to services.

Phase 3: Control

Service are controlled using the SOAP protocol. A service action is invoked by sending a HTTP POST request to the *control URL*, which is found in the device description. The response returned by the service includes either the result of the operation, with output arguments if any, or a SOAP-exception if an exception occurred while processing the request.

Phase 4: Eventing

For eventing the GENA is used, which is also layered on top of HTTP. Below we first look at how control points can subscribe and unsubscribe to events and later describe how events are delivered to the control point.

Subscribing to events

A control point sends a *SUBSCRIBE* request to the service's *event subscription URL*, which is specified in the device description. A general SUBSCRIBE-request is shown below in Table 9.4, followed by a description of it.

Table 9.4. SUBSCRIBE request headers.

Header	Description
CALLBACK	Location URL(s) where events should be delivered to.
NT	Notification type. Must be set to *upnp:event*
TIMEOUT	Requested duration for the subscription. Notice that this is only a proposed value, and the device has the final saying.

Listing 9.5. GENA SUBSCRIBE Request

```
    SUBSCRIBE (publisher path) HTTP/1.1
    HOST: (publisher host):(publisher port)
    CALLBACK: (delivery url)
    NT: upnp:event
 5  TIMEOUT: (requested duration of the subscription)
```

In the request-line the *publisher path* is retrieved from the *event subscriber URL* in the device description. Let us now have a look at the headers of the request.

A success response looks like the following:

Listing 9.6. GENA SUBSCRIBE Response

```
    HTTP/1.1 200 OK
    DATE: (when the response was generated)
    SERVER: OS/version UPnP/1.0 product/version
    SID: (uuid:subscription-UUID)
 5  TIMEOUT: (actual duration of the subscription)
```

Most of these headers have been described above in Listing 9.5 and 9.6. The SID header contains the subscription ID and must be included in requests to renew or cancel this subscription. In the Timeout-header the actual subscription length (in seconds) can be found. A control point must renew the request before the subscription period runs out, otherwise the device will terminate the subscription. How does a control point renew a subscription? A new SUBSCRIBE-request is sent, similar to the one shown above. But it must not include the Callback- and NT-headers. Instead it must include a SID-header with the value from the response to the original SUBSCRIBE-request. Finally, let us look at how control points can unsubscribe from an event, using the UNSUBSCRIBE-request (See Listing 9.7).

Listing 9.7. GENA UNSUBSCRIBE Response

```
    UNSUBSCRIBE (publisher path) HTTP/1.1
    HOST: (publisher host):(publisher port)
    SID: (uuid:subscription UUID)
```

As can be seen it is very similar to a re-SUBSCRIBE request. Please see above for a description of these headers. If successful the device should reply with a standard HTTP 200 OK response (see Listing 9.8).

Listing 9.8. HTTP OK Response

```
    HTTP/1.1 200 OK
```

Events

Above we have covered how control points subscribe to events. Now we will look at how services notify control points about events and what these events look like. When the state of an eventing state variable changes the service sends a NOTIFY-request to all subscribers. This can for example occur when a light switch changes state (on/off) or media playing stops (or starts playing, etc.). Let us have a look at the headers of a NOTIFY-request as given in Listing 9.9:

Listing 9.9. GENA NOTIFY Request

```
   NOTIFY (delivery path) HTTP/1.1
   HOST: (delivery host):(delivery port)
   CONTENT-TYPE: text/xml
4  CONTENT-LENGTH: (body length, in bytes)
   NT: upnp:event
   NTS: upnp:propchange
   SID: (uuid:subscription-UUID)
   SEQ: (event key)
```

The delivery-information, see request-line and HOST-header, is specified by the subscriber in the first SUBSCRIBE-request. Because events are encoded in XML the Content-Type-header is "text/xml". Similar for the subscription-requests the NT (notification type)-header is "upnp:event". The NTS-header, Notification subtype, is "upnp:propchange", which indicates that the reason for the request is that the state of variable has changed. In the SID-header we find the same Subscription-ID which was assigned to the control point in the response to the first SUBSCRIBE-request. Finally, there is the event key which identifies the events sent to a particular subscriber. The initial event message is 0 and then the number increases with 1 for each notification message, up to the maximum value of 4294967295 before it wraps to 1. This SEQ-header is useful for the control points to handle duplicate messages and also if messages are delayed. For example, sequence 2 arrives *after* sequence 3, then the control point will know that this is not the most recent state. A subscriber must reply with a standard HTTP 200 OK response, as shown in Listing 9.8, within 30 seconds of receiving the event. However, even if a subscriber does not send a response the service should not remove it from the subscriber list before the subscription time is up. As mentioned, the body of a NOTIFY-request includes event information and is encoded in XML. This XML-document follows the XML-schema shown in Listing 9.10.

Listing 9.10. UPnP Template Language for Eventing

```
   <?xml version="1.0" ?>
2  <Schema name="urn:schemas-upnp-org:event-1-0"
       xmlns="urn:schemas-microsoft-com:xml-data"
       xmlns:dt="urn:schemas-microsoft-com:datatypes">
       <ElementType name="propertyset" content="eltOnly">
           <element type="property" minOccurs="1" maxOccurs="*" />
7      </ElementType>
       <ElementType name="property" content="eltOnly" />
   </Schema>
```

An example event notification document using this schema can be seen in Listing 9.11.

Listing 9.11. Event notification

```
1   <?xml version=" 1.0 " ?>
    <e:propertyset xmlns:e=" urn:schemas−upnp−org:event −1−0">
        <e:property>
            <variableName>new value</variableName>
        </e:property>
6       <!−− More VariableName elements with their values may
            follow , see
                maxOccurs in the schema .
        −−>
    </e:propertyset>
```

As can be seen, the notification just contains one or more tuples of (variable name, value). Some state variables can change very rapidly, such as the current speed of a fan. For this reason the UPnP Forum standards are augmented with two values, maximumRate and minimumDelta, which describe how often events should be published.

Phase 5: Presentation

This phase is optional, so not all devices support it. If we go back to the device description, which was described in Section 9.3.2, it includes an element called

Table 9.5. UPnP libraries.

Library	Description	Environment(s) supported
Siemens SDK for UPnP technologies [18]	Also includes a generic control point useful for testing.	Java (Personal Java, Java 1.1, Java 1.3, and Java 1.4) and C++ (Microsoft Windows and Microsoft Windows CE)
Intel Authoring Tools for UPnP Technologies [9]	Intel also provides many tools useful for testing and validation of UPnP devices and services.	Microsoft .NET Framework
Microsoft Windows UPnP APIs [11]	Windows comes with native support for hosting of UPnP devices. See http://www. microsoft.com/whdc/device/ netattach/upnp/default.mspx for more information about these APIs.	Microsoft Windows (included in Platform SDK) and Microsoft Windows CE.
CyberGarage CyberLink [10]	Also a reference implementation of the UPnP AV MediaServer device is available. Available at: http: //www.cybergarage.org/net/.	Java, C, C++ and Perl

presentationURL. If this element is present in the device description then it supports the Presentation phase.

At the *presentationURL* a control point can retrieve a web page which provides a presentation of the device. Everything about the web page, such as how it looks like and functions, is up to the vendor to decide. Therefore the UDA does not give more details about it.

9.3.3 Libraries for UPnP Development

When describing the UPnP protocol stack in Section 9.3.1 the layer "Vendor-specific Application Programming Interfaces" was covered. In Table 9.5 a few of the available libraries are mentioned, which can be used for UPnP development.

References

1. N. Apte and T. Mehta. *UDDI: Building Registry-based Web Services Solutions.* Prentice-Hall PTR, December 2003.
2. S. Cheshire, B. Aboba, and E. Guttman. Dynamic Configuration of IPv4 Link-Local Addresses. RFC 3927, IETF, May 2005.
3. R. Droms. Dynamic Host Configuration Protocol. RFC 2131, IETF, March 1997.
4. G. Meredith E. Christensen, F. Curbera, and S. Weerawaran. Web services description language (wsdl) 1.1. W3C note, World Wide Web Consortium (W3C), March 2001.
5. M.W. Mutka F. Zhu, and L.M. Ni. Service discovery in pervasive computing environments. *IEEE Pervasive Computing*, 4(4):81–90, October/December 2005.
6. R. Fielding, J. Gettys, J. Mogul, H. Frystyk, L. Masinter, P. Leach, and T. Berners-Lee. Hypertext Transfer Protocol – HTTP/1.1. RFC 2616, IETF, June 1999.
7. UPnP^TM Forum. UPnP^TM Device Architecture 1.0. `http://www.upnp.org/resources/documents/CleanUPnPDA101-20031202s.pdf`, December 2003. Version 1.0.1.
8. E. Guttman, C. Perkins, J. Veizades, and M. Day. Service Location Protocol, Version 2. RFC 2608, IETF, June 1999.
9. Intel. Intel upnp tools, 2007. http://www.intel.com/cd/ids/developer/asmo-na/eng/downloads/upnp/overview/index.htm.
10. Satoshi Konno. Cybergarage cyberlink, 2007. http://www.cybergarage.org/.
11. Microsoft. Windows hardware developer central: Universal plug and play, 2007. http://www.microsoft.com/whdc/device/netattach/upnp/default.mspx.
12. H. Nielsen, P. Leach, and S. Lawrence. An HTTP Extension Framework. RFC 2774, IETF, February 2000.
13. K. Permanente P.V. Biron and A. Malhotra. XML schema part 2: Datatypes 2nd edition. W3C Recommendation REC-xmlschema-2-20041028, World Wide Web Consortium (W3C), October 2004. Available at http://www.w3.org/TR/xmlschema-2/.

14. J. Postel. User Datagram Protocol. RFC 0768, IETF, August 1980.
15. J. Postel. Internet Protocol. RFC 0791, IETF, September 1981.
16. J. Postel. Transmission Control Protocol. RFC 0793, IETF, September 1981.
17. J. Sermersheim and Ed. Lightweight Directory Access Protocol (LDAP): The Protocol. RFC 4511, IETF, June 2006.
18. Siemens. Siemens upnp toolkit, 2007. http://www.plug-n-play-technologies.com/.
19. S. Vinoski. Invocation styles. *IEEE Internet Computing*, 7(4):83–85, 2003.
20. S. Vinoski. Service discovery 101. *IEEE Internet Computing*, 7(1):69–71, 2003.

Part IV

Peer-to-Peer Communication

10

Digital Ownership for P2P Networks

Michael Stini[1], and Martin Mauve[1], and Frank H.P. Fitzek[2]

[1] University of Düsseldorf {stini|mauve}@cs.uni-duesseldorf.de
[2] Aalborg University ff@es.aau.dk

Summary. This chapter introduces the digital ownership idea and its need for peer-to-peer networking due to some shortcomings of digital right management. This chapter is the ground work for the SMARTEX application introduced in the following chapter.

10.1 Introduction

The providers of digital content are fighting a losing battle against piracy copies distributed via peer-to-peer file sharing in the Internet or on self-burned DVDs and CDs. Even though new forms of digital content (e.g., digital collectibles, ringtones or virtual items for on-line games) and portable multimedia players have significantly increased the demand of digital content, illegal distribution has dramatically reduced its overall market volume. In an attempt to counter this development content providers have turned towards Digital Rights Management (DRM) [3]. DRM seeks to address the problem of piracy copies by limiting the possibilities of those persons who have legally purchased the digital content. For example, the number of times the digital content can be burned on a CD or transferred to a portable player may be limited by DRM. Unfortunately, DRM in its current form does not aid the providers of digital content in any significant way. The reasoning for this is twofold. First, experience shows that DRM as well as all existing copy protection systems are not able to prevent that a piece of digital content is distributed in an unprotected fashion. There is a multitude of ways to achieve this, ranging from breaking the DRM scheme to resampling the content when it is played back. The second reason why DRM does not solve the problem of the content providers is the fact that it restricts the actions of those who have legally purchased the content. This means that a consumer who legally purchases DRM protected content not only has to pay for the content but also gets an inferior product compared to someone who obtained an illegal, unprotected copy for free. We consider that a radically different approach is much more promising than

259

F.H.P. Fitzek and F. Reichert (eds.), Mobile Phone Programming and its Application to Wireless Networking, 259–270.
© 2007 *Springer.*

trying to protect the digital content from being illegally distributed (which is technically impossible) and artificially restrict the usefulness of the content for those users who legally purchased it (which is decidedly unwise). We propose to securely manage the information on the legal owners of a given piece of digital content and call this *Digital Ownership Management*(DOM). Services, such as being able to access and download their collection of digital content at any time from any place they choose, are provided to the owners. If these *Digital Ownership Services* (DOS) are sufficiently attractive, customers will buy the ownership of the digital content, even if the content itself could be obtained for free via illegal actions. In contrast to protecting the content itself, the secure *management of ownership information* is technically feasible. In its most simple form, the management can be realized by a set of databases maintained by trusted authorities. Furthermore, customers who legally bought the digital content will gain a direct benefit compared to those users who got an illegal copy. The combination of those two facts will make the ownership of digital content valuable to users. This is in sharp contrast to all existing DRM approaches. In addition to the facts mentioned above, we claim that the users themselves are interested in having a clear notion of owning digital objects and of being able to distinguish legally owned content from pirated copies. The main reason for this is that the perceived value of an object directly depends on the effort required to obtain it. This holds true for digital and real-world objects alike. A rare stamp like the "Blue Mauritius" derives its value from the limited number of legitimate copies. Currently, a song or a video has almost no value at all due to its free availability via file-sharing. By introducing the notion of digital ownership associated with benefits for owners a very basic need of customers is satisfied: to own things. Therefore digital ownership management is a result of customer demands. This is another main difference to DRM which targets content provider interests only and massively violates the interests of the customers.

10.2 Digital Content and Its Value

Audio, specifically music, is currently the predominant digital content. Logos, pictures, electronic books, and movies are also quite common. However, there exist many more types of digital content as well. An example of rather unusual digital content are digital collectibles. They are the digital equivalent of real-world trading cards and stickers that are purchased and traded in order to assemble a collection. A complete collection may then be exchanged for a much higher price than the sum of all single collectibles. Besides that, it can be used as the basis for a game. Given the success of real-world collectibles [5], it is not unreasonable to assume that digital collectibles will become quite popular either as a marketing instrument or as a stand-alone product. One key attraction of collectibles is the strong feeling of owning something which is rare. Thus this application area would profit tremendously from the idea of

DOM as it is proposed in this position paper. Virtual items in on-line computer games are an example for content that has no counterpart in the real world. In some of these games, the player controls an avatar which can improve by hard training, virtual work, or by acquiring in-game items. The more powerful an avatar or an item is, the more valuable it is to the players of the game. Although the avatars and items are entirely virtual, they are currently sold for real money on eBay [2]. Thus they represent digital content which would profit from DOM. Sony recognized this and started its own exchange platform called "Station Exchange" [6] explicitly for this purpose. Digital content may also have a direct impact on the real world. Admission tickets, for example, can be represented as digital content, getting rid of the physical form that can be lost or being stolen. One-time usage for a movie theater or public transport, one year admission to an exhibition or even shared usage of these rights can be realized through DOM. By giving these examples we want to stress that the management of ownership for content which does not have a physical representation is a very fundamental problem. It is not limited to audio and video and it is likely to become increasingly important as consumers begin to assign value to a broad range of intangible products.

As long as there exists no perfect copy protection system and as long as there are people willing to share digital content, the data representing digital content (e.g., the file containing a song) has no significant monetary value. At a first glance this might seem to be a highly controversial statement. However, given the fact, that digital content can be copied without any loss of quality and given the presence of low-cost distribution channels (i.e., the Internet or self-burned CDs/DVDs), digital content can be obtained at a negligible cost if these two preconditions (no perfect copy protection and willingness to share) hold. While there have been attempts to establish new copy protection schemes and to legally prosecute those who share digital content, these efforts have been without any noticeable impact so far. The preconditions mentioned above are still valid. We propose to choose a radically different approach and to simply accept the fact that the digital content itself does not have a value. In contrast, being the rightful owner of digital content is indeed of value to a consumer. There are at least three factors that make ownership valuable: legal proof of ownership, convenience, and social status. Being able to prove the ownership of digital content enables consumers to use it without any fear of legal prosecution. Furthermore it is the basis for being able to trade or sell digital content. Both aspects have a monetary value. Proof of ownership is therefore the most fundamental service offered by DOM. If there are services which can only be used by the legal owner of digital content, then the value of ownership is increased further. For example, there could be a service enabling the owner of a song to conveniently download it at anytime, anywhere, and in any desired format. Or the owner may be granted preferred access to booking a live event of the artist who created the digital content. Finally ownership may increase the social status of the owner. Either by being able to show off the ownership or by demonstrating that the owner explicitly supports the artist who created the digital content. As a consequence we propose to be concerned with the management and protection of the ownership status rather than with the management and the protection of the content itself. In order to increase the convenience and the social status of digital content owners we furthermore propose to introduce the notion of *ownership services*, i.e., services that can only be used by the rightful owners of digital content.

10.3 Basic Elements of Digital Ownership

Digital ownership comprises two key elements: Digital Ownership Management (DOM) and Digital Ownership Services (DOS). Figure 10.1 provides a high level example of how users and content providers interact with DOM and DOS. It is important to realize that DOM is only concerned with associating user IDs with the content IDs in order to identify which user owns a copy of the content. DOM is not involved in handling content data, DOS will provide the copy itself. In a first step, content providers register their content with the digital ownership management to attract potential customers *(1)*. During this process a unique ID is assigned to each piece of digital content. In order to make ownership worthwhile to the user, the content providers offer digital ownership services *(2)*. One rather obvious service is the delivery of the content in a format desired by the user. Users interested in obtaining ownership of digital content register with the DOM and an unique user ID will be assigned *(3)*. To obtain ownership the user needs to purchase the ownership from the DOM *(4)*. This results in an association between the unique user ID and the unique content ID which is maintained by the DOM. At a later time the user wants to take advantage of the ownership by requesting ownership services *(5)*. For example, the user requests a download of the digital content in a format suitable for his current application, e.g., a media player. The DOS

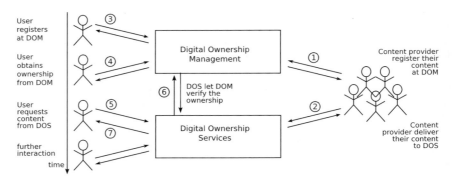

Figure 10.1. Building blocks and interaction within DOM and DOS.

checks with the DOM that the user actually owns the content he is requesting ownership services for *(6)*. Upon successful verification, the DOS provides the requested service and delivers the content to the user *(7)*.

10.4 Digital Ownership Management

This section provides an insight into the basic concepts of DOM and its building blocks. The interaction with the closely related DOS are explained. The most important task of DOM is the secure and trustful management of the information necessary to answer the question: *Does X own Y?* In its simplest form all information can be provided by a single system employing a database with just three relations: users, content, and ownership information, as depicted in Figure 10.2. This information is sufficient to provide the basic functionality of DOM: users may register with the DOM system. This will create an entry in the user table. Similarly, content providers may register their content. When a user acquires the ownership of a piece of content, the *(userID, contentID)* pair is simply added to the *ownership details* table. Given this set-up, the basic management of ownership information is simple and secure as long as the system managing the ownership information has not been compromised. This is fundamentally different from DRM where each system that has access to the content and the content itself must be secured. These are daunting tasks, considering that those systems are under the control of the users. In addition to the rather trivial functionality described so far, DOM will have to provide a broad range of services for the management of ownership. These include on-line and off-line transfer of ownership (trading and selling between users), temporary transfer of ownership (lending something to someone else), on-line and off-line proof of ownership. While some of these services can be realized in a straightforward fashion, others still require significant research effort. In the remainder of this section we will specify the more important services which should be provided by DOM and highlight possible solutions as well as open challenges.

On-line transfer of ownership enables users to trade digital ownership. Similar to trading the ownership of objects in the real world digital ownership

ID	User
1	Alice
2	Bob
4	David
...	...

(a) User details

ID	Content
11	trading card A
14	trading card D
16	trading card F
...	...

(b) Content details

User ID	Content ID
1	14
4	16
1	11
...	...

(c) Ownership details

Figure 10.2. Minimal DOM database.

can be given away, sold, or exchanged between users. During the negotiation both parties involved in the transfer are connected to the DOM. When both parties have agreed on a transaction, the ownership will be transfered by altering the ownership information stored in the database. As user authentication, the negotiation itself, and the final agreements are supervised by the DOM the on-line transfer of digital ownership does not pose significant new challenges.

Off-line transfer enables users to trade in an ad hoc manner, i.e., without the necessity of being connected to the DOM. Users may communicate via short-range communication and negotiate the trade based on the ownership information available off-line. After the negotiation has been concluded, both parties will agree on a trade contract. To become effective, the contract must be submitted to the DOM by at least one user. Off-line transfer poses a number of challenging problems, such as ensuring that the trade partner actually owns content that he offers for trade, guaranteeing that both trade partners will be able to submit the contract while the transaction may be performed only once even if the contract is submitted multiple times and avoiding situations where one contracting party is at a disadvantage compared to the other trade partner, e.g., by signing the contract earlier than the other contracting party.

Temporary transfer of ownership is equivalent to lending ownership to another user. Until the ownership is handed back, the borrower is treated as being the owner although some restrictions may apply (e.g., it may be forbidden to trade or sell the ownership). In particular ownership services may be used during that time. From the perspective of the DOM the negotiation process and submission is similar to a normal transfer. Additional information in the DOM is introduced, listing conferred ownership and date of restitution. Compared to real-world lending of ownership, there is the additional convenience that conferred ownership is given back automatically and the content cannot be harmed.

On-line proof of ownership is typically required when the user wants to access an ownership service that does have on-line access to the DOM. In its simplest form, on-line proof of ownership is a query to the DOM database to verify that the user is in fact a legal owner of the specified digital content. While this lookup and verification is rather straightforward, there remains one caveat: To support privacy, it must not be possible to verify the ownership status of a user unless he has given explicit consent to that check. This is not a trivial problem and requires careful investigation. *Off-line proof of ownership* is required for ownership services that are not able to connect to the DOM. An example would be an ownership service that allows a user to show off his collection of digital content to other users in radio range. Furthermore, off-line proof of ownership would solve one key problem of off-line trading: it would allow for a verification to verify that a trade partner actually owns the content he offers. One potential approach to off-line proof of ownership would be to let the DOM sign a certificate of ownership that the user can then carry around on his mobile device. The key problem with this idea is that the ownership can change after the DOM has issued a certificate of ownership. For example, the user requests a certificate of ownership and then trades the ownership to someone else. In this case the user would still be able to show the certificate even though he is no longer is the actual owner. One way to alleviate this would be to limit the lifetime of ownership certificates and disallow trading of the ownership while it is "checked out". Another idea could be to "punish" users who present certificates for content that they do no longer own. However, all these ideas imply significant drawbacks, thus off-line proof of ownership can currently be considered as a main challenge in the area of DOM.

10.5 Digital Ownership Services

The business models behind ownership services may be rather heterogeneous. One possibility would be to provide a given service free of charge to any owner. Thus the ownership service would be funded by the sale of digital ownership. Alternatively the ownership of digital content may be a mere prerequisite to use an ownership service for a charge. The specific ownership services that are made available to the customer will depend on the business model of the content provider. For free ownership services, this represents a trade-off between the cost required to provide the ownership services and the increase in value of the ownership as it is perceived by the consumer. In this section we introduce three examples of digital ownership services. They are not meant to be an exhaustive list. Rather we would like to demonstrate that it is possible to define ownership services that do add value to owning digital content. One prime ownership service has already been mentioned in the preceding sections: enabling a content's owner to use it on every platform, in any format, and at any time. Thus the content can be viewed on heterogeneous terminals, even if the terminal, such as a TV set in a hotel room, is not the property of the content owner. One example for content where this service would be applicable are digital movies. Once the ownership has been approved, the owner is able to display the movie on its mobile phone, at home, or on the TV set in a hotel room. The implementation of this service will require the setup of a digital content server, either hosting the content in all desired formats or transcoding it on the fly. The upcoming 4G wireless networks with heterogeneous terminals as well as the progressing fixed/mobile convergence will make this service particularly desirable for consumers in the near future. A second ownership service is the support of collecting and trading digital content. This includes functionality for offering items, bidding on offers, or bartering for items. The basic functionality for this kind of service is well known from trading real-world objects such as stamps, trading cards, or other collectibles. But as the items that are exchanged among users are in the digital domain, new types of actions become possible. One example is multiple description coded digital items as described in [4]. The idea behind multiple description coded digital items is digital content that can be aggregated. An example of multiple

Low class descriptor + Low class descriptor + Low class descriptor = High class descriptor

Figure 10.3. Digital Content able to be aggregated.

description coded digital item is given in Figure 10.3, where a given picture can be divided into three base parts referred to as low class descriptors. Each base part can be displayed in a stand alone fashion. The more base parts are collected the better the quality will be. Multiple description digital content is not limited to pictures, but is also possible for music and other examples. This concept is an example for collecting and trading functionality that is only available in the digital domain. A third example of an ownership service makes explicit use of mobile devices such as PDAs, phones, and laptops. The key idea is to allow owners to show off their ownership to other nearby users. This service can be realized in a straightforward fashion by letting the mobile devices announce the ownership of digital content to all other devices in radio range. These devices might alert their owners if the announcement matches their interests. A much more sophisticated approach will let users access and use the digital content announced by a friend either for a brief period of time after receiving the announcement, for as long as the announcing user is nearby or permanently at a low quality. This could be realized by enabling the owner of digital content to give out personalized vouchers that can be redeemed by the recipient for access to a reduced set of ownership services. As a consequence the owner who gives out these vouchers will earn the respect of his or her peers while, at the same time, promoting the digital content. Even rewards can be granted in the case of purchase due to his/her "advertisement".

10.6 First Steps

In order to investigate the concept of digital ownership in a realistic environment we have started the development of ownership management functionality and of ownership services. Our main goals are to identify key challenges and to take first steps towards solutions to the most important problems. The focus of the research group at the University of Aalborg is on making digital ownership attractive through ownership services, while the group at the University of Düsseldorf is mainly concerned with ownership management itself.

As a common platform for the client side of our work we have selected mobile phones. This allows us to experiment with functionality where the physical proximity of users can be exploited to increase the advantages of owning digital content. At the same time it is an environment where ownership management is particularly challenging because a direct connection to the server maintaining ownership information might not be available all the time.

In a first step we set up a single server with a database system to manage the ownership information shown in Figure 10.2. The client runs on mobile phones as well as regular PCs. Registration of users and content is currently done manually. Based on the information contained in the database it was straightforward to implement functionality for on-line transfer of ownership and on-line proof of ownership. Programming was done entirely in Java to provide platform-independent libraries and applications. Even this simple setup emphasized that for a real system a number of open issues need to be addressed. For example, it is unrealistic to assume that ownership information of a significant number of producers of heterogeneous digital content are managed by one single worldwide authority. Thus a distributed and scalable solution to secure storage and management of ownership information is required. Furthermore, it may then be desirable that the unique content ID is assigned in a structured way, potentially encoding some form of hierarchical information. In a second step we moved to off-line trading. We allowed clients to communicate directly with each other (using bluetooth on the mobile devices). Essentially a trade in this environment involves that both parties sign a binding trade contract. Either of the two involved parties may then turn in the contract at any time. Once the DOM is informed about the trade it will perform the appropriate database operations and log the contract so that it cannot be turned in again. Although off-line trading seems to be fairly simple, a number of challenges remain. Essentially each party may sign a contract that it is not able to fulfill. It could be argued that due to signing a binding contract a culprit can be identified and held responsible for any damage caused by violating the contract. However, the situation becomes much more complicated when the ownership acquired off-line is traded further. This can generate a chain of illegally signed contracts which is very hard to resolve. To overcome this, we currently disallow the trading of an ownership which has been acquired off-line as long as the trade has not been verified with the server. Certainly this is not an optimal solution. Summarizing, our first steps to realize ownership management functionality have made us confident that it is possible to develop the required functionality. At the same time we identified a number of very interesting problems waiting to be solved.

Our main focus in the area of ownership services is currently on exploring unusual ideas to make ownership of digital content more attractive. Thus in a first step we investigated methods allowing users to perceive and potentially appreciate the fact that someone else owns digital content. Essentially this

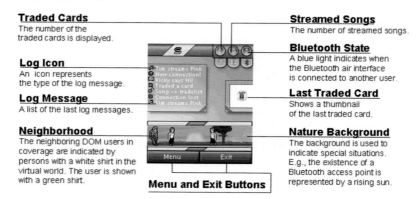

Figure 10.4. Screenshot of a Nokia 6600 with the Digital Ownership Virtual World.

is the basis for converting digital content into status symbols similar to real-world objects. The user interface of a collection of ownership services is shown in Figure 10.4. In this framework a virtual world is shown, where the user himself as well as other users located physically close are depicted at the bottom of the screen. This virtual world is also used to inform the user about ownership in his local vicinity. To achieve this, the application is constantly searching for other users that enter or leave the proximity of the phone. In order to facilitate trading, the local user is able to specify lists of digital content that he is looking for or is willing to sell or trade. The user can then explore the content offered or searched for by the other users and potentially negotiate a trade (or leave the negotiation to the program). If a trade is performed this action is represented in the virtual world and the last acquired object is displayed. In addition to trading, that virtual world allows users to explore the content owned by other users. If digital content is marked as *shared* by the user, other users may experience it as long as they stay in the proximity of the phone. For example the owner of a piece of music could grant the users nearby the temporary right to listen to it. In our prototype the digital content is then streamed from the phone of the owner to the phone of the user who is exploring the content. In the future an advanced interaction with the ownership management will be realized: users will be able to either grant other users some form of temporally limited ownership or the unlimited ownership of an inferior version of the digital content.

10.7 An Example for DRM Usage: iTunes

A full discussion of the vast variety of existing DRM technology would certainly exceed the scope of this position paper. Instead we focus on a single system that employs DRM: Apple's *iTunes* [1]. In April 2003 the *iTunes* application was introduced as a proprietary player and organizer for digital music

and video files purchased from the *iTunes Music Store iTMS*. iTunes is used
to access the iTMS and as an interface to manage Apple's mobile digital music
players, the *iPod* family introduced in late 2001. In contrast to all other digital music stores, the concept of *owning* music and videos, i.e., *digital content*,
has been adopted. Thus a customer owns the downloaded digital content, and
not just the license to use it. In 2006 downloads of songs were sold for 99
cents, music videos and episodes of TV shows for $1,99. Audio-books start
at $2,95 each. At the end of 2005 over 42 million iPods were sold. iTMS
sold its 1.000.000.000th song on February 23, 2006, thus on average about
24 songs were sold per iPod. Digital content sold in the iTMS is encoded in
the AAC open standard format and protected by Apple's proprietary DRM
system *Fairplay*. Playback is limited to compatible devices, which comprise
Apple's *iPod* family and their *iTunes* application. In contrast to earlier versions of iTunes, sharing music over the Internet is now restricted to the local
network using *Bonjour*, Apple's implementation of the *Zeroconf* open network
standard [7], to five authorized computers every 24 hours. Although the term
ownership implies that the owner has the control over the usage on the one
hand and is able to make use of the first-sale doctrine on the other hand this
anticipation is disappointed. In particular the following additional restrictions
are imposed on the owner: if the content is damaged or lost and no backups
have been made, the content must be purchased again. The rights granted
at the time of purchase can be unilaterally restricted by Apple afterwards.
Purchased content must stay with the account it was purchased with, which
means that no trading is possible. Although content can be stored on an unlimited number of computers, only five computers at a time can be authorized
for playback music or coping music to *iPods*. Music can be burned to CDs
an unlimited number of times, but only 7 (formerly 10) times with the same
playlist. Content cannot be edited, excerpted, or otherwise altered. Products
purchased with up to five different accounts can be stored simultaneously on
a single device, such as an iPod, at a time. Apple's *iTunes* concept uses DRM
and it is generally considered a huge success. Thus the question arises whether
this contradicts our line of argument that DRM is the wrong approach and
does not solve the problems of the content providers. In fact we believe that
the opposite is true: the success of iTunes supports our case for DOM rather
than weakening it. A large part of iTune's success is caused by the fact that
Fairplay is much less restrictive than the DRM approaches used by its competitors. Furthermore iTunes has introduced first mechanisms that we would
call ownership services. One example for this is the ability to share music with
authorized computers on the same subnetwork. Thus in our opinion iTunes
represent a first (small) step toward DOM. Furthermore it is important to
realize that the average number of songs sold per iPod is quite limited. Thus
the vast majority of songs played on iPods still comes from other – potentially
illegal – sources. We strongly believe that taking further steps toward DOM
is likely to change this and increase the demand for legally purchased content
dramatically.

10.8 Conclusions and Outlook

In this position paper we have made the case for digital ownership as a key technology to allow the sale and trade of digital content. We argue that digital ownership consist of two main components: digital ownership management allows the sale, trade, and proof of ownership, while digital ownership services provide added value to the rightful owners of digital content. In contrast to DRM our approach does not require a (perfect) copy protection and it is actually desirable for those who acquire the ownership of digital content. While we are convinced that the idea of digital ownership is the right approach to support the trading of digital content, many problems remain open at this time. Some of them, such as off-line trading and novel ownership services, have been mentioned. Others include an appropriate architecture for global DOM, because a flat database is unlikely to be sufficient, standardized formats for ownership information and active probing for potential threats that may target a real-world digital ownership system. However, we consider these problems solvable and we believe that digital ownership has the potential to become a very productive and challenging field of research.

References

1. Apple Computer Inc. Apple - iPod + iTunes. http://www.apple.com/itunes.
2. eBay Inc. eBay. http://www.ebay.com.
3. European Information, Communications and Consumer Electronics Technology Industry Association (EICTA). europe4DRM, DRM, Europe, Digital Rights Management. http://www.europe4drm.com,.
4. F.H.P. Fitzek, B. Can, R. Prasad, and M. Katz. Traffic Analysis and Video Quality Evaluation of Multiple Description Coded Video Services for Fourth Generation Wireless IP Networks. *Special Issue of the International Journal on Wireless Personal Communications*, 2005.
5. Panini Spa. Panini. http://www.paninionline.com/collectibles/institutional/ea/usa/.
6. Sony Online Entertainment Inc. Station Exchange: The Official Secure Marketplace for EverQuest II Players. http://stationexchange.station.sony.com.
7. Zeroconf. Zero Configuration Networking (Zeroconf). http://www.zeroconf.org.

11

SMARTEX: The SmartME Application

Morten V. Pedersen and Frank H.P. Fitzek

Aalborg University {mvpe|ff}@es.aau.dk

11.1 Introduction

In this chapter the SmartME application is introduced to demonstrate the idea presented in Chapter 10 about Digital Ownership Management on real mobile devices. The main goal of the SmartME application is to enable spontaneous collaboration between its users establishing ad hoc peer-to-peer networks using Bluetooth. Within these networks the users are able to provide each other with two different services: The CollectME service is derived from the idea of digital collectibles. When the application is initially started the user is given an incomplete collection of soccer cards (currently containing the logos of Danish soccer teams). The objective from a user's perspective is to complete an entire collection. A new user is given twelve randomly selected cards, where some of them are duplicates and the user can then start exchanging cards. To enable the exchange of cards the user can access his current collection of soccer cards, here he can configure his wish list and trade list. The wish list contains the cards that the user is currently looking for and the trade list the cards that the user wants to trade as shown in Figure 11.1.

The current exchange of cards is done in a one-for-one fashion and the user can configure the application to perform the transfer automatically if a match is found. However in later versions of the application more social aspects could be introduced, motivating the user to become more involved. This could be done by adding items with different values and allowing cards to be traded in, e.g., a two-for-one ratio introducing a negotiation element. A way of implementing this, is to let the user divide his or her items into the following three classes:

Class 1: Nontransferable item.
Class 2: Negotiable item.
Class 3: Available item.

When two users come within communication range of each other, they should exchange their collection lists. The users then know which items they

F.H.P. Fitzek and F. Reichert (eds.), Mobile Phone Programming and its Application to Wireless Networking, 271–274.
© 2007 *Springer.*

Figure 11.1. The user interface allowing the user to browse his card collection and configure his wish and trade lists.

Figure 11.2. The main user interface of the SmartMe application, the screen keeps the users informed about the activity in his/her "virtual neighborhood".

can exchange without negotiation, which they have to negotiate about, and which they cannot get. The StreamME service enables users to *loan* and play each others music while in the vicinity of each other. This kind of service would increase the *coolness factor* of the owner of the digital content. The service was implemented in an emulator as actual streaming of music, but on the mobile phones this failed due to restraints in the needed API. Therefore the StreamME service was redesigned so that the entire song was transferred before the user could play it. After the song was played it was deleted. In order to play the song again the source had to still be available in the network. To benefit from any of these two services, the users need to seek out other users using the application. After starting the application the user can see his/her *virtual neighborhood* as seen in Figure 11.2. The neighborhood shows the currently active users in the vicinity represented by the people wearing white shirts. The user himself is represented by the person wearing a green shirt. Future versions could allow users to modify and customize their avatars, making the characters more personal.

Furthermore indicators on the screen is all the time updating how active the user is, creating a *coolness factor*, aimed at motivating the users to be active in their neighborhoods. The success of the SmartME application is based on the users' willingness to collaborate and provide services to each other. Therefore the social motivation factor becomes an important issue to create a high density of users.

11.2 The Software

The software was during the design phase divided into several modules. These can be seen in Figure 11.3, as a layered model. In SmartME modularity was a desirable property due to a belief that it augments the testability and maintainability of the software.

The network and RMS (Record Managment System) functionality was grouped into one layer accessing and providing the application with network connectivity and persistent storage. A centralized layer called *Core* was placed on top of these to ensure initialization of all layers and providing connectivity between the higher and lower layers. In addition the *Core* should also safeguard a graceful shutdown and initialization of the application. On the left on Figure 11.3 the crosscutting domain layer is shown. This is used to encapsulate shared data in a proper way, which was achieved by using the Constant Data Manager pattern. On top of the Core are the application services namely the StreamME and SmartME layers. Implementing the specific functionality needed by the card-collection game and the music streaming. The topmost layer is the GUI, which enables data presentation and user interaction. During the project the application was tested on the Nokia 6600 smart phone. This device facilitates Bluetooth connectivity and Java 2 Micro Edition (J2ME) execution environment. The developed application is currently an alpha release, and the source code and executable can be found on the DVD.

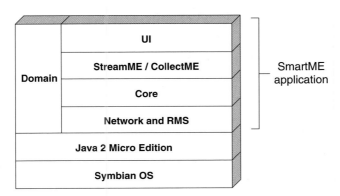

Figure 11.3. The software modules comprising the SmartMe application, running in the J2ME execution environment on the device.

11.3 Summary

The SmartME application successfully demonstrates how users can use the short-range communication technologies of their terminals to exchange digital content, a tendency that is becoming more and more prevalent in, e.g., MMORPG (Massive Multiplayer Online Role Playing Games). In some communities digital content is becoming more important than real physical items. They are used to enhance the properties of mobile terminals (e.g., ring tones), as collections (e.g., baseball cards), or for multiplayer role games (e.g., virtual tools). The application creates a *virtual marketplace* motivating users not only to interact through the application, but also in the real world.

12

The Walkie Talkie Application

Morten V. Pedersen and Frank H.P. Fitzek

Aalborg University {mvpe|ff}@kom.aau.dk

12.1 Introduction

As Voice-over IP (VoIP) services are dominating the telephony backbone already, the main focus of this project is to investigate the possibility of using VoIP over the last hop between the home IP router and the mobile phone. Most homes are already equipped with fixed IP communication lines, and many network providers plan to use VoIP to the IP router at the homes. The project investigates the possibility to use VoIP between the mobile phone, hosting the personal information, and the home router. As a first result a Walkie-Talkie has been developed allowing Push-to-Talk services in a Peer-to-Peer manner called the *Direct Link* scenario as shown in Figure 12.1.

In a second step the possibility to communicate between the mobile phone and home router is investigated referred to as the *Internet Link* scenario. The Internet link scenario describes an extension to the Direct Link scenario. However instead of connecting and communicating directly via Bluetooth, the mobile phones establish a Bluetooth connection to the home router, see Figure 12.2.

The home router forwards IP-packets with the voice payload injecting them onto the Internet. Using this architecture it is possible to exchange data between two mobile phones utilizing the existing VoIP infrastructure. The short-term goal of this project was to replace the PSTN phone by the mobile phone, the long-term goal, to make mobile VoIP in all-IP networks available. The short-term goal is extremely useful both in domestic and office environments, as the mobile phone is the most widespread terminal with about four mobile phones per PC in the world [1]. In addition to that, implementing the feature for existing handsets would not require customers to buy new handsets, they would only need the home router and the mobile phone client software. The project was carried out in Symbian C++ as it offered both the implementation flexibility and huge market penetration. The following will describe the implantation of the two scenarios.

F.H.P. Fitzek and F. Reichert (eds.), Mobile Phone Programming and its Application to Wireless Networking, 275–279.

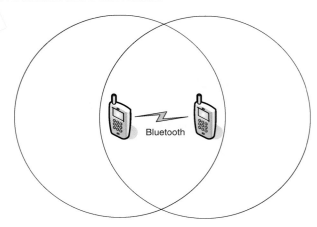

Figure 12.1. The Direct Link scenario describing two mobile phones, connecting via Bluetooth.

Figure 12.2. The Internet link scenario describing two mobile phones, connecting over the Internet using Bluetooth.

12.2 The Software

The first scenario describes a walkie-talkie application allowing voice data to be transferred between two mobile phones. To achieve this goal two software modules were developed, namely the Voice and Bluetooth modules. The two modules provide the core functionality of the walkie-talkie application. However, in order to correctly invoke the functionality of the two modules, two additional elements were added; a User Interface (UI) layer giving the user control of the application, and a control layer which directly invokes the functionality of the core modules through a small protocol. The logical organization of the modules is shown in Figure 12.3.

12.2.1 UI Layer

The purpose of the UI layer is to add a Graphical User Interface (GUI) to the walkie-talkie application. Additionally the UI layer will handle user interaction, which may be invoked through menu selection or by pressing the talk-key (5). This will allow the user to activate the features needed to operate

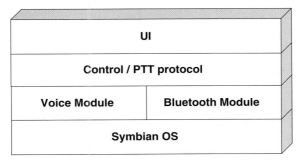

Figure 12.3. Layers composing the walkie-talkie application.

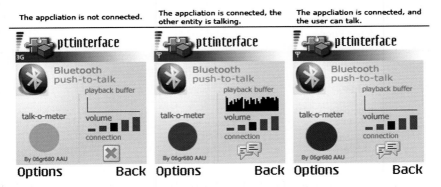

Figure 12.4. The GUI of the walkie-talkie application.

the application. Figure 12.4 shows a number of screen shots of the walkie-talkie GUI.

12.2.2 The Control/Push-To-Talk Protocol Layer

This layer acts as an intermediary between the UI, Voice, and Bluetooth Module. This is done by introducing a small protocol defining the flow of data between two users of the application. In order to invoke the functionality in the Bluetooth and Voice Module, the Control layer implements the observer classes defined in each module. Internally in the application this results in the flow seen in Figure 12.5, depicting how data is passed between the different layers.

12.2.3 Bluetooth and Audio Modules

The walkie-talkie application also serves as a first integration test of the Voice and Bluetooth Modules. The Voice Module enables control of the device's audio capabilities allowing 12.2 kbit/s streaming of voice data in AMR,

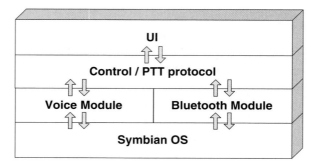

Figure 12.5. Message flow between the different layers of the walkie-talkie application.

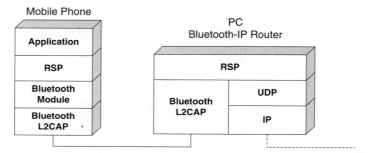

Figure 12.6. Remote Socket Protocol stack.

while the Bluetooth module exposes an extensive API for managing L2CAP connections. During the development of the Voice Module, a problem regarding the full-duplex capabilities of the module was encountered. According to Nokia [2], there is no public API allowing full-duplex audio operations. These are reserved for the system applications such as the video telephony server. Activating these audio operations would require setting a high audio priority, which would result in interference with system functionalities, such as clock alarms and incoming message alerts. For this reason no access to these API's have been granted by Nokia for use in 3rd-party applications. The final walkie-talkie application therefore only supported half-duplex audio.

12.3 Bluetooth-IP Integration

The second scenario allows a user of the walkie-talkie application to reach the Internet, thereby extending the reach of the application through the home IP router. To achieve this, a small glue protocol called RSP (Remote Socket Protocol) was introduced, allowing the mobile phone to access the socket API of the home router as shown in Figure 12.6.

Figure 12.7. Proof of concept setup of Bluetooth-IP.

In order for the walkie-talkie application to make use of the functionality provided by the server, an RSP client module needed be integrated into the application. Due to the modularized development approach, the integration of the RSP module only required changing the module layers directly above the RSP client module. The Bluetooth-IP connectivity was successfully demonstrated allowing two users to communicate using their mobiles over IP as shown in Figure 12.7.

12.4 Summary

The project successfully demonstrated how mobile devices in a relatively simple way could replace currently fixed phone solutions, allowing users to be more mobile in their home/office environments while using cheap VoIP techniques making voice calls. In the future additional services besides VoIP could be implemented on the home router, integrating it with, e.g., other devices in the home/office environment.

References

1. Wendy Holloway association manager of the dotMobi Advisory Group. 2007.
2. Nokia. 2nd Edition Platforms: Known Issues. November 2005.

## Part V

Cooperative Communication

13

Cooperative Wireless Networking: An Introductory Overview

Exploiting Cooperative Principles in the Wireless World: A Wireless Device Approach

Transport of the mails, 'transport of the human voice, transport of flickering pictures – in this century, as in others, our highest accomplishments still have the single aim of bringing men together. **Antoine de Saint-Exupéry 1939**

Cooperation is the thorough conviction that nobody can get there unless everybody gets there. **Virginia Burden**

Marcos D. Katz and Frank H.P. Fitzek

[1] VTT, Technical Research Centre of Finland, `marcos.katz@vtt.fi`
[2] Aalborg University, Denmark, `ff@es.aau.dk`

Summary. This chapter provides an introduction to cooperative techniques in wireless networks, with emphasis on the wireless devices. A motivating introduction is first presented, discussing the scope, scenarios, and potentials of cooperative techniques. Some key challenges are also identified and discussed. Cooperative approaches are then classified according to the employed principles, scope, and goals as communicational, operational, and social cooperation. Finally, a very promising paradigm for cooperation in heterogeneous wireless networks is proposed and explained. This approach is based on a composite centralized-distributed architecture of two combined access networks, a wide area cellular network and a local short-range network.

13.1 Introduction

In recent years, while developing future wireless and mobile communication systems, both industry and research community have identified key challenges as well as promising concepts and technologies with potential to tackle them. One of the key principles with strong potential to help system designers to solve many of the foreseen problems is cooperation. Indeed, we are today witnessing an increasing interest of applying cooperative principles in many areas of wireless communications. The purpose of this chapter is to provide a motivating introduction to this field, in particular from the wireless device

F.H.P. Fitzek and F. Reichert (eds.), Mobile Phone Programming and its Application to Wireless Networking, 283–297.
© 2007 *Springer.*

standpoint. Cooperation is not a new concept in wireless networks. In fact, it has always been used in a passive or tacit manner, for instance by agreeing among the involved entities on using some common formats for the signals, protocols, rules, etc. Clearly, without such a common agreement no communication would take place. In general we refer here to a more active and advanced form of cooperation, where rich interaction among the involved entities are purposely implemented by design. A diverse array of entities can in principle collaborate directly or being part of a wide cooperative interaction, namely users, terminals, OSI layers, complete functionalities, algorithms, signals, etc. There is not one but many reasons explaining the great interest in exploiting cooperative principles. First and most importantly, cooperation has the potential to improve not only network and link performance but also efficiency in the use of key radio resources, in general with a good complexity-performance trade-off. Moreover, wireless communication systems are increasingly becoming complex, with signalling schemes and protocols supporting richer and richer interactions, and with an increasingly fine-grain resource resolution (e.g., frequency, space, time). The emergence of distributed access architectures, such as those found in ad hoc networks, naturally advocates cooperative interaction among connected entities. There exists already a rather wide body of research showing the potential cooperative principles in wireless networks. An extensive and detailed account of the most important cooperative techniques applied to wireless communication networks can be found in [4].

A summary of the characteristics, scenarios, and potentials of cooperative techniques in wireless networks is shown in Figure 13.1. Cooperation can be actively applied in basically all practical scenarios, namely centralized access (e.g., cellular networks) and distributed access (e.g., ad hoc networks), and for every form of signal distribution, i.e., broadcast, multicast, and unicast. Also, the scope of cooperation covers pure altruism (aiming at bringing benefits to a third party like in the case of a repeating station), the actual cooperation (where all interacting entities gain by cooperating), and the autonomous or noncooperation situation as an extreme case. Figure 13.1 also shows the possible gains that can be obtained by the use of cooperative techniques, including achievable data throughput, QoS, network capacity, and coverage. In addition, the efficiency in the utilization of radio resources can be improved by these techniques, notably power (at the base station), energy (at the terminal), and spectrum (at network level). Moreover, terminal cooperation brings the possibility of having cooperative clusters where augmentation and sharing of capabilities can be exploited.

Future wireless and mobile communication systems, often known as the four generation (4G) systems, will integrate a number of different wireless networks providing ubiquitous and seamless connectivity regardless of the type and location of the wireless devices exchanging information. Coverage-wise, the main wireless network components are (a) wide area networks, (b)

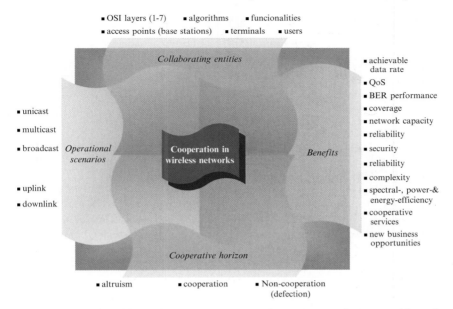

Figure 13.1. Cooperation in wireless networks: An overview of scope and benefits.

metropolitan area networks, and (c) short-range networks. The first type of networks encompass mostly the current second and third generations (2G, 3G) cellular systems as well as evolutionary developments of them. The secondly mentioned networks include mainly high-performance wireless networks optimized for urban use, like networks based on WiMAX technology (IEEE 802.16 standard). The third type of networks comprises in reality a large array of network technologies and approaches, with very different requirements and serving a wide range of applications. Examples of these are wireless local area networks (WLAN), wireless personal area networks (WPAN), wireless body area networks (WBAN), wireless sensor networks (WSN) car–to–car communications (C2C), Radio Frequency Identification (RFID), and Near Field Communications (NFC). Figure 13.2 depicts a classification of wireless communication networks according to the service coverage, showing the role of past, current, and future generations.

13.2 Challenges

We identify here some of the major challenges of future wireless communication systems. These challenges will serve as motivation for the use of cooperative principles as it will be discussed in the next subsection.

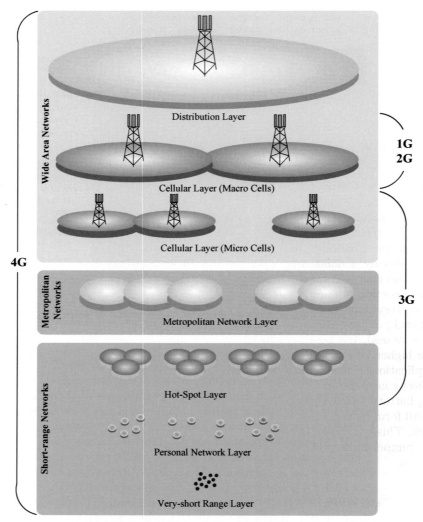

Figure 13.2. Classification of future wireless networks according to their coverage.

13.2.1 Increasing Demand of Energy/Power

Out of the many challenges of future wireless networks one stands out for its essential importance, namely the increasingly higher energy and power demands. The problem is less critical in power-limited systems like access points or base stations, where power emission is regulated and strictly limited by directives issued by the corresponding local communication administration. The immediate impact of power limitation is evident on the achievable data rate and coverage area. In energy-limited systems like wireless devices, higher and higher energy requirement becomes crucial for two main reasons. First, the operating time (i.e., capability of autonomous communications on a truly wireless mode) becomes critically reduced, compromising the use, effectivity, and ultimately the "appealing power" of the wireless device. Certainly this has an direct effect on the user and eventually it dictates his election when purchasing a new device, as shown in a recent study [5]. In [4] was shown that over the past, present, and future generations of mobile terminals roughly half of the energy requirements correspond to communicational functionalities like RF and baseband processing for wide and local connectivity and their associated application engines. Recent measurements, as present in this book in Chapter 25, show that the share of the communication-related function can be as high as two thirds of the total requirements. The second significant consequence of the rapidly increasing energy requirements is the fact that relatively large power consumptions (e.g., typically $P > 6$ Watts for current form factors) require the use of some form of active cooling systems, undoubtedly a nondesired option by terminal manufacturers. It is believed that this energy trap will become more pronounced in the future. Reasons for this assertion abound. We mention here the appreciably slow developments of battery capacity (i.e., energy density), the ever increasing required data throughput (requiring more power to keep the energy per bit within acceptable values), the possible widespread use of multi–antenna transceivers and the higher DSP processing power required on board to cope with advanced applications (imaging, positioning, etc.) Note that the need to keep limited the power consumption in terminals is important even though eventually, promising battery technologies like fuel cells will deliver enough energy. Indeed, the small form factors of handheld devices leads to limited heat transfer capabilities. This, together with significantly high power consumptions means that the temperature in the devices will without doubt rise to uncomfortably high values.

13.2.2 Spectrum Scarcity

The worldwide explosive growth of the number of wireless subscribers together with proliferation of advanced services requiring wider and wider bandwidths pose unique challenges on the use of the limited spectrum. However, it appears in practice that the spectrum is inefficiently utilized. Indeed, even though the allocated frequency bands seem quite congested, their temporal utilization factor is rather low, not more than 10-20 percent according to some recent measurements. New techniques need to be developed to opportunistically exploit the unused bands of the spectrum. Thus, these bands could be already allocated to some users but, if not currently in use, after identifying them they could be temporarily allocated to other users, resulting in better spectral efficiency.

13.2.3 Complexity

The implementation complexity of terminals and access points (base stations) has steadily increased from one generation to another and this trend is likely to accentuate in the future as more processing power and associated function-alities are required to cope with the requirements of high-speed connectivity and sophisticated capabilities. One of the consequences of the use of advanced multiple-antenna techniques is the need of parallel receiving and transmitting branches, which have a direct impact on complexity, which could be crucial at the terminal side. Complexity is also a key factor for terminal develop-ment, as it is directly related to terminal cost and hence, it has to be kept low to result in competitive terminal prices. In general terminal complexity and its associated cost can be traded at the expense of more complex (though expensive) infrastructure. However, as hinted in [4], it is in principle possible to exploit cooperation in a fashion that simple terminals will collaborate, sharing resources and capabilities, aiming at achieving virtual capabilities impossible to attain with single terminals but attainable through a cooperative terminal cluster.

13.2.4 Appealing Services and Applications

Future wireless networks should be designed in a user-centric manner, try-ing ultimately to fulfill the expectations of the users, and not developed for the sake of introducing new advanced technologies. We are already witnessing with 3G systems that the lack of attractive services has slowed down notably the penetration of these new technology. Novel appealing and useful services will guarantee a widespread acceptance of new technologies. In that respect, and following the rapidly emergent trends of social interaction across networks (e.g., Internet), it appears that services and applications exploiting coopera-tive interaction among wirelessly connected users is a very promising answer to this important challenge.

13.2.5 Demanding Communicational Capabilities

The next wireless and mobile communication systems will have stringent com-municational requirements. From the terminal perspective these can be sum-marized as (a) multimodality where several air interfaces will be integrated onboard providing concurrent access to several wireless networks, (b) support of very high data rates, of the order of tens of Mbps in wide area networks and hundreds of Mbps in local area access, (c) Quality of Service (QoS) provision-ing, (d) seamless connectivity along any given network (horizontal handover) as well as across heterogeneous networks (vertical handover), (e) all range of terminal mobility support, from fixed to very high speed (several hundred Km/h) scenarios, and (f) latency (short connection and transmission delays).

13.3 Cooperative Principles in Wireless Networks: An Introduction

We have witnessed during the past years an increasing interest in exploiting cooperative techniques in the field of wireless communications. A vast literature already exists showing the potentials of cooperation basically in each and every OSI layer of a communication system. A classification of cooperative techniques was proposed in [4] where the principal approaches are *communicational cooperation, operational cooperation,* and *social cooperation.* These three concepts are presented and discussed next. Figure 13.3 depicts this classification, highlighting the scope and goals as well.

13.3.1 Communicational Cooperation

The so-called *communicational cooperation* approaches are the most explored aspects of cooperation in wireless networks. Such methods encompass different techniques exploiting the joint collaborative efforts of multiple entities in the system aimed at bringing some advantages to the involved parts. In general, communicational cooperation is inherently embedded in the wireless network and therefore invisible to the user. The mutually interacting entities include signals, algorithms, processing elements, building blocks, and complete units. The goals of communicational cooperation are to enhance key performance figures at link and network level, to improve the utilization of basic

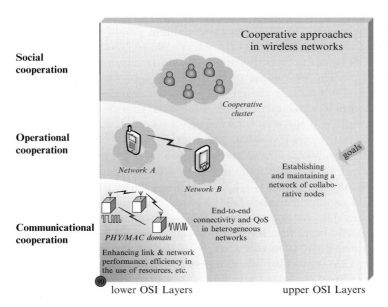

Figure 13.3. Cooperation in wireless networks: classification, scope, and goals.

radio resources, and to create virtual augmented capabilities of the interacting entities of a cooperative cluster. A further classification of communicational cooperation includes implicit and explicit cooperation. The former refers to cases where interactions take place without any preestablished cooperative framework, such as protocols (e.g., all network nodes tacitly agree to use a common protocol like TCP, ALOHA, etc). The latter form of cooperation refers to interactions established through a given framework, that is, cooperative behavior is allowed and supported by design, allowing counterpart entities to actively interact directly with each other. The core of communicational cooperation mostly takes place in lower OSI layers, particularly in the physical (PHY) and MAC layers, though link and network layers may also be involved.

The most representative cooperative techniques are briefly presented in the following paragraphs:

Multi-hop (Relaying) Techniques

Of all cooperative approaches multi-hop techniques have perhaps received the largest share of attention, probably because of their inherent simplicity. The basic idea is to have one or more intermediate nodes whose main function is to repeat or retransmit in a convenient manner the received signal. In this way it is possible to extend coverage (range), resulting also in a more uniform provisioning of QoS within the service area. Multi-hop techniques, a basic technology of ad hoc networks, is also being considered for cellular networks, where, in addition to enlarging the coverage area, they will help in guaranteeing high data throughput even at the edge of the cell. Multi-hop techniques usually exploit two basic approaches, namely amplify-and-forward (AF) [7] and decode-and forward (DF) [11]. AF is a simple method where, as its name implies, the signal is directly amplified and retransmitted to the next node in the chain. The processing station or device in the case of AF is usually denominated *repeater*. In the DF approach the signal is received and digitally regenerated before being transmitted forward, thus the noise is not amplified over the multi-hop chain as with the AF method. The processing unit in this case is referred to as the *relay*. There exist other schemes as estimate-and-forward and store-and-forward as well as some hybrid combinations of them. In general, a two-hop solution (single repeater) is a good engineering compromise in cellular networks, while in distributed networks a generic multi-hop approach is usually considered. Multi-hop techniques are presented in detail in [4]. Note, that particularly in cellular networks one of the two hops need not necessarily be over a wireless link but it can be implemented on a wireline or optical fiber, as in the concepts of distributed base stations and radio over fiber. One important advantage of multi-hop techniques lies in the fact that the deployed repeating/relaying stations reduce the average link distance of the active terminals, resulting in a relaxed link budget and ultimately in extended duration of the battery.

Cooperative Diversity

Cooperative diversity refers to several techniques exploiting the presence of one or mode nodes between the source and destination. These distributed nodes cooperatively helps the source to improve the overall capacity achieved between source and destination. The joint contribution of several nodes can be seen as a special form of spatial diversity and hence the end-to-end communication reliability can be also enhanced. These techniques are based on protocols allowing sequential reception and further retransmission in a time division fashion as simultaneous reception and transmission is not practically feasible (half-duplex constraint). Additional reading on cooperative diversity can be found in [6, 8] as well as in [4].

Cooperative/Distributed Antennas

Terminals in close proximity and forming a cooperative cluster can be used to form a virtual antenna array (VAA) by forming with single-antenna terminals a multiple input multiple output (MIMO) system. The VAA approach is also also known as distributed antenna or distributed MIMO. Each terminal contributes with a single antenna and by exploiting the short-range links these antennas can be interconnected making the cooperative cluster to appear either as a transmitting or receiving array. Such array, when communicating with a counterpart one in another cluster or in a base station form the basic MIMO structure. A comprehensive overview of cooperative/distributed antenna techniques is presented in [4]. The concept of space-time coding, that is, the joint coding over temporal (e.g., repetition) and spatial (e.g., antennas) domains can be also applied in such a distributed antenna scenario, as showed in [12]. A different approach with distributed antennas is the concept of distributed beamforming where each antenna element of the terminals forms a random array. Thus, assuming that nodes exchange their information and synchronization is achieved over the short-range links, it is possible to obtain array gain in both receiving and transmitting directions. In principle it is possible to achieve full array gain in a given desired direction as with conventional arrays (e.g., uniform linear arrays) but given the random distribution of nodes the side lobes cannot be controlled. Some insights of distributed beamforming are given in [4].

Network Coding

Another form of cooperation which has emerged rapidly in recent years is network coding. Network coding assumes several repeating/relaying nodes between the source and destination. Source and destination may not necessarily be punctual but multiple sources and destinations can be considered. Network coding in general comprises the joint design of routing and coding in such multi-node scenario. A node will not just retransmit its received information but in principle will combine information received from many nodes to then forward that joint information to other nodes. From the packet standpoint, network coding exploit the fact that each wireless node broadcasts its packets and a generic node thus receives and combines packets from many sources. Readers are referred to [4] for an in-depth discussion of network coding principles.

13.3.2 Operational Cooperation

An inherent characteristic of future wireless networks will be their heterogeneity, manifested by the presence of different access networks, a great variety of terminals with different capabilities and a large array of varied services. Clearly, providing connectivity in such an eclectic scenario is a far from trivial task. *Operational cooperation* can be defined as the interactional and negotiating procedures between entities required to establish and maintain communication between different networks. The main target is to ensure end-to-end seamless connectivity, where the main players could be different terminals operating in different networks. Research activities in this field are being carried out by academia and industry, we highlight here the work at the Wireless World Research Forum (WWRF) where network cooperation is mostly approached from the transport and network layers [10]. Also Ambient Networks project under the umbrella of EU's Sixth Framework Programme addresses the problem of cooperation in heterogeneous networks, particularly where networks belong to different providers or exploit different access technologies [1, 9]. The ultimate goal of these projects is to ensure seamless operation regarding the type and associated networks of both source and destination. In addition to developing cooperative procedures, these initiatives also explore architectures required to support provision of end-to-end connectivity.

13.3.3 Social Cooperation

One of the most interesting aspects of cooperation is perhaps its social perspective. In short, *social cooperation* can be defined as the dynamic process of establishing and maintaining a network of collaborative nodes, in our case, wireless terminals. The process of node engagement is important as each node needs to decide on its participation in a (ad hoc) network, each decision having an individual and collective impact on performance. Unlike the communicational and operational where the cooperative mechanisms are embedded in the system and therefore they remain mostly invisible to the user, here the user can take a key role as he will ultimately decide whether to cooperate or not. Since individuals are part of the social equation modeling cooperation, nontechnical aspects like trust, socio-geographical attitudes toward cooperation, incentives for cooperation, reputation, and others will eventually have an effect on system performance. It is worth mentioning the clear positive attitude toward cooperation that exists today, in particular though the virtual world created by Internet. A great deal of very popular initiatives exploit cooperative principles, including distributed ventures like eBay, Linux, OpenSource, Wikipedia. Collaborative efforts are applied in file sharing and distribution protocols like BitTorrent and many others. Distributed grid and computing efforts are also becoming increasingly popular lately. We expect that the rapid and pervasive proliferation of wireless networks will extend this phenomenon toward a wirelessly connected world. Already we are witnessing the rapid emerging phenomenon of wireless communities, where metropolitan- and rural-wide wireless networks connect cooperatively computers and users over short-range broadband networks. In a few years it is expected that several million users worldwide will be sharing resources and communicating through wireless community networks.

13.4 Cooperation in Heterogeneous Networks: The Emergence of a Novel Paradigm

We start this section by briefly discussing a couple of noteworthy trends that will likely dominate the wireless communication scene of the future. In the first place, the number of wireless terminals (or other devices with communicational capabilities) will continue to grow dramatically leading to a very high density of terminals. The WWRF predicts that by year 2017 there will be approximately one thousand wireless devices per each inhabitant of the globe. Most of these wireless devices will be for short-range communications. In other words, around every single terminal we can expect that there will always be a potential cluster to cooperate with. We also expect that the current trend of equipping mobile devices with several air interfaces will continue and even strengthen in the future. The two main interface types correspond to wide area cellular access and local access (short-range communications). Such air interface diversity can be realized by a multimodality implementation, integrating onboard the air interfaces with different chip sets, or by exploiting the flexible concept of software defined radio.

The above introduction to this section, together with the previous discussions on the emergent role of cooperative techniques served as a motivating background toward the introduction of a new paradigm in wireless networks, namely advocating cooperation between the two principal wireless networks, wide area (cellular) and local area (distributed). We strongly believe that the existing synergy between such a networks with highly complementary characteristics (e.g., licensed/unlicensed spectrum, high/low energy per transmitted bit, centralized/decentralized architecture) can be exploited, resulting in a number or advantages, as will be discussed below. Figure 13.4 illustrates the evolution of access architectures following wide area and local area developments. In the cellular networks the main evolutionary element is the

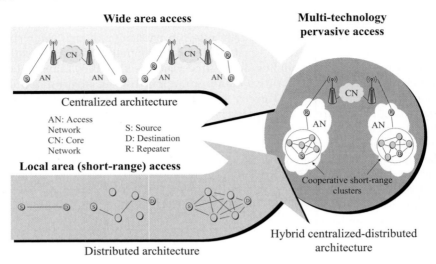

Figure 13.4. Evolution of wireless networks access architectures.

introduction of one or more relaying/repeating stations, while in short-range networks the development towards distributed ad hoc architectures is evident. We see that eventually these approaches will be merged into a hybrid (or composite) architecture. The composite (centralized-distributed) architecture of the last approach is sometimes referred to as "cellular-controlled peer-to-peer communications". The idea here is that the information is delivered to a particular terminal not directly from the base station (access point) or its serving repeaters, but through a short-range network surrounding the end terminal. The term "short–range network" can be replaced by several other possible equivalent designations, like "cooperative cluster", "wireless grid", "wireless sensor network", "ad hoc broadband network", etc. Figure 13.5 illustrates the discussed concept. Note that the clusters can have a variety of possible network configurations, with homogeneous or heterogeneous component nodes. The cellular network serves as a service entry point while the local short-range network cooperatively delivers the information to the target terminal. Clearly, terminals exploit dynamically both short-range and cellular air interfaces.

Three typical scenarios where users spend most of their time lend themselves perfectly to host the considered composite access architecture. In the *home environment*, the cooperative cluster can be formed by entities of the personal network, for instance other wireless devices, home appliances, etc., and by the terminals of other family members. In the *office environment* also wireless devices of coworkers can participate in the cooperative cluster, spontaneously and/or encouraged (or even required) by the employer. In *public places* cooperation can take place among terminals basically unknown to each

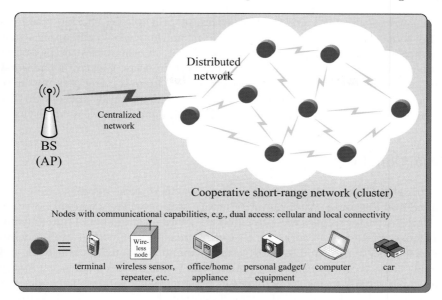

Figure 13.5. Composite (centralized-distributed) access architecture in heterogeneous wireless networks.

other. In such a case cooperation can emerge not only spontaneously but also it can be encouraged by operators by different incentives.

Among the advantages of the considered composite architecture we highlight a significant reduction in power consumption of the cooperating terminals [4]; cost attractive cooperative services (e.g., multicast); highly secure communications exploiting the fact that the conveyed information can be conveniently spread over the cellular and distributed network and combined at the target destination; increased capacity as the nodes of the cooperative cluster can cooperatively exchange signal components (e.g., video streams) over the short-range links, resulting for each terminal in higher data rates that would normally be achievable in a noncooperative manner, etc. Additional discussions on this composite architecture can be found in [2].

Another interesting aspect of the envisioned composite access architecture is to consider the cooperative cluster as a platform for sharing and augmenting capabilities of the collaborating terminals. For that purpose and in a generic approach we consider the cluster as a heterogeneous wireless grid where *user interfaces* (e.g., imaging devices like camera, displays; audio devices like microphones and speakers; and others like sensors, keyboards, etc.), as well as *built-in resources* (e.g., mass storage devices, CPUs, energy sources) will be exchanged and/or shared over the short-range links. The wireless grid then has a pool of different resources and according to the current needs or particular services associated with each terminal these resources. This can be seen as a *wirelessly scalable architecture*, where in principle by sharing resources every cooperative entity will gain. Depending on the scenario and applications being used, all users may simultaneously gain or, in other cases the pay-off of cooperation will temporarily be enjoyed by certain users. Clearly, in the worst case, where nodes do not exploit cooperation at all, every user gets not more or not less than the capabilities of his own terminal. In [3] authors further discuss the potential of this scalable architecture and its associated services.

13.5 Conclusion

In this chapter several aspects of cooperation in wireless networks were presented and discussed. After an introductory section where a general overview of cooperative techniques was presented we identified a number of fundamental challenges of future networks. The three main aspects of cooperation in wireless networks were then discussed, namely communicational, operational, and social. These techniques have the potential to tackle the key challenges. A very promising paradigm for cooperation in heterogeneous wireless networks was proposed and explained, highlighting its potentials. The considered approach is based on a composite or hybrid (centralized-distributed) architecture of two access networks, a wide area cellular network, and a local short-range network.

References

1. B. Ahlgren, L. Eggert, B. Ohlman, and A. Schieder. Ambient networks: Bridging heterogeneous network domains. *International Symposium on Personal Indoor and Mobile Radio Communications (PIMRC 2005)*, 2005.
2. F.H.P. Fitzek, M.D. Katz, and Q. Zhang. Cellular Controlled Short-Range Communication for Cooperative P2P Networking. *WWRF 17, WG5*, Heidelberg, Germany, 2006.
3. F.H.P. Fitzek, M. Pedersen, and M.D. Katz. A Scalable Cooperative Wireless Grid Architecture and Associated Services for Future Communications. *European Wireless Conference, EWC'07*, Paris, France, 2007.
4. F.H.P. Fitzek and M.D. Katz, editors. *Cooperation in Wireless Networks: Principles and Applications – Real Egoistic Behavior is to Cooperate!* ISBN 1-4020-4710-X. Springer, April 2006.
5. Taylor Nelson Sofres Consulting Group. Two-day batter life tops wish list for future all-in-one phone device. In *Technical report*, 2005.
6. J.N. Laneman, D. Tse, and G.W. Wornell. Cooperative diversity in wireless networks: Efficient protocols and outage behavior. *IEEE Trans. Inf. Theory*, 2004.
7. J.N. Laneman and G.W. Wornell. Energy-efficient antenna sharing and relaying for wireless networks. *IEEE WCNC*, Chicago, IL, 2000.
8. J.N. Laneman and G.W. Wornell. Distributed space-time-coded protocols for exploiting cooperative diversity in wireless networks. *IEEE Trans. Inf. Theory*, 2003.

9. N. Niebert, A. Schieder, H. Abramowicz, G. Malmgren, J. Sachs, U. Horn, C. Prehofer, and H. Karl. Ambient networks: An architecture for communication networks beyond 3G. *IEEE Communications Magazine*, 2004.
10. C. Politis, T. Oda, S. Dixit, A. Schieder, K.Y. Lach, M. Smirnov, S. Uskela, and R. Tafazolli. Cooperative networks for the future wireless world. *IEEE Communications Magazine*, 2004.
11. A. Sendonaris, E. Erkip, and B. Aazhang. Increasing uplink capacity via user cooperation diversity. *IEEE ISIT*, Cambridge, MA, 1998.
12. A. Stefanov and E. Erkip. Cooperative space-time coding for wireless networks. *Proc. IEEE ITW*, 2003.

14

The Medium is the Message

Anders Grauballe, Ulrik Wilken Rasmussen, Mikkel Gade Jensen, and Frank H.P. Fitzek

Aalborg University {agraubal|wilken|mgade|ff}@es.aau.dk

Summary. A simple Python program transmitting a short text message over the GSM network without causing any costs for the customer is presented. The application is intended to underline the strength of cooperation in cellular controlled peer-to-peer networks. Furthermore an example how to extend the Python library with Symbian/C++ is given.

14.1 Motivation

In the 1960s Marshall McLuhan coined the phrase *The Medium is the Message* claiming that the medium over which information is transported is sometimes more important than the information itself. He was referring to the upcoming importance of television, but still 40 years later this sentence has some importance as we have shown in [1], where we advocate to exploit channel descriptor information in packet data communication networks to gain transmission capacity. Besides the normal data transmission, the channel descriptor (or character of the channel) can also be used to convey data. Introducing a novel access technology, we would like to motivate our idea by a short example out of the GSM world, which we later will realize with a short Python program.

At the beginning of the GSM deployment, there was a phenomenon in Italy called *squillo*. People, mostly young, would just ring each other (hanging up before the other side could pick up) using their mobile phones to say hello or to convey some other predefined messages. This kind of communication was very popular as it was not billed (still in many countries it is not). Inspired by this idea, we envision a scheme where multiple phones could be used to convey data over existing wireless networks using the signaling plane without any additional costs for the user.

An example of a possible setup is given in Figure 14.1, which shows two groups of phones where each side contains one master and three slaves. The sending master has the information to be conveyed to the master of

F.H.P. Fitzek and F. Reichert (eds.), Mobile Phone Programming and its Application to Wireless Networking, 299–310.

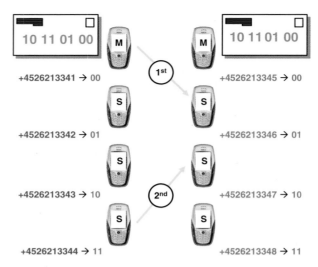

Figure 14.1. Data Transmission over the GSM signalling plane with two groups of four mobile phones. Both groups have one master (M) each and three slaves (S) communicating over Bluetooth among group members.

the receiving group. The eight sending and receiving phones are identified by their phone numbers and to each identity a two-bit address is assigned. This assignment is known to both masters of the group.

The sender master will read the digital message and by using the first two bits one phone of the sender group is chosen to call over GSM a phone of the receiver group, which is identified by the second two-bit tuple. In this example, first the master itself (**00**) will call the second phone of the receiving group (**01**). The receiving phone, by using the intragroup communication, informs the master about the received call (also which phone in the sending group made the call), which the master in turn can demap into four bits of information (**0100**). The second call from phone number four of the sender group (**11**) to the third phone of the receiver group (**10**) will be transformed into the information (**1011**). By each call four bits of information is transmitted.

The bits gained are of course paid for by the network provider and far fewer bits are conveyed than the network provider has to invest to make this transmission possible. Therefore it may be referred to as a *trick*. Leaving this example, we raise the question whether it is possible to convey data by exploiting the channel descriptor (in our example phone numbers) used instead of transmitting data (recall that we did not send any bits at all over the GSM bearer). In the following we would like to investigate this idea in more detail.

14.2 Identification of the Needed Functionality

To develop the application which uses this concept proposed in the previous section some functionalities are needed. The application must have access to basic functionality such as Bluetooth communication to communicate locally, set up a voice call, hang up on a voice call, see the actual call status, and pick up the caller ID. The prototype application must contain an editor where the user can enter a message to send.

14.2.1 Functionality Available in Python

Python for Symbian Series 60 (S60) offers many built-in modules which are useful and Python makes it easy to develop applications without any knowledge of the underlying hardware platform. Python for S60 does not include all standard modules, but many of them (some with modifications and extensions) plus special modules for using the native resources of the S60 mobile phones. Figure 14.1 shows the functionality needed for the described application and whether or not this is a built-in functionality in Python.

14.2.2 Adding Functionality to Python

Python is great for developing software prototypes, but it still misses some functionality (see Table 14.1). This problem can be solved through the Python/C API which allows a developer to write extensions in C/C++ to a standard library to gain access to, e.g., S60 functionalities. In the following a Call Status extension is explained to show how an extension to Python is made. The extension is developed using Symbian SDK 2nd edition FP2. To make the extension we need the following three files:

1. Source code (callstatus.cpp)
2. Project file (callstatus.mmp)
3. Build file (bld.inf)

Table 14.1. Table containing the functionality needed to make an application using this concept. The functionality marked is available in Python.

Functionality	Available in Python
Editor	Yes
Bluetooth	Yes
Bit mapping	Yes
Dialling	Yes
Hang Up	Caller Yes/Receiver No
Call Status	No
Caller ID	No

They must be saved in the directory:

`c:\Symbian\8.0a\S60_2nd_FP2\src\ext\callstatus`

Source code (callstatus.cpp)

The first file (callstatus.cpp) hosts the functionality we want to integrate in Python. It can be made from any function in C or C++. The C++ code example in Listing 14.1 shows the functionality of the module. The function makes a connection to the client handle through the **systemAgent**, and is then able to access the information from the state variables on the phone, among them the status of the phone.

Listing 14.1. Source File of the Extension of the Call Status in Python

```
1  #include "Python.h"
   #include "symbian_python_ext_util.h"
   #include <saclient.h>
   const TInt KUidCallValue = KUidCurrentCallValue;
   const TUid KUidCall ={KUidCallValue};
6
   extern "C" PyObject *
   callstatus(PyObject*){
      #ifdef __WINS__
      return Py_BuildValue("i", 0);
11    #else
      RSystemAgent systemAgent;
      TInt error = KErrNone;
      if ((error = systemAgent.Connect()) != KErrNone)
      return SPyErr_SetFromSymbianOSErr(error);
16    TInt CallValue = systemAgent.GetState(KUidCall);
      systemAgent.Close();
      return Py_BuildValue("i", CallValue);
      #endif
   }
21 extern "C" {
      static const PyMethodDef callstatus_methods[] = {
         {"GetStatus", (PyCFunction)callstatus,
         METH_NOARGS, NULL},
         {NULL,          NULL}
26       };
   DL_EXPORT(void) initcallstatus(void){
   PyObject *m;
   m = Py_InitModule("CallStatus", (PyMethodDef*)
      callstatus_methods);}
31 }
   GLDEF_C TInt E32Dll(TDllReason){
      return KErrNone;
   }
```

Lines 7–9 initialize the function and checks whether the program is running on an emulator, if this is the case the routine will always return 0. If the function is running on a phone line 12–14 initialize and connect to the **systemAgent**. In line 16 the variable **CallValue** is assigned to **KUidCall** that has a value between 0 and 9 depending on the actual call status, see Table 14.2. Lines 17 and 18 close the connection to the system agent and return the value from **KUidCall**.

Table 14.2. The meaning of the returned values from `CallStatus()`.

Values	Meaning
0	No call
1	Voice call
4	Calling
5	Incoming call
9	Call Disconnecting

Project File (callstatus.mmp)

This file tells the compiler which module to compile and what is needed for this module. The file is shown in Listing 14.2.

Listing 14.2. The Project File Which Shows the Libraries that have to be Included to Build the Python Extension File

```
1  targettype  dll
   TARGET          CallStatus.pyd
   TARGETPATH      \system\libs

   systeminclude \epoc32\include
6  systeminclude \epoc32\include\libc
   userinclude   ..\..\core\Symbian
   userinclude   ..\..\core\Include
   userinclude   ..\..\core\Python

11 USERINCLUDE    .
   LIBRARY python222.lib
   LIBRARY euser.lib
   LIBRARY sysagt.lib
   LIBRARY efsrv.lib
16 source  CallStatus.cpp
```

TARGET is the Python name of the module. It is very important that this is the same as the first argument in Py_InitModule() in the source code see Listing 14.1. Lines 5–9 give the paths of the various header files needed in the source code and lines 11–15 is a list of the libraries needed in the compilation. Source in line 16 specifies the name of the source code.

Build File (bld.inf)

This file specifies which platforms the module should be compiled to and what project file to use. **Wins** is for the PC emulator and **armi** is for the ARM processor of the phone. The file is shown in Listing 14.3.

Listing 14.3. The Build File Specifies the Platforms for the Extension

```
   PRJ_PLATFORMS
   wins winscw armi
   PRJ_MMPFILES
4  CallStatus.mmp
```

Compilation

To compile the module you have to open a DOS shell and make a virtual drive. This is done by:

```
subst t: c:\Symbian\8.0a\S60_2nd_FP2
set EPOCROOT=\
```

Go to the location of the module files, in this case:

`t:\src\ext\callstatus` Run:

```
bldmake bldfiles
```
This creates an abld.bat file needed for the compiler

Then run either:

```
abld build wins udeb
abld freeze
abld build wins udeb
```
This compiles a CallStaus.pyd file in your t:\epoc32\release\wins\udeb\ directory for the debug emulator.

Or:

```
abld build armi urel
abld freeze
abld build armi urel
```
This compiles a CallStatus.pyd file in your t:\epoc32\release\armi\urel\ directory for the phone.

The module works straight away for the emulator but to get the module to work in the phone you need to send CallStatus.pyd from the `t:\epoc32\release\armi\urel\` directory via Bluetooth or cable to your phone.

Usage in Python

To use the new extension module in Python you need to import it first. In the Python console this is done by:

```
import CallStatus
```

To use the function "GetStatus" type:

item `CallStatus.GetStatus()`

This should return one of the values from Table 14.2. As well as the Call-Status module a CallerID and HangUp module is made so all the needed functionalities are available in Python.

14.3 Realization of the Idea with Python

To demonstrate the idea on four phones, we developed an application in Python. The main purpose of the application is to send a message from one group of phones to another. This is done by making calls in a pattern specified by the bit composition of the message as given in Figure 14.1. The application developed to realize this, is composed of three different parts:

- The sender side
- The receiver side
- The GUI

The requirements to the application is that it should be able to send and receive a message. Thus the correct part of the application must be initialized when needed. On the sender side this is done when a user chooses to send a message in the user interface, and it is done on the receiver side when the phone receives the first call, which is made as an initialization call. The initialization takes place on the master phones (The sender phone which the user has written the message on and the receiver phone where the message is being delivered). The two masters must subsequently inform their slaves about the upcoming transmission by local communication. The message is then sent using a common protocol developed for the purpose, which is visualized in Figure 14.2. As we assume two mobile devices on each side, the message is sent in two-bits per call as both caller and receiver represents one bit each. The transmission is ended using a timeout value that is higher than the time it takes to establish the call between the phones.

The transmission of a message is controlled by a sequence of functions. To give the reader an understanding of how the application's sender side works, a message sequence flowchart is displayed in Figure 14.3.

The initial transmission is done when the user chooses to send the message. The application makes a call, which initializes the transmission on the receiver side and establishes a Bluetooth connection to the senders slave. The message is then converted into a list that can be sent by the communication protocol, and transmitted by establishing calls from either the master or slave to the receiver phones. On the receiver side the receiving starts when the master receives the initializing call. This starts the receiving sequence, which like the sender side controls the sequence of steps needed to complete the process, this can be seen in Figure 14.4.

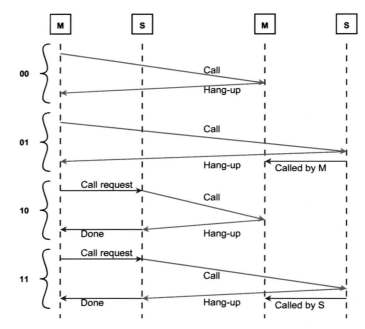

Figure 14.2. The protocol of the transmission with the sender on the left and the receiver on the right. Each side contains of a Master and a Slave. The green and red lines indicate phone calls and the blue lines indicate Bluetooth communication between the phones. The Bluetooth communication is used whenever a slave must make a call as it must be told to do so by its master and if a slave receives a call it must inform its master about it.

The receiving master will as initial run in a loop waiting for an initializing call, when this is received it establishes a Bluetooth connection to the receiver slave. The application will now record the sequence of the calls received, until it gets a timeout and then convert the received information to a message.

The User Interface

The GUI of the Python application is designed with a menu where the user can navigate through all the possibilities in the application. There are two different menus, one shows when the phone is set up to be a master and the other when the phone is a slave. The GUI is made similar to other phones SMS menus so the end user is familiar with it.

The most interesting menu is on the master phone it contains the following:

- Write new (Where to write and send a message)
- Inbox (Viewing incoming messages)
- Settings (Specifying the settings of the system)

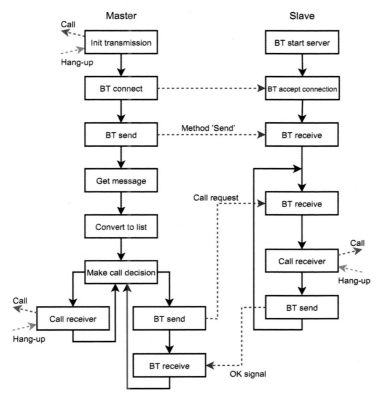

Figure 14.3. Flowchart of the applications sender part including the Bluetooth communication between the master and slave.

In Figure 14.5 the GUI of the master is shown in full, and in Figure 14.6 the GUI of the slave is shown.

To send a message with the application, all phones have to be turned on and configured. The user should select 'Write new' and in the editor write a new message. After this is done the user should select 'Send SMS', then the application will start the sending and receiving sequence and transmit the message.

14.4 Measurement Results

Several measurements have been carried out to measure how long it takes to send one character. The average time of sending a successful message of 20 characters was 31,46 minutes which is 1 minute and 34 seconds per character.

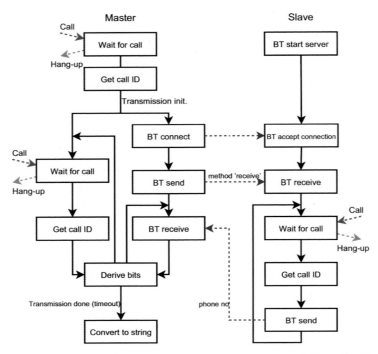

Figure 14.4. Flowchart of the applications receiver part including the Bluetooth communication between the master and slave.

This time is an average of 20 tests. This result is for two phones on each side with German to Danish SIM cards. This was needed as all tests were carried out in Denmark where the network operator bills even for setup of the voice call. By using foreign SIM cards, the network operator was not able to charge, but the voice call setup did take much more time. As we were more focused on the proof of concept, we did not consider the time aspect important because this is a proof of concept. If one character is sent in 1 minute and 34 seconds (94 seconds) with two phones on each side, this means that it takes 23,5 sec. to set up a call (t). The following equation can be used to calculate the time (T) to send one character if more phones are available:

$$T = t \times \frac{8}{bits/call}$$

Table 14.3 gives the time needed to transmit a single eight-bit character for different number of phones on each side.

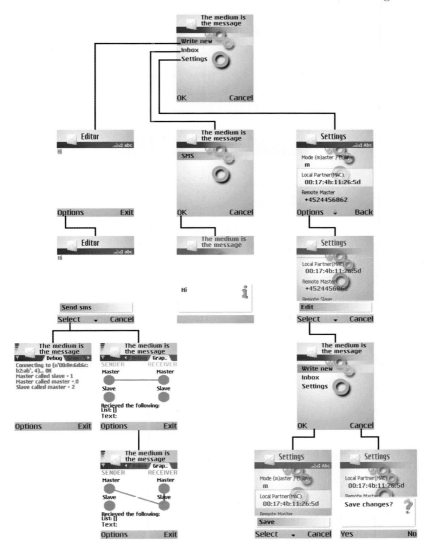

Figure 14.5. The GUI of the phone in master mode.

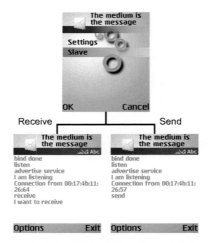

Figure 14.6. The GUI of the phone in slave mode

Table 14.3. The calculated time to send a character.

Phones/side	Time to transmit a character
2	94 seconds
4	47 seconds
8	31 seconds
16	23 seconds

14.5 Conclusion

In this chapter we have shown the proof of concept of a cooperative communication cluster to avoid any costs transmitting a SMS message. Furthermore the example shows the beauty of Python as a quick and flexible way to implement ideas on a mobile device.

References

1. F.H.P. Fitzek. The Medium Is The Message. *IEEE International Conference on Communication (ICC)*, June 2006.

15

Peer-to-Peer File Sharing for Mobile Devices

Imre Kelényi, Gergely Csúcs, Bertalan Forstner, and Hassan Charaf

BUTE {imre.kelenyi|gergely.csucs|bertalan.forstner|
hassan@aut.bme.hu}@aut.bme.hu

Summary. Peer-to-peer (P2P) file sharing is a relatively new addition to the growing list of mobile-based technologies. The advanced connectivity features and powerful hardware of new generation mobile devices facilitates the implementation of even the most complex P2P technologies. The characteristics of the P2P architecture, such as reliability, easy one-to-one communication, and extensible distribution of resources, make it one of the most suitable networking technologies for mobile systems. In this chapter, we give a summary of our experience in implementing the Gnutella and the BitTorrent file sharing technologies on a mobile platform. We examine the particular protocols and emphasize the points relevant to the mobile implementation.

15.1 Peer-to-Peer File Sharing in General

For the last couple of years, peer-to-peer (p2p) file sharing has been the focus of interest in the world of information technology. While new or refined protocols and technologies are released almost every month, the basics of decentralized file sharing have remained the same since the birth of the original Napster service. The *recipe* is well known: download a file from multiple locations simultaneously and you have better bandwidth utilization, and thus, increased transfer speed. Form a network of individual peers and give them the ability to search the network for shared resources. These simple ideas may seem trivial, but simplicity should not be underestimated in computer science, and this is also one of the reasons why P2P file sharing is so popular nowadays. This chapter is not intended to give a detailed view of the topic, but we briefly summarize the basic ideas and techniques behind file sharing before moving on to topics involving the mobile implementation. P2P file sharing systems aim to share information among a large number of users with minimal assistance of explicit servers or no servers at all, thus, the network relies primarily on the bandwidth and computing power of the participating nodes rather than on the scarce resources of a limited number of central servers [5]. Although

F.H.P. Fitzek and F. Reichert (eds.), Mobile Phone Programming and its Application to Wireless Networking, 311–324.

with the spread of 3G networks and high-speed communication technologies, bandwidth is becoming less and less expensive, multimedia applications also have the tendency to demand a large amount of network resources by transferring documents, music files, or even videos over the Internet. If we did not have peer-to-peer technologies in hand we would still have to direct all traffic through some central servers and invest into hardware upgrades each time we run out of bandwidth or computing power, which is clearly not a long-term option. Many content providers have already recognized the possibilities behind these concepts and started serving their content in a P2P nature. Good examples could be TV shows distributed by BBC or the ingame patching system of Blizzard Entertainment's highly acclaimed World of Warcraft video game. Not only the content providers but also the users benefit from the increased transfer speeds and better response times which are the results of the reliable architecture. The most important characteristics which make P2P file sharing so attractive when it comes to large scale content distribution, especially in a mobile network, could be summarized as follows:

- Efficient bandwidth utilization
- Scalability and adaptability
- Reliability
- Cost-effectiveness

Despite the many advantages P2P has to offer, we should not forget to mention that there are also disadvantages including the legal issues raised by the distribution of copyrighted materials. The more decentralized a network is, the harder it is to trace the shared content back to its source. File sharing has been in the headlines several times described as a *source of piracy* or a *tool for software pirates*. In fact, nobody can claim that sharing copyrighted materials is not one of the main reasons behind the popularity of P2P, but it should be emphasized that this is not the responsibility of the technology itself, and many recent protocols have features which help network operators to track down illegal users [6]. Previously, P2P file sharing and mobile devices were considered quite incompatible technologies but the appearance of mobile phones equipped with advanced connectivity features and considerable computing power has changed the situation. The amount of information handled and generated by applications running on mobile phones increases every year. Transferring this large amount of content over a network of nodes can be done in several ways but the key characteristics of mobile networks, such as the adhoc nature of network topology, the limited bandwidth of the participating nodes and the typically unreliable connections make P2P file sharing an ideal candidate for the task. Enabling P2P technologies on mobile devices could have a large impact on how we think of content providing and sharing. By the end of the year, the number of mobile phones used worldwide is estimated to exceed 2.7 billions. Imagine that we have a billion tiny computers connected to the Internet; all of them operating as a file sharing client. With the spread of 3G, flat rates, and WLAN coverage, it is not just a vision any more. In spite of the diversity of P2P file sharing solutions, the main concepts could be well demonstrated on two well-known and widely used protocols: Gnutella and BitTorrent. These two systems are designed for different purposes and they take a quite different approach when it comes to the distribution of the shared files. We will discuss these issues in more detail in the following sections.

15.1.1 Gnutella

The birth of the Gnutella Network could be tracked back to 2000 when two developers of Nullsoft (actually, they belonged to AOL then) released the first client application. This first version of the protocol soon became very popular among the former users of Napster. Gnutella promised an *indestructible* network, which in theory could not be shut down by an external authority simply because there were no central servers, all peers in the network had completely identical functionality. Generally speaking, Gnutella is an unstructured, pure P2P system, which mainly aims at managing a network of nodes and searching among the shared resources (in most cases files), the actual file transfers are handled using HTTP [4]. What does all this mean in practice? Consider a network of randomly connected nodes, all of them run the same client application. The nodes communicate with standard messages. To join the network and start searching for some content, first we need to access a node which is already connected. This initial step is called *bootstrapping*. In the first version of the protocol, users had to manually specify a connected peer, but this method was soon replaced by more sophisticated technologies like the Gnutella Web Caching System or the currently recommended UDP Host Cache scheme. Both methods involve the use of some dedicated servers which host a dynamically updated list of peer addresses. Establishing a connection with a Gnutella node also involves the exchange of some introductory messages which is often referred to as *handshaking*. We can start searching for files as soon as the client has managed to connect to the network. Search queries, in the form of Gnutella messages, are sent to the neighboring nodes which then forward the requests to the nodes they are connected to. If a search request finds a result, a *Query Hit* message containing the address of the matching peer is propagated back through the chain of nodes to the source of the request. Although this type of message broadcasting works well with a relatively small number of peers, the increasing number of messages can easily overload the nodes with slower connections in a larger network. There have been many important changes and improvements since the initial release of the Gnutella protocol, but clearly one of the most significant was the introduction of the Ultrapeer system. The problems of randomly connected nodes are alleviated by organizing the network in a more structured form. Instead of treating each node equally, some nodes are now treated as *ultrapeers*, routing search requests and responses for nodes connected to them. This new dedicated group of nodes

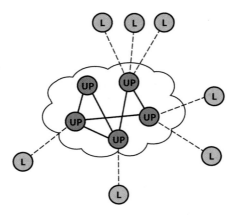

Figure 15.1. The network topology of the Ultrapeer System. Ultrapeers (UP) form the backbone of the network. Leaves (L) only communicate with each other via the layer of ultrapeers.

forms a new layer the main purpose of which is to relieve casual nodes of routing tasks. These ordinary nodes connected to ultrapeers are referred to as *leaves*. The concept is demonstrated in Figure 15.1.

As we mentioned earlier, file transfers are handled outside the Gnutella network, through HTTP. The Gnutella protocol participates only in the searching process. Since all clients operate also as tiny HTTP servers, they can request files from each other by using standard HTTP GET messages. The particular files are identified by their names and a dedicated index number which comes from the received *Query Hit* message. The Gnutella protocol can be implemented relatively easily on mobile devices due in no small part to the Ultrapeer system. If we do not want to be involved in the routing tasks, we can implement the client to work only as a leaf. Leaves are full-featured clients in terms of file sharing, however, they only need to implement the following features:

- Some sort of bootstrapping algorithm (GWebCache or UDP Host Cache)
- The handshaking process
- Processing of the five standard Gnutella messages (Ping, Pong, Query, Query Hit, and optionally, Push)
- HTTP-based file transfer (GET messages)

Of course, there are many more extensions we could add to the client, but a basic implementation requires only the listed features. One serious drawback of this approach is that it cannot be used in homogeneous mobile networks. Since all mobile clients are leaves, at least one ultrapeer is needed to route the messages. However, if our goal is to connect to the standard Gnutella network, this is not an issue. We show you the details of implementing the protocol for Symbian OS in Section 15.3.

15.1.2 BitTorrent

BitTorrent is a relatively new P2P file sharing protocol originally designed and created by programmer Bram Cohen. With BitTorrent, when several peers are downloading the same file at the same time, they upload pieces of the file to each other. This redistributes the cost of upload to downloads [2]. The authors of BitTorrent describes it as a *free speech tool* since it enables everyone to publish whatever they like without spending lots of money on costly hosting services. The more popular the files become, the more *free* bandwidth is gained. The BitTorrent protocol has an official client application which is also referred to as *BitTorrent*. However, since the specifications of the protocol are freely available, anyone can create applications based on the BitTorrent technology. Opera Software integrated BitTorrent into Opera Web Browser and a Taiwan-based company has recently announced that it is releasing a chip that can process BitTorrent feeds through hardware rather than software. These are just a few examples which reflect that BitTorrent is quickly gaining industry support. BitTorrent takes a different approach from Gnutella by concentrating only on distributed file transfer. It does not have any built-in service for searching. Generally speaking, BitTorrent is designed to distribute large amounts of data without incurring the corresponding consumption in expensive server and bandwidth resources.

Sharing files over BitTorrent needs at least one dedicated peer in the network which is called the *tracker*. The tracker coordinates the file distribution and can be queried for the shared resources which are under its supervision. A network can contain several trackers. A particular shared file, however, can be associated with only one tracker. This means that when a peer starts sharing a file (or group of files), a specific tracker must be chosen, and cannot be changed afterward. This mechanism could cause some issues in a transient network environment. The process of sharing a file or a group of files begins with the creation of a *torrent file* which contains metadata about the the chosen tracker and the files to be shared. The data in the torrent file is stored in an encoded form which is referred to as *Bencode* or *Bencoding*. The torrent file must be registered with the tracker; afterward, any client which obtains the torrent file can connect to the swarm and download or upload the files. The torrent file itself is relatively small in size, transferring it over the network does not consume significant amount of bandwidth. It is usually hosted on a web server. Peers are required to periodically check in with the tracker (this process is called *announcing*); thus, the tracker can maintain an up-to-date list of the participating clients. The tracker can also offer several additional features such as *leech resistance* which encourages clients to not just download, but upload file fragments as well. When two clients establish a connection they communicate over the *Peer Wire* protocol afterward. This protocol facilitates the exchange of file fragments, or *pieces*, as described in the torrent file. It includes messages for requesting and uploading a piece and can regulate data traffic by preventing the choking of clients.

Concerning legal issues, BitTorrent, similarly to any other file transfer protocol, can be used to distribute files without the permission of the copyright holder. However, a user who wishes to make a file available must run a tracker on a specific host and distribute the tracker's address in the torrent file. This feature of the protocol does imply some degree of vulnerability that other protocols lack. It is far easier to request that the server's ISP shut the site down than it is to find and identify every user sharing a file on a traditional peer-to-peer network. It should be noted that there is also a trackerless version of the BitTorrent protocol [1], which relies heavily on DHTs (Distributed Hash Tables). The maintenance of DHTs needs considerable computing power, especially in a very transient environment, and also results in great increase in network traffic which is highly undesirable [3]. A desktop computer can easily handle the overhead of this extension but our experiences has shown that the traditional tracker-coordinated version of the protocol is more appropriate for mobile devices. In this section, you could get a brief overview of how BitTorrent works and what it can be used for. Mobile phones were excluded from the BitTorrent community until now, but things will definitely change in the near future. There is no reason why portable devices with Internet connectivity capabilities should not participate in BitTorrent swarms. To demonstrate the concepts, we created SymTorrent, the first BitTorrent client for Symbian OS, which is covered in detail in Section 15.4.

15.2 Thoughts on the Mobile Implementation

Implementing a P2P file sharing system on an advanced mobile device is somewhat different from doing it on a desktop computer. Although the application is still written in C++, Java, etc., mobile operating systems often restrict the use of the APIs which could be considered *harmful*. If the target platform is native Symbian OS, we even have to accommodate to the use of its restricted version of C++. Furthermore, we have to deal with the simplistic user interface along with the lower available computing resources. The realization of a P2P application should begin with the thorough examination of the particular protocol. We should focus on the general concepts rather than the details. It is essential to achieve a comprehensive picture of the protocol's properties before designing the architecture and starting the implementation.

15.2.1 Separating the User Interface from the Application Engine

The concept of separating the user interface (UI) from the business logic has been used for a long time. The Model-View-Controller (MVC) architecture is used in the vast majority of modern applications. Although these are quite general concepts and can be applied on almost any platforms, they are especially important in the case of mobile devices where so many different screen sizes and control schemes exist. If we implement the engine of our application separately, we can save ourselves from doing unnecessary work when porting the application. Although Mobile Java tries to ease the situation by providing standard APIs for all kinds of mobile UI layouts, there is still no reason why we should refuse the other benefits of the architecture. In case of native Symbian OS, if we want to support the largest set of phones, it is indispensable to use a separate UI layer. Symbian OS forms only the basis of modern devices, and a separate UI layer is responsible for handling the user input and output. The APIs of this UI layer is somewhat different in the currently available three Symbian UI platforms: S60, UIQ, and Series 80. This is the main reason why we implemented the engine as a separate DLL in all of our Symbian-based P2P file sharing applications.

15.2.2 Accessing the Mobile Nodes

A significant issue in cell phone networks is that mobile operators usually use network access translation (NAT) and provide only *local IP addresses* for network users. Furthermore, hosts behind a NAT-enabled router do not have true end-to-end connectivity which is one of the key principles of P2P file sharing. Although there are no technical obstacles that would prevent operators from providing *public IP addresses* for the users, this is not a common practice in the current cellular phone networks. And even if a public address is provided, we still have to deal with the lack of fixed IP addresses, which is another problem. Gnutella has introduced *Push* messages to ease the problem. A Gnutella client may send a *Push* message if it receives a *Query Hit* message from another node that does not support incoming connections. Upon receiving the request, the client that cannot accept incoming connections establishes a new connection to the initiator of the *Push*. The main shortcoming of this approach is that it only works when at least one of the two parties can accept incoming connections. Another possible workaround could be some kind of proxy server which forwards incoming connections toward the mobile devices. The proxy must have a *public IP addresses* which is accessible to anyone in the network. Our particular solution is referred to as *Connector Server* and is developed to run on a separate desktop computer. The main task of the proxy is to provide public and semipermanent (or long-term) IP addresses for the connecting clients. Figure 15.2 shows an architectural overview of the proxy-based network. Without going into details, we discuss the principal features that the proxy should implement to support the concept.

The proxy maintains two dedicated ports for supporting the process. The one responsible for service messages is called *control port*. Clients connect to this port to apply for a public IP address. This *control connection* remains active during the whole session. Each client obtains the (same) IP address of the proxy as a public address but with a different port number. The proxy stores the IMEI number (International Mobile Equipment Identity, a number unique to every GSM and UMTS mobile phone) of the connecting device, thus,

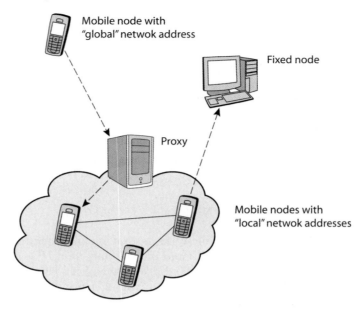

Mobile node with
"global" netwok address

Fixed node

Proxy

Mobile nodes with
"local" netwok addresses

Figure 15.2. Basic architecture of the proxy-based network.

next time when a client connects to the proxy, the same port can be supplied. Generally speaking, the address and port uniquely identifies the client. When the proxy receives an incoming connection to one of the assigned public addresses, it notifies the corresponding client, which then opens a new connection to the proxy's other dedicated port, called *forward port*. As the name suggests, this port is responsible for forwarding incoming connections toward the mobile clients. The *forwarder connection* is initialized by the client, and acts as transmitter between the host connecting to the public address (which is handled by the proxy) and the mobile client. This architecture provides full transparency for the connecting party. Connecting to the public address and port through standard TCP seems as if it were connected directly to the phone. The architecture is semitransparent from the phone point of view. After the forwarder connection is initialized, the socket can be used in a normal way as if it were connected directly to the connecting party. If the proxy is implemented in a platform-independent language, such as Java, then it can be used under different operating systems. Actually, in an extreme case, even a mobile device can run the proxy if it has a permanent IP address and enough bandwidth resources. It should also be noted that connecting through the proxy is optional in the sense that mobile devices can still listen on their own IP address if it is accessible to other hosts.

15.3 Symella

Symella is the first Gnutella client for Symbian OS [7]. It was released in the summer of 2005 as an open source application under the terms of the GNU General Public Licence. It has a massive user basis which consists of around ten thousand users worldwide. Currently, Symella supports mobile phones based on the S60 platform 3rd or 2nd edition (it covers more than thirty mobile phone models at the time of writing). Figure 15.3 shows a screenshot of Symella.

Symella can search the global Gnutella network and support multi-threaded downloading. It works only as a leaf, however, it does not affect searching or downloading from the user's point of view. Many advanced Gnutella extensions were left out of the application in order to save memory and computing power but the results are still more than sufficient, especially for a mobile phone. After bootstrapping Symella initiates connections to only a small number of ultrapeers. Peer addresses are continually collected in a host cache which classify the peers based on their uptime and some further properties. The host cache is saved on exiting the application thus the stored peers can be reused next time at the bootstrapping. Searching involves sending a standard Gnutella Query message to the connected ultrapeers. The returned query hit messages contain the name, the size, and optionally a hashed value of a matching file along with the peers address which hosts the file. Based on the hash value, or if it is not available then the size and the name, the client groups together with the hits for the same file. Multithreaded downloading is one of the key features of Symella. The application tries to download the particular file from more sources simultaneously. Since Gnutella uses standard HTTP for file transfers, it is possible to request only a given part of a particular file. Each file is divided into equally sized pieces which can be downloaded separately. After the completion of the download, the application reassembles the pieces to obtain the complete file.

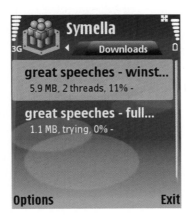

Figure 15.3. The download view in Symella.

15.4 SymTorrent

SymTorrent is the first, and at the point of writing, the only BitTorrent client for Symbian OS. Our main goal was to transfer the BitTorrent technology to a mobile platform and demonstrate the possible use cases of BitTorrent-based file sharing on a real device. In addition, we developed some new concepts during the development which resulted in an integrated client-tracker application. SymTorrent not only works as a standard BitTorrent client, it also has its own built-in tracker. Running a tracker on a mobile phone may seem a bit bizarre at first but it can have several interesting use cases. Sharing files instantly between a small group of users without depending on external servers is just one example. A screenshot of SymTorrent is shown in Figure 15.4.

During the last three months, SymTorrent has been downloaded more than 10,000 times. Most users employ it as standard BitTorrent client for downloading files with their mobile phones through GPRS or WLAN. While we have reports of users downloading torrents with SymTorrent during the night just to avoid the noise of their desktop computers, others prefer to download even while driving to work or waiting in a supermarket checkout line. Since the first version of the application, which was released in October 2006, SymTorrent has grown in features. Now it is a full-featured and complete implementation of the BitTorrent protocol which can operate either as a standard client or even as a tracker. The main features are as follows:

- Handling any standard torrent file
- Downloading or sharing multiple torrents simultaneously
- Built-in tracker
- Built-in torrent maker

Similarly to Symella, SymTorrent is available for mobile phones based on the S60 platform 3rd or 2nd edition. However, it can be easily ported to any

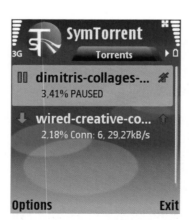

Figure 15.4. The main view in SymTorrent's user interface

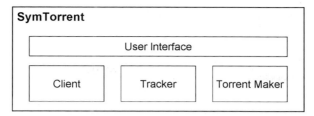

Figure 15.5. The general architecture of SymTorrent.

other Symbian UI platform. Since the source codes are freely available under the terms of the GNU General Public Licence, any developer can access them through the official website [8].

The general architecture of SymTorrent is shown in Figure 15.5. We cover each part in more detail in the following subsections.

15.4.1 The Client

The main and largest part of SymTorrent is the client engine. The client implements the standard BitTorrent protocol which consists of the following:

- Parsing Bencoded torrent files
- Communicating with the tracker over HTTP
- Communicating with the peers over the Peer Wire protocol

Without going too much into details, we would like to describe the client's architecture and summarize how it works. Although most class names are prefixed with some kind of abbreviation, we will use only the main part of the label (e.g., `PeerConnection` instead of `CSTPeerConnection`). The core of the engine is implemented in the class named `TorrentManager`. All other entities including the loaded torrents are supervised by this class. In normal circumstances, it operates as a singleton thus only one instance exists. The event detection method is polling-based: a timer triggers the `TorrentManager` in every second then the status of the torrents are updated according to the recent changes. This approach has a small overhead compared to more sophisticated event models, but we have the advantage of handling all changes together. Moreover, this class functions as the connection point between the UI and the engine. Further entities of the BitTorrent protocol are implemented as separate classes like `Torrent`, `Peer`, `Piece`, and `TrackerConnection`. The settings of the client can be accessed through the `Preferences`. Between starting to download a torrent and receiving the first piece, the following steps are involved:

1. The `TorrentManager` parses the given *torrent file* which results in a new `Torrent` instance
2. The `Torrent` creates a `TrackerConnection` instance which announces to the tracker and processes the reply

3. Each peer address returned by the tracker is encapsulated in a separate `Peer` instance
4. The `Torrent` initiates connections to a number of peers by creating `PeerConnection` instances
5. `PeerConnection` requests a piece from the connected peer

15.4.2 Using the Class Library

The engine of the client was implemented as a separate DLL which can be reused in the form of a class library by any other application. It enables to download and share torrents with only a few lines of codes. In this section we will give a couple of examples to demonstrate the basic APIs. First of all, we must obtain a `TorrentManager` instance in order to access the client's services. This could be achieved either by creating a new instance or using the *singleton wrapper*. Then we only need to set the basic preferences before the transfer of the torrent can start. The settings can be loaded from a stored file or can be set manually. The following code demonstrates how to initialize the engine and start downloading a given torrent.

Listing 15.1. Initializing the Engine and Starting to Download a Torrent

```
1  CSTTorrentManager* torrentManager =
           CSTTorrentManager::NewL();

   torrentMgr->Preferences()->
           SetPreferencesFileL(_L("C:\\symtorrent.cfg"));
6  torrentMgr->Preferences()->LoadSettingsL();

   torrentMgr->Preferences()->
           SetDownloadPathL(_L("E:\\Downloads"));

11 CSTTorrent* torrent = NULL;
   TInt err = iTorrentMgr->
           OpenTorrentL(_L("c:\\x.torrent"), torrent);

   if (err == KErrNone)
16 {
           if (torrent->HasEnoughDiskSpaceToDownload())
                   torrent->StartL();
   }
```

The engine can send notifications on the status of the torrents. To sign up for a specific event, we must first implement the interface `MSTTorrentObserver`.

Listing 15.2. The Torrent Observer Interface

```
1  class MSTTorrentObserver
   {
   public:
     virtual void TorrentChangedL(CSTTorrent* aTorrent,
                                  TInt aIndex,
6                                 TUint32 aEventFlags) = 0;
   };
```

In the enumeration `TSTTorrentObserverEventFlags` the available event types are defined. There are many events specified; we can receive notifications when the torrent is started, stopped, or finished. The following example code shows how to subscribe for a notification in case the torrent is opened or the download speed changed.

Listing 15.3. Signing Up for Events

```
iTorrentMgr->AddTorrentObserverL (
  this ,
  torrent ,
  EEventTorrentOpen   |   EEventDownloadSpeedChanged ) ;
```

15.4.3 The Tracker

Since a basic tracker operates as a simple peer address cache, its architecture is very simple. After accepting an incoming HTTP request and parsing the received announce data, it returns the list of peers connected to the particular torrent. Having a tracker on hand enables us to quickly share files without depending on an external server. SymTorrent also includes a torrent maker application which can create torrent files for any specified group of files. The particular steps of sharing a file are as follows:

1. Create a torrent file
2. Register it with the tracker
3. Start sharing with the client
4. Publish the torrent file by, for instance, sending it in MMS or over Bluetooth

If these steps are automated then the user needs only to select the files to be shared; all other operations are carried out by the system.

15.5 Summary and Conclusion

In this chapter, we have covered the basics of peer-to-peer file sharing along with the examination of two particular protocol: Gnutella and BitTorrent. We have introduced our Symbian-based mobile implementation of both Gnutella and BitTorrent, and briefly described the architecture of these applications. We have also given some code examples on how to use the engine of SymTorrent. Building a completely new application for handheld devices could take a lot of time and effort but it becomes much easier if you have a base to build on. Before you start developing your own software, it is worth studying the source code of Symella and SymTorrent. The more protocols and platforms are supported, the better the network becomes and that is what peer-to-peer is intended to achieve.

References

1. The bittorrent protocol homepage. http://www.bittorrent.org.
2. B. Cohen. Incentives build robustness in bittorrent. *Proceedings of the 1st Workshop on Economics of Peer-to-Peer Systems*, 2003.
3. G. Ding and B. Bhargava. Peer-to-peer file-sharing over mobile ad hoc networks. *Proceedings of the 2nd IEEE Annual Conference on Pervasive Computing and Communications Workshops (PERCOMW '04)*, March 2004.
4. The gnutella protocol homepage. http://www.the-gdf.org.
5. A. Oram, editor. *Peer-to-Peer: Harnessing the benefits of a distruptive technology.* ISBN: 0-5960-0110-X. O'Reilly and Associates, 2001.
6. P. Rodriguez, See-Mong Tan, and C. Gkantsidis. On the feasibility of commercial, legal p2p content distribution. *SIGCOMM Comput. Commun. Rev.*, 36(1):75–78, January 2006.
7. The symella application homepage. http://symella.aut.bme.hu.
8. The symtorrent application homepage. http://symtorrent.aut.bme.hu.

16

Energy Saving Aspects and Services for Cooperative Wireless Networks
First Step out of the Energy Trap!

Leonardo Militano[1], Gergely Csúcs[2], and Frank H.P. Fitzek[3]

[1] Università Mediterranea di Reggio Calabria `leonardo.militano@unirc.it`
[2] BUTE `gergely.csucs@aut.bme.hu`
[3] Aalborg University `ff@es.aau.dk`

Summary. This chapter describes the implementation of wireless cooperative networks. The application was developed to prove energy savings for cooperative mobile devices.

16.1 Motivation

As previously introduced, cooperation among mobile devices controlled by a cellular network is a novel wireless architecture capable of overcoming problems and limits encountered by conventional wireless networks. To prove these claims, a very simple application has been implemented to advocate the use of cooperative behavior in wireless networks. Furthermore the application can act in a stand-alone fashion too, without cooperation. Both approaches are compared with each other. The application itself underlies the same concept as BitTorrent in the wired world. In the first step we are not focusing on a commercial application. Larger interest lies in the understanding of the cooperative concepts and its related performance compared to state-of-the-art technologies. In the end of the chapter we will implement the new cooperative concept in the SymTorrent application explained before.

16.2 Test-bed Setup

Even if cooperative techniques are not limited to multicast services, those kind of service types illustrate the benefit of cooperation in the best way. In case all users are interested in the same service, for instance, to download a movie, the news, or a sport event to be displayed on their mobile phone, the mobile devices should team up and cooperate among each other. Conventionally they would set up a cellular link, for instance a GPRS or a UMTS connection, to

F.H.P. Fitzek and F. Reichert (eds.), Mobile Phone Programming and its Application to Wireless Networking, 325–339.

download the data stream from the web server offering the service. Considering instead the situation where the devices are in coverage for a short-range connection, they could think about cooperating with each other. By setting up a short-range link they can exchange partial or total information regarding the service. Very challenging is the case where each user is in charge of downloading only a portion of the complete data stream and exchange it with the other users over the short-range connection. At the end, each device gets the complete data, but uses the cellular connection for a shorter time and a smaller data amount. As we will describe throughout this chapter, the devices will save energy of their mobile battery and will experience a higher virtual data rate.

Having in mind the commercial goal, a simplified scenario has been built. A two-devices scenario has been considered which foresees a web server offering an mp4 movie file for downloading. It is assumed that the two mobile devices interested in downloading the file from the server are in coverage for a short-range connection. The server is capable of supporting cooperating and noncooperating terminals. The initial mp4 file is split in two shares of exactly the same size and installed on the server hosting also the complete movie file. When the two devices cooperate with each other, each of them will download over the cellular link only a half of the video file. The downloaded data will simultaneously be exchanged over the short-range connection with the peer device. Consequently the two devices receive the two halves of the video needed to build the complete file to be displayed. In the described scenario, we loaded on the server an mp4 file of 2.546 Kbytes and two files of 1.273 Kbytes representing, respectively, the first and the second half of the complete file. In order to set up the introduced scenario, a multimode mobile device is needed, equipped with a cellular and a short-range connection capability. Moreover the device is required to be capable of working simultaneously on the two network links. Suitable candidates for a short-range communication system are for example Bluetooth and WLAN (Wireless Local Area Network), while GPRS and UMTS are both possible for the cellular link. A good choice for a test-bed implementation, is to use a GPRS connection for the cellular link and Bluetooth for the short-range connection. Bluetooth has been chosen as short-range communication system for the following simple reasons:

- Bluetooth gives the opportunity to extend the cooperative short-range connection to more than two devices (this is important for future extensions of the work).
- Not many mobile devices are equipped with WLAN access.
- Many mobile devices are nowadays equipped with Bluetooth technology.

The phone used in the test-bed is a Nokia N70, which belongs to the Series 60 models. The phone's characteristics of interest for the test-bed setup are the following:

- Bluetooth connectivity.
- GPRS class B, multi-slot class 10.

Series 60 phones support both Java and Symbian implemented applications. The complexity of the application development and the need to efficiently debug time after time the code, require to set up an emulated system on computer. Nokia provides a very useful software development kit. This includes an emulator of an Series 60 Nokia phone, a plug-in for connecting the emulator to internet through an Ethernet connection, and a useful tool to set up a Bluetooth connection between two emulated devices. Once the application works correctly on the emulated system it can be installed on the mobile phones. These can then be used in the reference scenario to analyze the real micro cooperation benefits.

16.3 Description of the Source Code

The application has been developed for Series 60 Nokia phones using Symbian C++ programming language as it offers the required flexibility needed for this project such as full access to Bluetooth and networking functions. As the application is part of the DVD provided with this book, the reader may test the application beforehand. The application provides four different views, which are selected by pressing the left or right arrow button on the mobile device. The first view shows ongoing network activities. During the download process data rates of the involved networks are displayed, as given in Figure 16.1(a), divided into four values: *Bluetooth In*, *Bluetooth Out*, *GPRS In*, and *Total In*. The latter one is the sum of the incoming cellular and short-range communication. Once the transaction is completed, mean data rate and overall time used are displayed in this first view. The second view displays information about Bluetooth operations, whereas the third view is dedicated to the operations on the GPRS connection. The last of the four views lists the downloaded files, which can be displayed once the transaction is completed. The coming part of this section goes more into detail of the source code, showing the implementation issues concerning Bluetooth, GPRS, and the other important parts of the application. During the implementation of a Bluetooth connection the following points have to be taken into account:

Bluetooth sockets: used to discover other devices and to read and write data over a Bluetooth connection. The APIs allow the socket to work as a client connected to a remote server, or as the server of the socket connection. Once a connection is made the device can send and receive on the socket until disconnection.

Bluetooth Service Discovery Database: is a local database for all local services. Through this a Bluetooth device can discover whether a service is available or not. Together with the Service Discovery Agent it implements the SDP (Service Discovery Protocol).

Bluetooth Service Discovery Agent: enables Bluetooth services and attributes discovery available on a remote device.

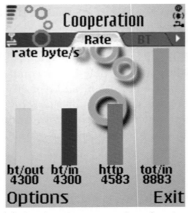

(a) Application screenshot showing instantaneous data rates on short-range and cellular link and virtual data rate (total instantaneous data rate incoming at the device).

(b) Application screenshot showing the final mp4 file displayed on the mobile device (Lecture room for *Mobile Phone Programming* course Spring 2006).

Figure 16.1. Screenshots for Cooperative Wireless Networking.

Bluetooth Security Manager: enables security-requirements setting for incoming connections requiring a Bluetooth service. These requirements could be *authentication*, *authorization*, and *encryption* and are set only for incoming connections.

UUID (Universally Unique Identifier): used by the SDP to identify a service univocally.

The Bluetooth APIs offer two different libraries to set up a connection. The first one needs user interaction in selecting the device wished to be connected to. The user is asked to select a device among those found during the device discovery procedure and listed in a dialog box. This first procedure exploits the *RNotifier* class. The second option is to let the application select the device automatically. By means of the *RHostResolver* Symbian class, it is possible to filter the devices which advertise the desired service and a connection to one of them is automatically started. This second possibility takes usually longer because all the found devices are asked about their services and attributes before filtering can take place. A Bluetooth module for Symbian, based on the *RNotifier* class and developed by Aalborg University has been integrated in the test-bed application. A standard version of Bluetooth implementation has been used, simplifying further future implementations involving the Bluetooth interface. In the reference scenario it is assumed that the two mobile phones, before starting to download any data from the server, already decided whether to cooperate or not. In case they are going to cooperate, they know which share of data to download from the server and which share to receive from the other device (related to the master or slave role in the Bluetooth link). Depending on the choice made by the mobile user to be master or slave of the Bluetooth connection, different function calls are made. A *HandleCommandL* function processes, by means of a *switch*, the commands that the user selects from the option menu. The user can choose to start a Bluetooth piconet as a master, or to advertise himself as a possible slave in the piconet as given in Listing 16.1.

Listing 16.1. The *HandleCommandL* Function Processes User's Choice to be Master or Slave in the Bluetooth Piconet

```
1  void  CCooperationAppUi :: HandleCommandL (TInt  aCommand)
       {
       switch  ( aCommand )
         {
         case  EBTPointToPointStartReceiver :
6          iGlobalVariables ->iServerBt ->StartL () ;
           break ;
         case  EBTPointToPointConnect :
           iGlobalVariables ->iClientBt ->ConnectL () ;
           break ;
11         . . .
         }
       }
```

When an *EBTPointToPointStartReceiver* call is made, the device starts as a slave of the Bluetooth connection. Differently an *EBTPointToPointConnect* call, causes the device to start as a master of the Bluetooth piconet. The integrated Bluetooth module offered the opportunity to configure some parameters of the possible connection. In Listing 16.2 it is shown how it is possible to set up a Logical Link Control and Adaptation Protocol (L2CAP) Bluetooth link and to configure it with a 7170 bytes Maximum Transmission Unit (MTU).

Listing 16.2. These Code Lines Show the Setup and the Configuration of a L2CAP Bluetooth Link

```
   void  CMessageServer :: StartL ()
2  {
   TBTConfiguration  conf(171, _L ("L2CAP") );
   conf.iReceiveMTU =7170;
   iLog.LogL ( _L (" Server_Starting") );
   if (iModule  ==  0)
7    {
     iModule  = CBTModule :: NewL (this ) ;
     TRAPD( result ,  iModule->
         StartL (CBTModule :: EServer , conf) ) ;
     if  (result  != KErrNone)
12     {
       iLog.LogL ( _L (" Error_starting _bt_server _%d") ,
           result );
       }
     iState  = Estarted ;
17   }
   }
```

Once the connection is set up the two mobile devices, after a short *handshake*, can start exchanging data over the link. The receiving and sending operations on the Bluetooth sockets do not block the application because master and slave are both implemented using *active objects*. The aforementioned operations are processed by the *active scheduler* which is informed about the active object's state. A correct implementation of the different states simplifies all events processing. A device is enabled to a send operation only when the outgoing buffer contains enough data. The *Send-Stuff()* function is called, which takes care of the operations on the buffer and of checking whether some additional information has to be sent. For instance the other device could be informed whether the sent data is the last portion of the specific file. Anyhow the main operation of the *Send-Stuff()* function is to build a Bluetooth packet as given in Listing 16.3. The user is informed whether the packet is sent successfully or a sending error occurred. When no error occurs the sent data is removed from the outgoing buffer.

Listing 16.3. These Code Lines Show how a Bluetooth Packet is Built and the Correspondent Errors are Processed

```
   CBTPacket* packet = CBTPacket::NewL(iMessage->Des());
2  TInt result = iModule->WriteData(packet);
   if(result != KErrNone)
      {
      CAknInformationNote* note = new
         (ELeave) CAknInformationNote;
7     TBuf<25>  errMsg;
      errMsg.Format(_L("WriteData_Error_%d"), result);
      note->ExecuteLD(errMsg);
      }
   else
12    {
      iTotalSent=0;
      iGlobalVar.iBtDataSent+=iByteRemaining;
      HBufC8* temp = HBufC8::
         NewL(iGlobalVar.iBufOutSize-iByteRemaining);
17    temp->Des().Copy(iGlobalVar.iBufOut->Des().
         Right(iGlobalVar.iBufOutSize-iByteRemaining));
      delete iGlobalVar.iBufOut;
      iglobalvar.iBufOut = NULL;
      iGlobalVar.iBufOut = HBufC8::
22       NewL(temp->Des().Length());
      iGlobalVar.iBufOut->Des().Copy(temp->Des());
      delete temp;
      temp = NULL;
      iGlobalVar.iBufOutSize-=iByteRemaining;
27    HBufC* textResource = StringLoader::
         LoadLC( R_BTPO_SENT_MESSAGE );
      iLog.LogL( *textResource );
      CleanupStack::PopAndDestroy( textResource );
      }
```

In case the *SendStuff()* function is called and no data is found ready to be sent, a timer is started. This in order to check again after one second whether the outgoing buffer is filled with new data.

Listing 16.4. Timer Call Back Value is Set to One Second as Shown in this Code

```
iTimeWaster.Cancel();
TInt delay = 1 * 1000000;
iTimeWaster.After(iStatus, delay);
iState=EWaitingActivate;
SetActive();
```

As given in Listing 16.4, when the timer calls the scheduler the active object's state is set to *EWaitingActivate*. This causes a new call of the *SendStuff()* function. Every time the *SendStuff()* function is called, a maximum amount of data equal to the MTU size can be sent. Before sending new data again, the device expects to receive the same amount of data from the cooperating peer device, unless the whole file has already been received. Two main reasons justify this implementation choice. The first is to avoid "cheaters" in cooperation, in the sense that none of the cooperating devices is allowed to receive the whole data amount without corresponding his share of data. The second reason is that a Bluetooth connection offers a much higher data rate compared to the GPRS connection. Instead of sending many small packets over the Bluetooth connection, the device sends a relatively big amount of data all at once saving time and energy. When the active scheduler receives a call from the Bluetooth module about data received on the Bluetooth socket, the *DataReceived* function is invoked. This function is able to distinguish whether the received data is a handshaking message or a data message. In the later case, the data is written on a temporary file on the file system. This is used at the end of the cooperation process to build the final mp4 file. When the complete data is exchanged, the two devices inform each other about the successful cooperative transaction. The active objects' state is set to *EDisconnected* and, as given in Listing 16.5 the Bluetooth connection is closed consequently.

Listing 16.5. These Code Lines Show How Bluetooth Module is Stopped When Active Objects' State is Set on *EDisconnected*

```
void CMessageServer::RunL()
{
switch ( iState )
    {
    case EDisconnected:
        {
        if(iModule != NULL)
            {
            iModule->Stop();
            delete iModule;
            iModule = 0;
            }
        }
        break;
    ...
    }
}
```

As mentioned previously, the GPRS network has been chosen for the cellular link. In the reference scenario, it is assumed that the mobile device takes the decision whether or not to cooperate with another device before downloading any data over GPRS. This makes it quite simple to distinguish the different files to be downloaded in the different cases. To start the HTTP (Hypertext Transfer Protocol) transaction as given in Listing 16.6, the user can select the *EClientGet* option from the menu.

Listing 16.6. These Code Lines Show How the GET Request is Handled by the *HandleCommandL* Function in Order to Start Up the HTTP Transaction

```
     void CCooperationAppUi::HandleCommandL(TInt aCommand)
     {
3    switch ( aCommand )
       {
       case EClientGet:
         {
         iClientHttp->CancelTransaction();
8        TBuf<256> uri;
         if(iGlobalVariables->iServerBt->iConnectedToDevice)
           {
           _LIT(HttpAddress,
               "http://kom.aau.dk/~leomil/FilePart2");
13         uri.Append(HttpAddress);
           }
         else if(iGlobalVariables->iClientBt->
               iConnectedToDevice)
               {
18             _LIT(HttpAddress,
                   "http://kom.aau.dk/~leomil/FilePart1");
               uri.Append(HttpAddress);
               }
             else
23             {
               _LIT(HttpAddress,
                   "http://kom.aau.dk/~leomil/File.mp4");
               uri.Append(HttpAddress);
               }
28       CAknTextQueryDialog* dlg = new (ELeave)
         CAknTextQueryDialog(uri, CAknQueryDialog::ENoTone);
         if (! dlg->ExecuteLD(R\_DIALOG\_URI\_QUERY))
             break;
         iHttpView->Reset();
33       TBuf8<256> uri8;
         uri.LowerCase();
         if(uri.Find(KHttpPrefix) == KErrNotFound &&
             uri.Find(KHttpsPrefix) == KErrNotFound)
           {
38         uri8.Append(KHttpPrefix8);
           uri8.Append(uri);
           }
         else
           {
43         uri8.Copy(uri);
           }
         iClientHttp->IssueHTTPGetL(uri8);
         }
         break;
48     ...
       }
     }
```

The main function the *EClientGet* call invokes is *IssueHTTPGetL(uri8)*, which receives the only URI parameter. This value is different for the master and for the slave of the Bluetooth connection and for the noncooperating device.

The *IssueHTTPGetL* function parses the received URI and submits the method string for the HTTP GET transaction to the framework as given in Listing 16.7.

Listing 16.7. The *IssueHTTPGetLReceived* Function Parses the Received URI and Submits it to the Framework

```
void CEngine::IssueHTTPGetL(const TDesC8& aUri)
{
TUriParser8 uri;
uri.Parse(aUri);
5   RStringF method = iSession.StringPool().
        StringF(HTTP::EGET,RHTTPSession::GetTable());
iTransaction = iSession.
        OpenTransactionL(uri, *this, method);
RHTTPHeaders hdr = iTransaction.
10      Request().GetHeaderCollection();
SetHeaderL(hdr, HTTP::EUserAgent, KUserAgent);
SetHeaderL(hdr, HTTP::EAccept, KAccept);
iTransaction.SubmitL();
}
```

The *IssueHTTPGetL* function works like an active object. The framework calls the *MHFRunL* and *MHFRunError* functions, to inform them about transaction events. As given in Listing 16.8, the *MHFRunL* function by means of a *switch* command manages all the different events received from the framework. When the body of the data is received, it is written on a file on the file system. This is the final complete mp4 movie in the noncooperative case. In the cooperative behaving terminals instead, it is a temporary file needed to build the complete mp4 movie at the end of the transaction. When the device is cooperating, the received data is also added to the outgoing buffer. This buffer contains the data to be sent to the peer device over the Bluetooth link. The aforementioned *MHFRunError* function instead, manages all error messages received from the framework.

Listing 16.8. HTTP Transaction Events Handling Code. For Clearness Sake Not Too Many Details are Reported.

```
1   void CEngine::MHFRunL(RHTTPTransaction aTransaction,
        const THTTPEvent& aEvent)
{
switch (aEvent.iStatus)
    {
6       case THTTPEvent::EGotResponseHeaders:
            {...}
            break;
        case THTTPEvent::EGotResponseBodyData:
            {...}
11          break;
        case THTTPEvent::EResponseComplete:
            {...}
            break;
        case THTTPEvent::ESucceeded:
16          {...}
            break;
        case THTTPEvent::EFailed:
            {...}
            break;
21      }
}
```

The two main parts of the application have now been described. To exploit the application as a test-bed for micro cooperation gain measurements, however, some other code parts need to be implemented. For instance, this test-bed aims to point out the virtual data rate gain obtainable through cooperation. Therefore two timers have been implemented in the code to measure the elapsed time and consequently the data rate values. These values refer to send and receive operations on the GPRS and Bluetooth links in the cooperative scenario or only on the GPRS link in the noncooperative scenario. In the later case, a timer is started upon reception of the first data chunk over the GPRS link. This is afterwards stopped upon reception of the last data chunk. In the cooperative situation another similar behaving timer has been implemented for the Bluetooth link. For evaluating the virtual data rate, the highest value of the two timers is taken as transaction time for the whole file in the cooperative scenario. The *e32svr.h* library offers the opportunity to use timers for profiling purposes. As given in Listing 16.9, when the first data is received the timer is started.

Listing 16.9. These Code Lines Show How the Timer is Started for Profiling Purposes

```
RDebug :: ProfileReset (0 ,21);
RDebug :: ProfileStart (20);
```

Afterwards when the transaction is completed the timer is stopped and the information is used for transfer time and virtual data rate purposes as given in Listing 16.10.

Listing 16.10. These Code Lines Show How the Timer is Stopped and the Information is Saved for Further Use

```
RDebug :: ProfileEnd (20);
TProfile   result;
RDebug :: ProfileResult(&result ,20 ,1);
TReal  time  =  result . iTime /1000000.00;
```

Among others, the obtained values are used in the Bluetooth and GPRS dedicated views. All the gathered information regarding the received data, the elapsed time, and the state of the framework is reported on the display to inform the user. As previously mentioned, the first view is used for instantaneous and final results. The instantaneous incoming and outgoing Bluetooth rate, the instantaneous incoming GPRS rate and the total at device incoming rate (virtual data rate) are updated on a time interval of 10 seconds. Every 10 seconds, the instantaneous bytes per second value is calculated and a drawing with colored bars and exact numerical values is displayed as shown in Figure 16.1(a).

When all the data is received, the view displays the total results in terms of bytes received, transfer time, and data rate. These values refer to the GPRS link, the Bluetooth link, and the total cooperative architecture. For this purpose the *CPeriodic* class has been used. This class implements a timer which offers the possibility to call back one or more functions after a given amount of time (see Listing 16.11).

Listing 16.11. A *CPeriodic* Timer is Implemented as Shown in the Code Lines

```
1  TTimeIntervalMicroSeconds32 tic (10 * 1000000);
   iPeriodic = CPeriodic::NewL(0);
   iPeriodic->Start(tic, tic, TCallBack(Tick, this));
```

A ten seconds interval has been considered, after which the *Tick* function is invoked. This function is used to cast and call the nonstatic *DoTick()* function. All the information regarding elapsed time and the received data, is used here to correctly set the variables for the *Draw* function which is called with the *DrawNow()* command. Every time the *Draw* function is invoked, the transaction is checked for completeness and to verify whether all the data has been received on the active links. In this case, the final information is reported. Otherwise the previously calculated instantaneous rate and the data amount values on the different links are used to draw graphical information for the user. Another important operation is to build the final file. In the cooperative case, this needs to be done immediately after the complete reception of the two halves of the mp4 file. For this purpose, the *CComposeCoopFile* class has been implemented. This class uses an *active object* to read from the temporary files and to copy the content in the correct order into a new file. The Symbian *cooperative multitasking* has been a very useful feature in this class for allowing to manage the asynchronous behavior of the file system read and write operations. When *active objects* are not used, big files especially cause the application to be blocked for a long time because it is busy with in the asynchronous file system tasks. When the complete file is built, as given Listing 16.12 new choices are available in the option menu, in order to allow the user to display the obtained complete movie file. The movie is displayed in the fourth view as shown in Figure 16.1(b).

Listing 16.12. Menu Options for Video Displaying and Handling are Displayed When the Video File is Complete

```
   void CCooperationAppUi::HandleCommandL(TInt aCommand)
2  {
   switch ( aCommand )
       {
       case EVideoCmdAppPlay:
           {
7          DoPlayL();
           break;
           }
       case EVideoCmdAppStop:
           {
12         DoStopL();
           break;
           }
       case EVideoCmdAppPause:
           {
17         DoPauseL();
           break;
```

```
          }
      case  EVideoCmdAppContinue :
          {
22        DoResumeL ( ) ;
          break ;
          }
      . . .
      }
27  }
```

16.4 Measurement Results

Even though the main focus lies on the programming aspect, we shortly report on the measurement results of the previously described application. The measurements focus on the *energy savings*, the *virtual data rate*, and the *transmission time*. Cooperative as well as noncooperative download scenarios were carried out and compared with each other. The measurements were carried out at the same time of the day and at the same location to guarantee similar channel qualities. In terms of *virtual data rate*, in the noncooperative behavior an average download rate of 42 Kbps can be achieved, which is reasonable for a stand-alone GPRS connection. In the cooperative case the *virtual data rate* is 82.8 Kbps. Comparing the two values, it is easily deductible that the virtual data rate is almost doubled in the cooperative scenario. This is not surprising, since we accumulate the cellular links. The transmission time, defined as the average time to download the complete mp4 file from the server, compared to the cooperative case shows a reduction of about 50%. This is an expected result as the accumulated data rate has doubled. The downloading time for the cooperative mobile devices is now only a half compared to the stand-alone approach. The most interesting result is the *energy saving*. It could be shown that in the cooperative behavior a value of 44.33% energy saving is achieved. This means that even though we use an additional air interface, in this case the Bluetooth interface, the mobile devices save energy. The reason is that the mobile devices use the GPRS air interface for a shorter time and therefore use less energy on the cellular link. As we have already shown this behavior analytically in [1], the test-bed proved the analytical results. Two more cooperation benefits can be listed here. From the customers' point of view, the service cost is reduced by half and from the network operator's point of view, the bandwidth usage is reduced by half as well.

16.5 The BitTorrent Experiment

After the first proof of concept for wireless cooperative networking, it seems to be an exciting challenge to migrate the cooperative concept into existing applications. One possible candidate for this is the SymTorrent application introduced in Chapter 15. SymTorrent is based on the BitTorrent technology

Figure 16.2. Such a small download as the BitTorrent client itself already consists of approximately 100 (6.9 MB / 64 kB) pieces.

applied to the mobile world. BitTorrent employs a balanced downloading pattern referred to as *swarming*. In a nutshell, swarming means that downloadable content is split into equally sized pieces, and pieces are treated (shared and downloaded) independently from each other. This concept enables serving partially available content, thus reducing the load on participants and entities. In the case of the BitTorrent protocol, typical piece size is some tens or hundreds of kilobytes (usually a power of two such as 64 kB or 512 kB – see Figure 16.2 for an example), and the downloadable content is described by a *torrent* file, referencing one or more actual files.

Torrent downloading involves numerous independent stream connections to other BitTorrent clients and the exchange of torrent pieces with them. The introduction of another, cooperative short-range, connection is therefore a logical and easy extension. The implemented local connection behaves as follows:

1. The local connection uses Bluetooth serial port profile to exchange data locally. The number of cooperative entities is currently restricted to two mobile devices.
2. The original CSTTorrent class (representing a torrent) has been extended with a bitfield indicating availability of pieces over the local connection.
3. The new local-availability bitfield gets modified in three ways:
   a) when the local connection is set up: participants exchange data on already downloaded pieces of their torrent set
   b) when downloading a piece (of any torrent) is completed: an availability update is sent to the local peer
   c) when the local connection is closed: data on local-availability is removed
4. CSTTorrent::GetPieceToDownloadL is updated
   a) pieces available on the local peer are not eligible for downloading over the Internet

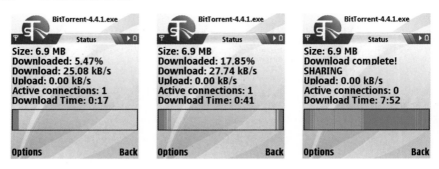

Figure 16.3. A cooperative downloading session.

   b) the connector device starts choosing pieces from the last piece and downward, the acceptor device starts choosing pieces from the first piece and upward

5. missing pieces available on the local peer are downloaded one by one over the local connection (when all available pieces are synchronized, incoming availability-updates initiate checking for new pieces)

That is more or less everything that needs to be done to realize the first mass market cooperative mobile application. The pattern itself can be applied to other file transfer protocols as well. Screenshots of the first application are shown for a two-device demo in Figure 16.3.

The first demonstration application actually works in a very robust manner. Users can initiate and drop cooperation at any time. This shows the robustness of the application and users can decide when and how long to cooperate. This means that also cooperating over a shorter time will lead to the benefits of larger throughput and less energy. Nevertheless, the largest gain is achieved when cooperation is established and maintained over the full download period. The benefits in cooperating is listed below:

- Reduced download time for all cooperating mobile devices. Measurements with the actual SymTorrent application has shown that two mobile devices have the possibility to get the download time halved, but due to some implementation issues, the current version takes a little bit more time. Note, SymTorrent is still under development, but these issues will be solved soon.
- Reduced download traffic (even if downloading lasts longer than expected, it is true that approximately half of a torrent is downloaded from the Internet).
- multicolored progress bar (in addition to the default green bar, yellow color represents pieces downloaded from the local peer).
- energy conservation can be also expected (as in the Cooperation application), however no measurements have yet been made regarding this.

Figure 16.4. "Local" menu items.

From the user's perspective, all he/she has to do is to use the new menu items *Connect Local* and *Accept Local* to enable cooperation within the short-range cluster as given in Figure 16.4. The manual preparation for cooperation will be replaced in our future work by automatic cooperation establishment.

16.6 Conclusion

In this chapter we have proven the feasibility of implementing cooperative wireless networking on mobile devices. With a first demonstrator we could show that cooperation among two mobile devices results in an energy saving of nearly 44% using GPRS and Bluetooth connections. Energy savings in that order may be considered as significant and could be a first step out of the energy trap. Despite the energy savings, also the service quality in terms of download time has been improved. In order to show that we are not limited to the very first demonstrator, the cooperative idea of local short-range exchange has been added to the SymTorrent application. This crossover has been started and more applications in this area are expected. Especially the user friendliness to support cooperative services will be part of our future research activities.

References

1. F.H.P. Fitzek and M.D. Katz, editors. *Cooperation in Wireless Networks: Principles and Applications – Real Egoistic Behavior is to Cooperate!* ISBN 1-4020-4710-X. Springer, April 2006.

## Part VI

Cross-Layer Communication

17

Cross-Layer Protocol Design for Wireless Communication

Thomas Arildsen and Frank H.P. Fitzek

Aalborg University {tha|ff}@es.aau.dk

Summary. This chapter provides an introduction to cross-layer protocol design. Cross-layer design is needed in order to enhance data transmission across especially wireless network connections. This enhancement is achievable by being able to optimize certain parameters jointly across multiple layers instead of considering each layer separately. Cross-layer protocol design is a principle that provides the possibility of enhancing the architecture and operation of the layered protocol stack by allowing communication between nonadjacent layers as an extension of what is possible in the OSI model. We briefly summarize the layered network protocol stack concept of the ISO OSI model and explain cross-layer communication in this context. We provide an overview of different categories of exchanging data across layers of the network protocol stack and point out advantages and drawbacks of different approaches. We provide examples from the literature of such approaches. Similarly, we review a number of recent examples from research literature of how cross-layer protocol design is being put to use. The examples provide an overview of which layers are involved in different cross-layer optimization approaches. The examples also provide an overview of which technological areas are currently considered in – and which optimization techniques employ – cross-layer design.

17.1 Introduction

Recent years' development within wireless communication means that wireless data transmission has found its way into a wide range of mobile or portable consumer electronics. The use of mobile phones is gradually shifting its focus from voice-only applications to multimedia streaming, Internet browsing, file downloading, etc. Network access from computers is now commonly wireless, e.g., 802.11, UMTS, WiMAXX.

The various network protocols generally use the layered architecture known from the ISO open systems interconnection (OSI) model [11]. The OSI model is a descriptive network scheme designed to facilitate interoperation between various network technologies. The model describes how data is transferred from one application on a computer system to another application on another

F.H.P. Fitzek and F. Reichert (eds.), Mobile Phone Programming and its Application to Wireless Networking, 343–362.

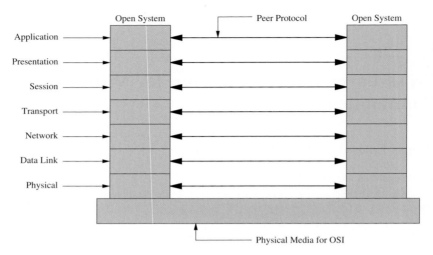

Figure 17.1. The layers of the OSI model.

computer system through a network medium. The OSI model deals with this data transfer by dividing the involved tasks into seven separate layers as shown in Figure 17.1.

The division into separate layers reduces the complexity of the protocol stack, because it makes it possible to restrict one's attention to one specific layer at a time while other layers can be abstracted from that layer by standardized interfaces. Each layer communicates with the corresponding layer at the other end of the network through the layers below it. Each layer uses functionality of the layer below it and provides its services to the layer above it. Thus network layers in the OSI model only communicate directly with their immediate neighbors. The seven layers are the following:

Application Layer provides network functionality to applications outside the protocol stack (e.g., user applications such as office applications, and games). The application layer establishes the availability of intended communication partners, synchronies, and establishes agreement on procedures for error recovery and control of data integrity.

Presentation Layer ensures that information sent by the application layer of one system is readable by the application layer of another system. If necessary, the presentation layer translates between multiple data formats by using a common format. It provides encryption and compression of data.

Session Layer defines how to start, control, and end conversations (called sessions) between applications. It also synchronies dialogue between two hosts' presentation layers and manages their data exchange. The session layer provides efficient data transfer.

Transport Layer regulates information flow to ensure end-to-end connectivity between host applications reliably and accurately. The transport layer segments data from the sending host's system and reassembles the data into a data stream on the receiving host's system.

The boundary between the session layer and the transport layer can be thought of as the boundary between application protocols and data-flow protocols. The application, presentation, and session layers are concerned with application issues, whereas the lower four layers are concerned with data transport issues.

Network Layer provides end-to-end delivery of packets across the network. It defines logical addressing so that any endpoint can be identified, how routing works and how routes are learned so that the packets can be delivered. The network layer also defines how to fragment a packet into smaller packets to accommodate different media.

Data Link Layer provides access to the networking medium and physical transmission across the medium and this enables the data to locate its intended destination on a network. The data link layer provides reliable transmission of data across a physical link by using the Medium Access Control (MAC) addresses. The data link layer uses the MAC address to define a hardware or data link address in order for multiple stations to share the same medium and still uniquely identify each other. The data link layer is concerned with network topology, network access, error notification, ordered delivery of frames, and flow control.

Physical Layer deals with the physical characteristics of the transmission medium. It defines the electrical, mechanical, procedural, and functional specifications for activating, maintaining, and deactivating the physical link between end systems.

Layer-N entities in a network communicate with layer-N peer entities at the other end(s) of the link/network using a specific layer-N protocol. They communicate through the facilities offered to them by layer $N - 1$. Thus information transmitted from a network node propagates from the top to the bottom of its protocol stack before finally being transmitted across the physical medium. Each layer encapsulates the data from its higher adjacent layer and adds its own control information. Each layer may partition the data from the higher layer into several parts or collect multiple parts into one before handing it to the its lower adjacent layer. The corresponding inverse operations take place in reverse order at the receiving network node(s). An example of data propagating through the protocol of a transmitting network node is shown in Figure 17.2.

The layered architecture facilitates development of protocol components by abstraction such that a particular layer only has to concern itself with the interfaces to the layer above it and to the layer below it. This modularity facilitates development of protocols, because individual layers can be tested separately. It also allows developers of network protocols to contribute with a particular layer instead of an entire protocol encompassing functionality corresponding to all layers. The layered approach facilitates standardization as well, again through division into smaller separate parts.

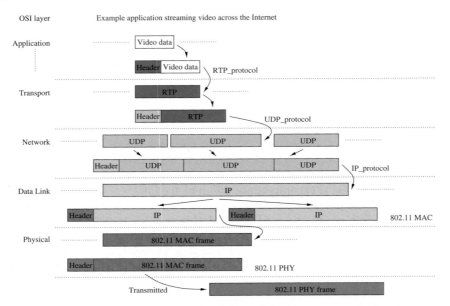

Figure 17.2. Data transmission in the OSI model with an example showing a scenario in which an application transmits video across a wireless Internet connection.

The layered architecture allows different applications to transport data in several different ways across the same networks and it allows networks to be connected across a heterogeneous variety of physical media – e.g., the Internet.

The layered network protocol architecture has enabled the evolution of networks into what they are today. However, many protocols were mainly developed for cabled networks and work very well with these. The fast growth of wireless network technologies means that wireless network access is becoming more and more common, especially on the last hop to the users. The wireless medium has very different properties in terms of, for example, channel fading and interference. Due to these differences, the existing layered protocols have several drawbacks and there is room for improvement in the way protocols at different layers cooperate.

17.2 Cross-Layer Protocol Design

17.2.1 The Principle

The principle of cross-layer protocol design is to extend the architecture of the layered protocol stack to allow communication between nonneighboring layers in addition to what is already possible as described in Section 17.1 and to allow reading and controlling parameters of one layer from other layers. By this definition, cross-layer protocol design is a very broad subject. Therefore, we will in the following describe what we need it for, how to organize the actual communication across the protocol stack, and how cross-layer design is currently being employed in the research literature.

At each layer in the protocol stack, various choices exist for transmission of the data units passing through them. This could be different speech codecs for a VoIP application, different transport protocols such as TCP or UDP at the transport layer, or for example, at the physical layer, different modulation types for transmission on a wireless channel. Such possibilities at the different layers constitute a flexibility in the overall protocol stack. Cross-layer design plays an important role in relation to this flexibility. Cross-layer design is a means by which one can get specific knowledge across the protocol stack between separate layers and thus exploit the flexibility through making the protocol stack adaptive. Cross-layer design makes it possible to control features of different protocol layers jointly across the network protocol stack.

Protocols at different layers of the protocol stack may implement similar functionality. This could introduce redundant operations in the protocol stack, e.g., forward error-correction (FEC) being applied at two different layers. Instead, such redundant operations could, for example, be undertaken at only one of the layers by appropriate adjustments or jointly controlled for overall optimal operation of the functionality in question at the involved layers. Similarly, different protocols could implement complementary operations at different layers which could be jointly optimized in order to exploit the resulting collective operations more efficiently. An example could be to coordinate the operation of automatic repeat request (ARQ) at one layer with a FEC mechanism at another layer such that in case a very strong FEC mechanism is currently applied at one layer, it might not be necessary to actually use ARQ at the other layer because FEC will be able to compensate most of the data losses – or vice versa.

Considering how to approach cross-layer design, it should be a "nondestructive" approach. The layered protocol architecture is fundamentally a good idea and should be kept intact – it still provides us with for example the flexibility of being able to adapt the protocol stack to different radio technologies by replacing some of the lower layers or to use different applications at the highest layers. Cross-layer design should be used to enhance the existing architecture by exploiting opportunities of jointly optimizing parameters/behavior of the protocol layers that would not otherwise be possible.

Cross-layer design enables performance gains in a multitude of different aspects of wireless networking. However, it also potentially goes against some important benefits of the original layered architecture. A given cross-layer design implementation may interweave two or more layers such that these layers cannot be separated. It may introduce dependencies in the protocol such that one protocol cannot simply replace another one at a certain layer. This means that it potentially degrades the modularity and freedom to 'compose' the protocol stack. Another thing to consider is that some cross-layer optimizations may drastically increase the computational demands of running the protocol stack due to the additional degrees of freedom that are introduced in optimizing the protocol stack's performance.

Cross-layer designs need to be considered carefully. In any design implemented, one should consider its possible impact on the existing protocol stack. This is, for example, pointed out in a somewhat pessimistic way in [15].

In the following sections we first describe different types of communication across network protocol layers in Section 17.2.2 and then review some of the existing ideas for cross-layer optimizations in Section 17.2.3.

17.2.2 Communication Across Protocol Layers

One can choose to see cross-layer design from different viewpoints. One viewpoint that needs to be considered for practical deployment of cross-layer optimizations is how to integrate the cross-layer design into the protocol stack – how is the communication between different layers going to be realized?

We consider the following two categories of cross-layer communication: Using existing protocols, with the subcategories implicit and explicit, or using dedicated signaling mechanisms, with the subcategories of signaling pipes, direct communication, and external cross-layer management. These categories are explained in the following:

1. **Communication using existing protocols** We divide this category into implicit/inherent and explicit communication using existing protocols:
   a) **Implicit/Inherent** Cross-layer communication here simply consists of lower layers reading and/or perhaps altering data within the data units passing through them, belonging to higher layers. In this type of cross-layer communication, no additional data is transferred between layers compared to what is already the case within the traditional OSI architecture. The only difference lies in the fact that lower layers snoop into higher layers' packets to gain knowledge of what is taking place there and to exploit this. We call this kind of communication inherent or implicit because the data exploited across layers is already available.
   One clear advantage to this type of communication is that nothing needs to be changed except at the layer(s) that will be snooping into the data of other layers. The obvious drawback is that the data that can be exploited is limited to whatever data is transmitted by the higher layers. Another drawback is that data can only be exploited at lower layers relative to the layer to which the particular data belongs. This is due to the fact that the data needs to pass through the layer interested in that data, so for example, the transport layer cannot gain knowledge of the link layer using this type of communication. However, lower layers can possibly alter data passing through them to higher layers and thus manipulate the operation of these higher layers. The higher layers in question will generally be unaware of this interaction.

b) **Explicit** Cross-layer communication in this category is an extension of the above-mentioned category. The communication here is explicit since involved layers are aware of it and actively participating in the communication. However, the data is transferred by means of already existing protocols through the interfaces between layers defined by the OSI model.

Using this method, additional information can be transferred between layers compared to what is transferred between layers in the traditional approach. This is accomplished because the interacting layers are aware of the communication and can be designed to exchange specific additional information between them. However, there are still limitations as to where data can be sent to and from due to the use of existing protocols.

In addition to the above-mentioned drawbacks and benefits of using existing protocol formats, one could also mention the following, common to both implicit and explicit cross-layer communication:

- The information from higher layers may be difficult to access by lower layers due to, for example, segmentation, blocking, and concatenation of the higher layers' data units as illustrated in Figure 17.2. However, information flowing downward is easier to convey than information flowing upward since a particular layer's data units will pass through lower layers.
- Information from lower layers to higher layers is difficult to convey by these mechanisms since, for example, inserting extra information in packets to higher layers passing through a particular lower layer requires alteration of the packets including check sums and other content-dependent parameters, and the concerned packets may be segmented, concatenated, etc., across multiple of the lower layer's data units, as illustrated in Figure 17.2. In addition, lower layers would have to rely on data units, which they do not initiate, passing through them to upper layers.
- There is also a clear advantage of exchanging information through the mechanisms of existing protocols. The information exchange is transparent to intermediate layers so it does not require changing the protocols of intermediate layers in order get information through them.

2. **Explicit communication using dedicated mechanisms** Introducing dedicated mechanisms for communication across the layers of the protocol stack gives the ultimate freedom since the communication is not bound by restrictions of existing protocols that were not designed for this kind of mechanisms. Within these dedicated mechanisms we consider three different kinds:

a) The first mechanism arranges exchange of data across the layers as a 'signaling pipe' traversing all layers through which any layer can send data to or receive data from any other layer. This mechanism is illustrated in Figure 17.3. This provides a general framework under which

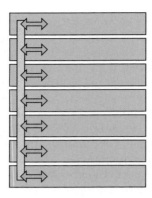

Figure 17.3. Dedicated signaling mechanism: a signaling pipe across the entire protocol stack.

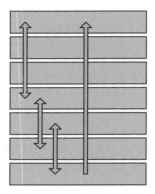

Figure 17.4. Dedicated signaling mechanism: specific interfaces between interacting layers.

cross-layer optimizations can be introduced at any layer, taking advantage of the available communication mechanism. However, all layers of the protocol stack must be modified to implement the signaling mechanism. Using this type of cross-layer communication also implies that any cross-layer optimization must be implemented inside one or more of the layers.

b) The second mechanism is a more specialized approach where signaling interfaces are introduced specifically for direct communication between interacting layers. It is illustrated in Figure 17.4. This allows introduction of dedicated signaling mechanisms only where needed. This could, for example, also provide benefits related to timing considerations where the above-mentioned signaling pipe as a general framework would be too slow a mechanism for very time-critical signaling.

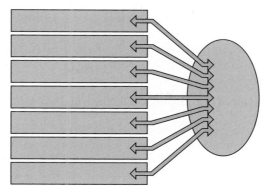

Figure 17.5. Dedicated signaling mechanism: communication through an external cross-layer manager.

Cross-layer optimizations also still need to be implemented inside one or more of the interacting layers.

c) The third mechanism is a general mechanism as the first one, but instead of data being exchanged through the layers of the protocol stack, an external management mechanism is introduced as illustrated in Figure 17.5. The layers of the protocol stack communicate individually with the external cross-layer manager. In this way, each layer only needs to consider cross-layer communication with one other party. All cross-layer optimization operations can be collected in the cross-layer manager which can take data from several layers into account and control these jointly. However, this type of mechanism may be considerably more computationally demanding and difficult to implement since this mechanism, among the three mentioned here, will require the most extra functionality compared to a traditional OSI architecture.

The approach of letting layers communicate directly with each other or with some management middle-ware external to the protocol layers allows the designer to tailor the framework exactly to the cross-layer optimizations in question. Especially if the optimization is of a kind that incorporates information from and/or control of more than two layers, the centralized management approach may be beneficial. However, it should be noted that there are also considerable drawbacks to this approach. It requires much more customization of the involved protocols in order to enable the communication mechanisms since the traditional protocols' mechanisms cannot be used in this context. Furthermore, the added communication mechanisms between the layers will be likely to increase the computational requirements of the system which may have very limited resources, especially in the case of mobile phones and similar devices.

In the following sections, existing work is covered that deals with cross-layer design concerning how to realize communication across the different layers. Since cross-layer design is a relatively young research area, there is not yet any standardization in the area concerning actual frameworks for realizing cross-layer designs. There have, however, been some attempts at defining such frameworks.

Explicit Communication Based on Existing Protocols

An example of the approach described as method 1a is [17]. In this example, a lower layer (MAC) simply reads priority information set in a header from a higher layer (application – the video coder's network adaptation layer). This is data already set at the application layer and all that is needed in order to introduce the cross-layer data exchange is to modify the link (MAC) layer to be able to read it.

Several examples exist of the approach described in the preceding section as method 1b. An early example of such an approach is Wu et. al.'s article from 1999 where they suggest Interlayer Signaling Pipes (ISP) which consists of exchanging the relevant information in the IPv6 header field Wireless Extension Header (WEH) [39]. This approach can actually be said to be an implementation of the category 2a approach by category 1b mechanisms. In [32] from 2001, Sudame and Badrinath suggest a framework, complete with API for Linux, that exchanges cross-layer information based on ICMP packets. Examples are provided with information from the driver (layers below network layer) exploited in application and transport layers. Mérigeault and Lamy's article from 2003 suggests a concept for signaling information between application and network access[1] layers [23]. They do so by piping the extra information through RTP packets. It is accomplished by means of network adaptation layers (NAL) in the form a Source Adaptation Layer (SAL) between application and transport layers and a Channel Adaptation Layer (CAL) between network and network access layers. The SAL and CAL handle the cross-layer information in the extra packets in relation to their protocol layers and filter out RTCP packets generated by the transport layer in response to the "artificial" extra packets. A similar concept is pursued much more extensively by the PHOENIX IST project that attempts to establish a framework for adapting video transmission in a cross-layer manner across an IPv6 network [22].

Dedicated Signaling Mechanisms

Some suggestions for cross-layer signaling frameworks approach the problem by introducing mechanisms to let any layers communicate directly with each other through new interfaces that circumvent the standardized interfaces of the OSI protocol stack (category 2b). One such example is [37]. The example

---

[1] Network access layer is used as a common name for DLC/MAC + PHY in this paper.

is not very detailed, but rather describes the general idea and points out its suggested framework, CLASS, as light-weight with high flexibility and fast signaling but with high complexity.

The bulk of this family of solutions however suggests some sort of management entity external to all the layers with which all of the involved layers will communicate individually and which will handle the considered cross-layer optimizations in a centralized manner, such as described in category 2c. One of the earliest examples is by Inouye et. al. from 1997 who suggest a framework for mobile computers for adaptation to the availability of different network interfaces based on several parameters that define this availability [10]. Recent examples of cross-layer frameworks include [3] which arranges cross-layer optimizations in so-called coordination planes and handles them through a cross-layer manager. There is the quite extensive ECLAIR framework [27] which introduces so-called tuning layers for access to the individual layers' parameters and a collection of "protocol optimizers" for handling the individual optimizations. The approach in [38] has a so-called local view for storing parameters from individual layers for access from all layers as well as a corresponding global view that serves the same purpose in a networkwide scope for cross-layer optimizations that span multiple network devices.

17.2.3 State of the Art

In this section, we go through some of the most recent work in the area to characterize which parts of the protocol stack are involved, what sort of information is exploited, and what purposes the optimizations serve. The overview is organized by most significant involved layer from the bottom up.

Considering what to optimize at the different layers, Raisinghani and Iyer's 2004 article provides a brief overview of what topics could be considered [26]. They go through the layers approximately corresponding to the OSI model and suggest parameters that could be interesting at the particular layer and what sort of interaction could be relevant with lower and higher layers, respectively. This is a starting point to getting an overview of cross-layer protocol design.

This section provides a more extensive overview of existing research in the field and attempts to bring it up to date.

Physical/Link Layers

While quite a lot of research work includes the physical layer in cross-layer optimizations, there is little focus mainly on this layer. Much of the work involving the physical layer revolves around the data link layer, especially MAC. For example, Alonso and Agustí focus on MAC-PHY interaction in [1, 2]. In the former they present a method using distributed queueing random access protocol (DQRAP) in a CDMA system to increase overall throughput and minimize power consumption and thus intercell interference. Cross-layer information consists of channel state information (CSI) and target spreading factor to reach a desired bit-error rate transmitted from receiver PHY to transmitter MAC[2]. In the latter they consider scheduling/prioritization in a CDMA base station MAC based on CSI from mobile nodes. They also cover somewhat the same as in the former paper and extend the principle to WLAN. In [1] they do not clearly specify how the information is exchanged between layers and in [2] it is stated that the layers interchange explicit control information by means of specific control channels – what we define as dedicated signaling mechanisms. In [14], which Alonso among others coauthored, they explore prioritization between real-time (VoIP) and nonreal-time traffic in a distributed-queueing MAC. Channel quality obtained from the physical layer is used to dynamically prioritize users with good conditions in order to improve overall throughput. It is merely stated that the MAC (link) layer acquires information through a "cross-layer dialogue" with the physical layer which could be implemented either through existing protocols or through a dedicated signaling mechanism.

Toufik and Knopp in [33] consider subcarrier allocation and antenna selection in a MIMO OFDMA system taking CSI from PHY into account in the allocation at MAC level. Song and Li's work is related in the sense that they consider subcarrier and power allocation in an OFDMA system. In [29] they optimize the system by maximizing utility functions based on transmission rates of the users, taking CSI into account. In [30] they extend it to base the utility functions on waiting time in addition to rates. They take it further in [31] where they mix the different utility maximizations from the two former papers in their simulations of a scenario consisting of users with different delay requirements, categorized as voice, streaming, and best-effort traffic users. Delay is very important to voice users, somewhat less important to streaming users, and has much lower importance to best-effort users. Filin et. al.'s paper [5] also considers resource allocation in an OFDMA system, but they focus on minimizing the time-frequency resource usage through a more heuristic approach. Their work includes MAC through scheduling of data flow segments as well as the physical layer through the control of transmission power, modulation, and coding as well as the use of SNR estimates. Compared to the previously mentioned approaches in this section, which all use information from the physical layer to change parameters in the link layer, [5] is interesting because they also control parameters in the physical layer based on information from the link layer. The above-mentioned five papers concentrate particularly on the optimizations, using the necessary parameters without considering how to exchange these between layers.

Zhang et. al.'s paper from 2006, [40], integrates physical and link layers in a model of the impact of physical layer Adaptive Modulation and Coding (AMC) and MIMO on link layer QoS provisioning. They model the physical layer service process as a finite-state Markov chain. As such, neither does this paper cover how to exchange the information between layers.

---

[2] In this case, information is exchanged not only across layers within the protocol stack of one mobile device, but between separate mobile devices as well.

Fawal et. al. in [4] also look at MAC-PHY interaction. Their work concentrates on impulse radio UWB (IR-UWB) where they describe a selection of aspects and how to address them with a PHY-aware MAC implementation. Their objective is mainly to achieve energy efficiency. Energy efficiency is also an issue in [24] where Mišić et. al. attempt to increase the lifetime of sensor networks by managing sensor activity taking MAC-layer congestion and interference as well as noise from PHY into consideration. None of these two articles seem to be concerned with how to exchange the actual information between layers.

Physical/Application Layers

[21] provides an example of an optimization involving the physical layer and the application layer. It is an extensive framework developed under the PHOENIX project [25]. It incorporates the mentioned layers at both transmitter and receiver side in optimizing transmission of a video stream. Thus the information exchange takes place both across layers in the individual mobile devices as well as across the network between individual devices. Parameters involved in the optimization are source significance information (SSI), source a priori information, source a posteriori information, and a video quality measure, all from the application layer, network state information from intermediate layers, as well as CSI and decision reliability information (DRI) from the physical layer. This is a very extensive and advanced piece of work. As mentioned earlier, an important aspect of the PHOENIX project, e.g., in [21, 28], is the framework for exchanging cross-layer information through already present protocols, e.g., via IPv6 extension headers and ICMPv6. Thus, the information exchange considered here is explicit, utilizing the existing protocols' capabilities.

Link/Network Layers

Tseng et al. address optimization of hand-off in a Mobile IP/802.11 scenario [34]. Instead of reading/adjusting parameters across layers, as most other examples do, the authors here signal events across layers. As such, this work is interesting because it is not a matter of reading parameters from other layers. Rather, it is a matter of the timing of specific signals introduced between the layers which constitutes the difference compared to the traditional OSI approach. Hand-off-related events are signaled between the link layer of the 802.11 protocol and the Mobile IP network layer in order to speed up the hand-off process. It is not stated how the information is exchanged between layers, but due to the timing-critical nature of the signaling it is likely that one cannot rely on the existing protocols and that dedicated signaling mechanisms between the two layers must be employed. The IEEE 802.21 working group is, among other things, dealing specifically with this type of signaling [9].

Link/Transport Layers

The examples presented for this combination of protocol layers are focused on taking action in the transport layer based on information from the link layer, i.e., an upward information flow.

One example of link/transport layer interaction is Wu et al.'s [39]. They investigate a cross-layer scheme involving TCP at the transport layer and Radio Link Protocol (RLP) at the link layer in the IS-707[3] standard. What they do is to enable TCP to exploit radio link parameters such as data rate, radio link round-trip delay, and fading conditions. This is done in order to mitigate unfortunate effects of TCP in a wireless environment. Wu et. al. exchange cross-layer information through existing protocol mechanisms.

Sudame and Badrinath's [32] is another example of link/transport layer interaction although the details of which lower layer is involved in the setup is not defined very precisely in OSI-terms. They test their proposed framework using modifications to UDP and TCP to decrease packet losses and delays, respectively, in case of WLAN handovers. The handovers are signaled from the network interface driver by means of ICMP messages. This approach is thus related to the previously mentioned [34] by their attempts to mitigate unwanted effects of handovers in the network. This is also an example of exchanging information through mechanisms of existing protocols.

Link/Application Layers

An example of a link-/application-cross-layer design is Liebl et al.'s [19] from 2004 where they consider VoIP over a packet-switched connection in a GERAN scenario with AMR speech coding. They attempt to optimize the utilization ratio of radio link control (RLC) data segments. This approach is centered around the application layer exploiting information (current segment size and utilization) from the link layer (upward information flow). The AMR speech coder's mode is selected to fit the resulting speech frames into a number of RLC segments minimizing the bit-stuffing of these. It is not mentioned how the relevant information should be exchanged between the involved layers.

Ksentini et al.'s recent work published in 2006 combines layered video coding at the application layer with classification of the video layers[4] into different QoS classes at the MAC layer [17], so this can be considered a downward information flow. Here, an implicit cross-layer information exchange through existing protocols is employed since they use a priority field set by the NAL of the video coder to classify the importance of the video packets at the MAC layer. One more example is found in Liu et al.'s 2006 paper [20]. They employ information on the importance of video frames from the application layer at

---

[3] A data service standard for a wideband spread spectrum system.

[4] This is not layers in the network stack sense; this is a way of partitioning the data produced by the video coder.

the link layer. As such, it can be compared to [17], but it is slightly different in the sense that Liu et al. consider differentially encoded video frames in a video group-of-pictures (GOP) to have decreasing importance according to their sequence number within the GOP. Based on this importance they use an ARQ scheme in which retransmission attempts are spent on video frames in decreasing order of importance in order to minimize the impact of packet losses on video quality. The link layer needs to know which position in a GOP each video frame has, but the paper does not specify how this information is obtained from the application layer. Haratcherev et al. present a somewhat more advanced scheme in which both video coding rate and link layer transmission rate are adapted based on CSI and link throughput provided by a so-called channel state predictor and a medium sharing predictor [6]. This employs mainly an upward information flow. The paper does not specify exactly how the information exchange between application and link layer was achieved.

Jenkac et al. explore FEC in [12]. They consider several error correction suggestions at different layers, but their most significant contribution in this article is their so-called permeable layer receiver in which error correction is performed at the application layer. The transport layer is also involved to some degree since the FEC produces extra parity data packaged in extra RTP packets associated with data RTP packets. The most important cross-layer aspect is in the receiver where erroneously received link layer segments (destroyed due to one or more radio bursts with errors) do not cause the entire corresponding RTP packet to be discarded. Instead, the data is passed up from the link layer with erasure symbols in the missing segments and the FEC mechanism attempts to reconstruct the erroneous data. It will require the link layer to be aware of the mechanism, but the cross-layer data exchange as such is implicit since it merely requires passing packets with erasure symbols up the stack in case of errors in the same way as with correctly received packets. This also illustrates the advantages of using the existing protocols since this is completely transparent to intermediate layers and utilizes mechanisms that are already available.

Physical/Link/Application Layers

All so far mentioned examples of specific cross-layer optimizations have only incorporated two layers. There are also recent examples of research involving three layers.

Jiang et al. explore a concept in which video is classified into different priority classes [13], similarly to other examples mentioned in the previous section. Here, transmission is adapted at link layer according to both the video data importance and the users' CSI on a cellwide basis in a CDMA cellular system. This is accomplished by dynamic-weight generalized processor sharing (DWGPS). It is not directly addressed how to exchange the required information between the layers. Likewise, Khan et al. focus on transmission of video in [16]. They do so by jointly adjusting video source rate at application layer, time slot allocation at data link layer, and modulation scheme at physical layer through observation of abstracted layer parameters. The cross-layer information exchange in this work is explicit and accomplished by letting the involved layers communicate with a common cross-layer optimizer through so-called layer abstractions. These serve the purpose of reducing the amount of parameters involved in the optimization in order to reduce the computational demands of the optimization operation.

Kwon et al. consider subcarrier allocation in an OFDMA system – 802.16e (WiBro) [18]. They do so according to users' CSI obtained from PHY and they furthermore, control adaptive modulation and coding and prioritize users' data streams according to QoS demands from the application layer. In addition, they suggest a protocol for implementing uplink channel sounding and downlink Channel Quality Information (CQI) feedback. Cross-layer information is exchanged between the involved layers through dedicated communication mechanisms between their "control information controller" at the physical layer and "MAC-c controller" at the link layer which also gathers application parameters. Hui et. al. also consider OFDMA resource allocation in [8] where they maximize average total system throughput as a subcarrier and power allocation problem under delay and queueing constraints, taking CSI and application layer source data rate into account. However, this paper does not describe how to exchange the required information between layers.

Schaar and Shankar explore different aspects in a wireless LAN setting in [35] where they incorporate the three layers and look at selecting the optimal modulation scheme, optimizing power consumption, and optimizing fairness among users, respectively. In [36], Schaar and Tekalp address the complexity of optimizing many layers' parameters jointly and suggest an off-line learning-based approach to the joint minimization of distortion, delay, rate, and complexity. The suggested scheme is simulated in a scenario where selection of application layer priorities and MAC layer retransmission limits is based on a training-based classification of low-level video content features, channel condition, and maximum available bit-rate. None of these two mentioned papers directly address the method for exchanging the required parameters between the involved layers.

Network/Transport/Application Layers

So far the only work we have seen centered around user input is the very recent [7] by Hasswa et al. Their proposal is to manage primarily vertical handovers[5] based on user-defined preferences from the application layer. These preferences concern cost of service, security, power consumption, network conditions, and network performance. The proposal handles handovers at the

---

[5] Handovers between networks with different wireless technologies, e.g., WLAN ↔ UMTS.

Table 17.1. Mechanisms for cross-layer information exchange in covered literature.

Existing	Dedicated	Not considered
[12, 17, 21, 28, 32, 39]	[2, 7, 14, 16, 18, 34]	[1, 4–6, 8, 13, 19, 20, 24, 29–31, 33, 35, 36, 40]

transport layer by using SCTP with mobility extensions. This deals with handovers through IP multihoming, i.e., the mobile device has multiple IP addresses registered at which it can be reached (one in each network), performing handover by redefining which address is its primary one. The network layer is involved in evaluation of some of the parameters for which the user has defined preferences. It is also involved in detection of the current availability of the different networks and acquisition of addresses within these. The whole framework consists of an application layer part called the "Handover Manager" and a transport layer part called the "Connection Manager". The former deals with the user preferences and controls handover decisions and the latter interworks with the SCTP protocol modified for this purpose which in turn interacts with the network layer. The two parts of the management framework apparently communicate with each other through an explicit mechanism developed for this purpose.

Overview

The preceding sections, of course, merely provide a taste of currently ongoing work. There are naturally hundreds of additional references that could not all be covered here. For example, the presented literature does not cover very much of the contributions within sensor and ad -hoc networks.

In relation to Section 17.2.2, Table 17.1 gives an overview of the described works on cross-layer design in terms of the type of cross-layer information exchange the respective articles employ. As the table shows, a large portion of the existing and very recent work on cross-layer protocol design does not consider how the actual exchange of information and setting of parameters are to be accomplished. This is mainly due to the fact that many of them have simulated their suggested concepts and not yet considered this aspect. This however underlines that there is a need for research in these practical mechanisms. However, while it may be debatable how to actually implement cross-layer design, researchers should definitely think cross-layer-wise.

17.3 Acknowledgments

This work was partially financed by the Danish government on behalf of the FTP activities within the X3MP project.

References

1. L. Alonso and R. Agustí. Automatic rate adaptation and energy-saving mechanisms based on cross-layer information for packet-switched data networks. *Communications Magazine, IEEE*, 42(3):S15–S20, 2004.
2. Luis Alonso and Ramon Agustí. Optimization of wireless communication systems using cross-layer information. *Signal Processing*, 86(8):1755–1772, August 2006.
3. G. Carneiro, J. Ruela, and M. Ricardo. Cross-layer design in 4g wireless terminals. *Wireless Communications, IEEE [see also IEEE Personal Communications]*, 11(2):7–13, 2004.
4. A. El Fawal, J.Y. Le Boudec, R. Merz, B. Radunović, J. Widmer, and G.M. Maggio. Trade-off analysis of phy-aware mac in low-rate low-power uwb networks. *Communications Magazine, IEEE*, 43(12):147–155, 2005.
5. Stanislav A. Filin, Sergey N. Moiseev, Mikhail S. Kondakov, Alexandre V. Garmonov, Do Hyon Yim, Jaeho Lee, Sunny Chang, and Yun Sang Park. Qos-guaranteed cross-layer transmission algorithms with adaptive frequency sub-channels allocation in the ieee 802.16 ofdma system. In *Communications, 2006. ICC '06. IEEE International Conference on*, 2006.
6. L. Haratcherev, J. Taal, K. Langendoen, R. Lagendijk, and H. Sips. Optimized video streaming over 802.11 by cross-layer signaling. *Communications Magazine, IEEE*, 44(1):115–121, 2006.
7. Ahmed Hasswa, Nidal Nasser, and Hossam Hassanein. Tramcar: A context-aware cross-layer architecture for next generation heterogeneous wireless networks. *Communications, 2006. ICC '06. IEEE International Conference on*, 2006.
8. David Shui Wing Hui, Vincent Kin Nang Lau, and Wong Hing Lam. Cross layer designs for ofdma wireless systems with heterogeneous delay requirements. *Communications, 2006. ICC '06. IEEE International Conference on*, 2006.
9. IEEE. Media independent handover services. `http://www.ieee802.org/21/`.
10. Jon Inouye, Jim Binkley, and Jonathan Walpole. Dynamic network reconfiguration support for mobile computers. *MobiCom '97: Proceedings of the 3rd annual ACM/IEEE international conference on Mobile computing and networking*, pages 13–22, New York, NY, USA, 1997. ACM Press.
11. ISO/IEC. Information technology – open systems interconnection – basic reference model: The basic model. Technical Report 7498-1, ISO/IEC, November 1994.
12. Hrvoje Jenkač, Thomas Stockhammer, and Wen Xu. Cross-layer assisted reliability design for wireless multimedia broadcast. *Signal Processing*, 86(8):1933–1949, August 2006.
13. Hai Jiang, Weihua Zhuang, and Xuemin Shen. Cross-layer design for resource allocation in 3g wireless networks and beyond. *Communications Magazine, IEEE*, 43(12):120–126, 2005.
14. E. Kartsakli, A. Cateura, C. Verikoukis, and L. Alonso. A cross-layer scheduling algorithm for dqca-based wlan systems with heterogeneous voice-data traffic. *Local and Metropolitan Area Networks, 2005. LANMAN 2005. The 14th IEEE Workshop on*, pages 1–6, 2005.
15. V. Kawadia and P.R. Kumar. A cautionary perspective on cross-layer design. *Wireless Communications, IEEE [see also IEEE Personal Communications]*, 12(1):3–11, 2005.

16. S. Khan, Y. Peng, E. Steinbach, M. Sgroi, and W. Kellerer. Application-driven cross-layer optimization for video streaming over wireless networks. *Communications Magazine, IEEE*, 44(1):122–130, 2006.

17. A. Ksentini, M. Naimi, and A. Gueroui. Toward an improvement of h.264 video transmission over ieee 802.11e through a cross-layer architecture. *Communications Magazine, IEEE*, 44(1):107–114, 2006.

18. Taesoo Kwon, Howon Lee, Sik Choi, Juyeop Kim, Dong-Ho Cho, Sunghyun Cho, Sangboh Yun, Won-Hyoung Park, and Kiho Kim. Design and implementation of a simulator based on a cross-layer protocol between mac and phy layers in a wibro compatible.ieee 802.16e ofdma system. *Communications Magazine, IEEE*, 43(12):136–146, 2005.

19. G. Liebl, M. Kaindl, and W. Xu. Enhanced packet-based transmission of multirate signals over geran. *Personal, Indoor and Mobile Radio Communications, 2004. PIMRC 2004. 15th IEEE International Symposium on*, volume 3, pages 1812–1816 Vol.3, 2004.

20. Hao Liu, Wenjun Zhang, Songyu Yu, and Xiaokang Yang. Channel-aware frame dropping for cellular video streaming. *International Conference on Acoustics, Speech, and Signal Processing, 2006. ICASSP. 2006 IEEE*, pages V–409–V–412, 2006.

21. Maria G. Martini, Matteo Mazzotti, Marco Chiani, Gianmarco Panza, Catherine Lamy-Bergot, Jyrki Huusko, Gabor Jeney, Gabor Feher, and Soon X. Ng. Controlling joint optimization of wireless video transmission: the phoenix basic demonstration platform. *14th IST Mobile & Wireless Communication Summit*, 2005.

22. Maria G. Martini, Matteo Mazzotti, Catherine Lamy-Bergot, Peter Amon, Gianmarco Panza, Jyrki Huusko, Johannes Peltola, Gábor Jeney, Gábor Feher, and Soon Xin Ng. A demonstration platform for network aware joint optimization of wireless video transmission. *IST Mobile Summit 2006*, June 2006.

23. S. Mérigeault and C. Lamy. Concepts for exchanging extra information between protocol layers transparently for the standard protocol stack. *Telecommunications, 2003. ICT 2003. 10th International Conference on*, 2: 981–985, 2003.

24. J. Mišić, S. Shafi, and V.B. Mišić. Cross-layer activity management in an 802-15.4 sensor network. *Communications Magazine, IEEE*, 44(1):131–136, 2006.

25. PHOENIX. Jointly optimising multimedia transmissions in ip based wireless networks. http://www.ist-phoenix.org/index.html.

26. V.T. Raisinghani and S. Iyer. Cross-layer design optimizations in wireless protocol stacks. *Computer Communications*, 27(8):720–724, May 2004.

27. V.T. Raisinghani and S. Iyer. Cross-layer feedback architecture for mobile device protocol stacks. *Communications Magazine, IEEE*, 44(1):85–92, 2006.

28. E. Roddolo, G. Panza, C. Lamy-Bergot, Peter Amon, Maria Martini, Gabor Jeney, Lajos Hanzo, and Jyrki Huusko. Joint source and channel (de)coding in 4g networks: the phoenix project. *Wireless personal multimedia communications, The Seventh International Symposium on*, 2004.

29. G. Song and Y. Li. Adaptive resource allocation based on utility optimization in ofdm. *Global Telecommunications Conference, 2003. GLOBECOM '03. IEEE*, 2: 586–590, 2003.

30. G. Song, Y. Li, Jr. Cimini, L.J., and H. Zheng. Joint channel-aware and queue-aware data scheduling in multiple shared wireless channels. *Wireless Communications and Networking Conference, 2004. WCNC. 2004 IEEE*, 3: 1939–1944, 2004.

362   T. Arildsen and F.H.P. Fitzek

31. Guocong Song and Ye Li. Utility-based resource allocation and scheduling in ofdm-based wireless broadband networks. *Communications Magazine, IEEE*, 43(12):127–134, 2005.
32. P. Sudame and B.R. Badrinath. On providing support for protocol adaptation in mobile wireless networks. *Mobile Networks and Applications*, 6(1):43–55, January 2001.
33. I. Toufik and R. Knopp. Channel allocation algorithms for multi-carrier multiple-antenna systems. *Signal Processing*, 86(8):1864–1878, August 2006.
34. Chien-Chao Tseng, Li-Hsing Yen, Hung-Hsin Chang, and Kai-Cheng Hsu. Topology-aided cross-layer fast handoff designs for ieee 802.11/mobile ip environments. *Communications Magazine, IEEE*, 43(12):156–163, 2005.
35. M. van der Schaar and Sai Shankar N. Cross-layer wireless multimedia transmission: challenges, principles, and new paradigms. *Wireless Communications, IEEE [see also IEEE Personal Communications]*, 12(4):50–58, 2005.
36. M. van der Schaar and M. Tekalp. Integrated multi-objective cross-layer optimization for wireless multimedia transmission. *Circuits and Systems, 2005. ISCAS 2005. IEEE International Symposium on*, 4: 3543–3546, 2005.
37. Qi Wang and M.A. Abu-Rgheff. Cross-layer signalling for next-generation wireless systems. *Wireless Communications and Networking, 2003. WCNC 2003. 2003 IEEE*, 2: 1084–1089, 2003.
38. R. Winter, J.H. Schiller, N. Nikaein, and C. Bonnet. Crosstalk: cross-layer decision support based on global knowledge. *Communications Magazine, IEEE*, 44(1):93–99, 2006.
39. Gang Wu, Yong Bai, Jie Lai, and A. Ogielski. Interactions between tcp and rlp in wireless internet. *Global Telecommunications Conference, 1999. GLOBECOM '99*, 1B:, 661–666, 1999.
40. Xi Zhang, Jia Tang, Hsiao-Hwa Chen, Song Ci, and M. Guizani. Cross-layer-based modeling for quality of service guarantees in mobile wireless networks. *Communications Magazine, IEEE*, 44(1):100–106, 2006.

18

Cross-Layer Example for Multimedia Services over Bluetooth

Morten V. Pedersen, Gian Paolo Perrucci, Thomas Arildsen, Tatiana Kozlova Madsen, and Frank H.P. Fitzek

Aalborg University {mvpe|gpp|tha|tatiana|ff}@es.aau.dk

Summary. This chapter gives a first idea how cross-layer design could be carried on mobile devices improving the overall performance of the communication system. As an example we refer to multimedia services that are conveyed over the Bluetooth technology. To increase the bandwidth efficiency on the wireless medium IP header compression schemes are used.

18.1 Introduction

The wireless delivery of multimedia services is one of the goals of the next generation's mobile communication systems. Users are expecting that the same services will be available in wireless networks as in wired ones. However, IP-based multimedia applications, including audio- and video-streaming and gaming, require more bandwidth than traditional voice services in circuit-switched networks. In order to use the limited resource, bandwidth, in the most efficient way, both IP packet header and packet payload should be compressed. A number of IP header compression schemes have been proposed over the last 15 years, e.g., Van Jacobsen HC (RFC 1114), Compressed RTP (RFC 2508), Enhanced Compressed RTP (RFC 3545), Robust Header Compression (RFC 3095). These techniques allow compression of a 40-byte IPv4 header down to some bytes. All these techniques apply compression of an IP packet header, but without taking into account the requirements of the link layer. A number of wireless technologies available today use fixed packet sizes at the link level, e.g., GPRS, UMTS, and Bluetooth.

Due to the specifics of packetizing of data implemented in the above-mentioned technologies, IP header compression schemes do not perform efficiently compared to technologies where the payload size of link layer packeta can freely vary, such as in WLAN IEEE 802.11 standard. Speaking about the header compression for packets with variable length, a useful parameter to consider is *bandwidth savings*. In this case, bandwidth savings are calculated as the ratio of the amount of data transmitted using a header compression

F.H.P. Fitzek and F. Reichert (eds.), Mobile Phone Programming and its Application to Wireless Networking, 363–371.

scheme to the amount of data required to send the same information without applying header compression. Then even a small compression of a header will lead to some savings. This is not the case when packets of fixed sizes are used on the link layer. In this situation, the communication is often based on some kind of time division access method: the channel is slotted and a packet transmission can start only at the beginning of the slot. Even if the packet length is smaller than maximally allowed by this packet type, no other packets can be sent in this slot and the bandwidth is wasted. Therefore, if applying header compression, the size of the packet cannot be reduced significantly (such that it will fit in another packet type), no bandwidth savings can be achieved. This problem of header compression inefficiency for fixed packet types has been addressed in [1].

18.2 Adaptive Header Compression for Bluetooth

We illustrate the novel header compression approach by considering Bluetooth technology as an example. The proposed scheme is called *Adaptive Header Compression for Bluetooth*. Bluetooth is a low-cost low-power wireless technology that provides connectivity among devices placed within short range. Bluetooth uses controlled scheduled transmissions. The Bluetooth system provides duplex transmission based on slotted time-division duplex (TDD) where the duration of each slot is 625 μs. Asynchronous links support payloads with or without a 2/3-rate FEC coding scheme. In addition, single-slot, three-slot, and five-slot packets are available. The Bluetooth packet types are summarized in Table 18.1. From the table one can notice the big difference between user payloads of 1-, 3-, and 5-slot packets. The payload length is variable and depends on the available user data. However, the maximum length is specified for each packet type. The whole slot (or 3 or 5 slots) is reserved for a transmission, even if the packet length is smaller than maximally available.

Following the proposed approach, we decide to compress IP headers or not, depending on the payload size. Table 18.2 shows when to apply header

Table 18.1. Bluetooth packet types.

Type	Payload Header (bytes)	User payload (bytes)	FEC
DM1	1	0-17	YES
DH1	1	0-27	NO
DM3	2	0-121	YES
DH3	2	0-183	NO
DM5	2	0-224	YES
DH5	2	0-339	NO

Table 18.2. Proposed Adaptive Scheme: switching HC mechanism ON/ OFF depending on the payload size (compressed or uncompressed header has been subtracted already).

Payload [Byte]	Header Compression	Packet Type	FEC	Pkt Type Number
0-13	ON	DM1	YES	1
14-23	ON	DH1	NO	2
24-81	OFF	DM3	YES	3
82-117	ON	DM3	YES	4
118-143	OFF	DH3	NO	5
144-179	ON	DH3	NO	6
180-184	OFF	DM5	YES	7
185-299	OFF	DH5	NO	8

compression (assuming that uncompressed header is 40 bytes, compressed header is 4 bytes). One should note that if payload is more than 339 bytes, segmentation is required.

In [1] we have considered two different scenarios: (i) broadcasting of data by a Bluetooth access point and (ii) data exchange between two Bluetooth devices. The robustness of the scheme wass shown for both scenarios. For the first scenario the proposed scheme performs better than the header compression scheme introduced in RFC2508 in terms of packet loss, and equally well for the bandwidth saving. The uncompressed transmission does not achieve any bandwidth savings, but lower packet losses than all compression schemes. For the second scenario, the proposed scheme shows better performance than the uncompressed and the RFC2508 with large bandwidth savings, that was only outperformed by RFC2508 for very good channel conditions.

18.3 Groundwork for Cross-Layer Design

Implementing the idea presented above, we can use any multimedia application and we need to implement only a header compression scheme. Fortunately, the mobile device platform gives us the flexibility to implement the header compression scheme. The more difficult part, and this is what we are presenting here, is to set the packet type of the link layer technology according to the higher layer needs. Fortunately mobile devices, in this case based on the Symbian OS, offer the flexibility to set the Bluetooth packet type. In the following, the basic understanding how to implement the packet selection is presented. Furthermore we present a small tool that is able to set the packet types and transmit a certain number of packets with the chosen packet type. This tool will be used to present measurements to show the impact of the packet type on the delivery time.

18.3.1 Symbian Basics

To access and use the Bluetooth stack on the mobile devices we need to utilize the Symbian OS Bluetooth v2 Sockets API introduced in Symbian OS v8.1b, but we experienced that not all mobile phones were responsive to baseband changes. So the problem seems to be solved for Symbian OS v9. The CBluetoothSocket class exposes the API needed to create asynchronous links. The API provides a BSD-like socket interface to the RFCOMM and L2CAP protocol layers, and baseband modification capabilities. In this section we will describe how to set up and configure a Bluetooth connection.

Setting up the Connection

Before we can start transferring data between two devices we need to set up the connection. One device will act as server, listening for incoming connections, and the other will act as client, initiating the connection. The following shows the steps involved in configuring the server socket using L2CAP and a fixed CID (channel identifier):

- Open connection to the Symbian OS socket server.
- Create a new Bluetooth socket object, using the L2CAP protocol and selected CID.
- Select the desired security settings.
- Allow other devices to connect to our device by:
  - Binding our socket to the selected CID.
  - Setting up a listening queue for incoming connections.
  - Accepting new connections.

The following listing shows the source code to set up a server socket:

```
1  // Setup connection
   TProtocolDesc protocol;
   TInt res = SockServ(aPacketTest).FindProtocol(KL2CAPDesC(),
       protocol);
   LOGLEAVE(res, "iSocketServ.FindProtocol");

5  TRAP(res, ListenSocket(aPacketTest) = CBluetoothSocket::NewL( *
       aPacketTest, SockServ(aPacketTest), protocol.iSockType,
       protocol.iProtocol));
   LOGLEAVE(res, "CBluetoothSocket::NewL()");

   // Set the wanted security settings
10 TBTServiceSecurity security;
   security.SetAuthentication(EFalse);
   security.SetAuthorisation(EFalse);
   security.SetEncryption(EFalse);
   security.SetDenied(EFalse);
15
   // Configure the
   TBTSockAddr addr;
   addr.SetPort(KDefaultCID);
   addr.SetSecurity(security);
20
   // Bind the socket
```

```
     res = ListenSocket(aPacketTest)->Bind(addr);
     LOGLEAVE(res, "iLisenerSocket->Bind()");

25   // Setup a listening queue
     res = ListenSocket(aPacketTest)->Listen(1);
     LOGLEAVE(res, "iLisenerSocket->Listen()");

     // Create a blank socket and start accepting connections
30   TRAP(res, ServiceSocket(aPacketTest) = CBluetoothSocket::NewL( *
         aPacketTest, SockServ(aPacketTest) ));
     LOGLEAVE(res, "CBluetoothSocket::NewL()");

     // Start accepting incomming connections
     res = ListenSocket(aPacketTest)->Accept(*ServiceSocket(aPacketTest)
         );
35   LOGLEAVE(res, "iLisenerSocket->Accept()");
```

The accept function call will complete asynchronously. To receive notification of completion, our application must inherit from the **MBluetoothSocket Notifier** class. The **MBluetoothSocketNotifier** class also defines a number of other callback functions that are used to notify our application of other occurring events. The following functions are defined in the **MBluetoothSocket Notifier** interface and need to be implemented by our application:

```
   void HandleConnectCompleteL(TInt);
   void HandleAcceptCompleteL(TInt);
   void HandleShutdownCompleteL(TInt);
   void HandleSendCompleteL(TInt) ;
5  void HandleReceiveCompleteL(TInt);
   void HandleIoctlCompleteL(TInt);
   void HandleActivateBasebandEventNotifierCompleteL(TInt,
       TBTBasebandEventNotification&);
```

Once a client connects to our device, we will receive notification through the **HandleAcceptCompleteL(TInt)** function. While the server is running and listening for incoming connections, we need to setup the client side of the communication. This can be done in the following steps:

- Find the Bluetooth address of our server device.
- Create a new Bluetooth socket object, using the L2CAP protocol and selected CID.
- Connect to the server device.

These steps are shown in the following listing:

```
1   // Using the RNotifier class to find other devices
    RNotifier notifier;
    TInt res = notifier.Connect();
    LOGLEAVE(res, "notifier.Connect()");
5   CleanupClosePushL(notifier);

    TBTDeviceSelectionParamsPckg filter;
    TBTDeviceResponseParamsPckg response;

10  TRequestStatus status;

    // Start the notifier and allow the user to select a device
    notifier.StartNotifierAndGetResponse(status,
        KDeviceSelectionNotifierUid, filter, response);
    User::WaitForRequest(status);
```

```
15  LOGLEAVE(status.Int(), "notifier.StartNotifierAndGetResponse()");

    CleanupStack::PopAndDestroy(1); // RNotifier

    // Setup connection
20  TProtocolDesc protocol;
    res = SockServ(aPacketTest).FindProtocol(KL2CAPDesC(), protocol);
    LOGLEAVE(res, "iSocketServ.FindProtocol");

    // Opening a socket using the protocol parameters
25  TRAP(res, ServiceSocket(aPacketTest) = CBluetoothSocket::NewL( *
        aPacketTest, SockServ(aPacketTest), protocol.iSockType,
        protocol.iProtocol));
    LOGLEAVE(res, "CBluetoothSocket::NewL()");

    // Set the address and CID of the remote device
    TBTSockAddr addr;
30  addr.SetBTAddr(response().BDAddr());
    addr.SetPort(KDefaultCID);

    // Connect to the remote device
    res = ServiceSocket(aPacketTest)->Connect(addr);
35  LOGLEAVE(res, "iServiceSocket->Connect()");
```

When the `CBluetoothSocket::Connect(TBTSockAddr)` call completes, our application will receive notification through the function `MBluetoothSocketNotifier::HandleConnectComplete(TInt)`. The connection is now fully established and we can start sending and receiving data. Also after establishing the connection we can start making changes to the baseband configuration.

Activating Baseband Notifications

To monitor the changes occurring at the baseband of the Bluetooth connection we need to enable baseband notifications using the `CBluetoothSocket::ActivateBasebandNotifications(TUint32)` function which as parameter takes a bitmask defining the events that we are interested in. To allow construction of the bit mask, the available values are defined in the bttypes.h header in two enumerations: `TBTPhysicalLinkStateNotifier`, `TBTPhysicalLinkStateNotifierCombinations`. The enumeration value `ENotifyAnyPhysicalLinkState` combines all possibilities to activate notification for all events. Any changes such as master/slave role switching or packet type changes will now result in a notification event through the `MBluetoothSocketNotifier::HandleActivateBasebandEventNotifierCompleteL(TInt,TBTBasebandEventNotification)` callback.

Changing the Packet Types

Changing the packet type on the link can be done through the `CBluetoothSocket::RequestChangeSupportedPackettypes(TUint16 aPacketTypes)` function, taking the packet types we wish to use as parameter. An enumeration `TBTPackettype` in the sdfs.h header file defines values needed to select a particular packet type. Note, initial tests performed with the packet switcher tool described in the following showed that changing the packet type had no effect on pre-Symbian v9 phones.

18.3.2 Packet Selection Tool

The first version of the Bluetooth packet selection tool seen in Figure 18.1 is currently used to verify whether the packet selection actually works, and to benchmark the performance of the different packet types.

Future versions of the tool will at a later stage also support the adaptive packet selection approach described previously. Currently the tool supports selection of the following parameters: Number of tests to run, number of packets to send, packet type to use, payload size, and reliability of the link. These configurations can be selected through the tool configuration view shown in Figure 18.2. The results from the different test runs are saved in a file for later analysis.

18.3.3 Measurements

To prove the application and to understand the impact of Bluetooth packing switching, we have used the packet selection tool to perform some measurements. The measurement includes two mobile devices, namely the Nokia N71 (3rd edition is a must here). The application is installed on both devices and tests are carried out for three different packet sizes such as 10, 100, and 200 bytes. A fixed number of 10,000 packets with one given packet size are transmitted using one packet type. The packet types used are DH1, DM1,

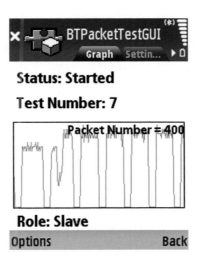

Figure 18.1. The main view of the application showing the current transfer speed and test progress.

Figure 18.2. The configuration view of the application.

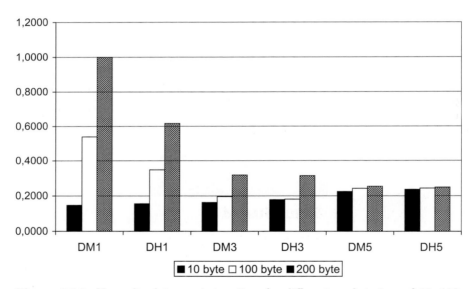

Figure 18.3. Normalized transmission time for different packet sizes of 10, 100, and 200 and Bluetooth packet types.

DH3, DM3, DH5, and DM5. For a better understanding of the packet types, please refer to Chapter 1. The tool allows us to repeat the measurement several times to get better confidence in our measurements. In Figure 18.3 the time needed to transmit the 10,000 packets with sizes of 10, 100, and 200 bytes using the different packet types is shown.

For packet size 10 it can be seen that the delivery time is increasing when we change the packet type from DH1/DM1 to DH5/DM5. The reason is that the type 5 packets allocate five Bluetooth slots, where only one Bluetooth slot would be enough to carry the 10 bytes. We could conclude that the Bluetooth link is not used efficiently in case type 3 and even more if type 5 packets are used. Choosing a larger packet size of 200 bytes results in larger delivery times for the type 1 packets than the type 3, while the delivery time for type 3 is also larger than type 5. The reason is that for type 1 and type 3 the 200 byte packet needs to be segmented into several Bluetooth packets and each packet is acknowledged by the counterpart communication entity. Of course, more segmentation is needed if type 1 packets are used instead of type 3 packets. Type 5 packets do not need any segmentation. For a packet size of 100 bytes, we notice the overlay of the two effects, namely efficiency and segmentation resulting in the shortest delivery time for the type 3 packet. While type 1 packets need to segment the 100 bytes, the type 5 packets suffer again from the inefficient use of the Bluetooth link.

18.4 Conclusion

The main contribution of this chapter is the introduction of cross-layer design on mobile phones. Cross-layer design is based on the flexibility in different protocol layers. While the flexibility in the network layer or higher is always given for the mobile device, here we have shown one possible approach to activating the flexibility even on the lower layers. Cross-layer design has the potential to improve the service quality and to reduce the energy consumption on the mobile platform.

References

1. T.K. Madsen, F.H.P. Fitzek, G.P. Perrucci, T. Arildsen, and S. Nethi. Novel IP Header Compression Technique for Wireless Technologies with Fixed Link Layer Packet Types. *GLOBECOM 2006*, December 2006.

Part VII

Sensor Networks

19

Convergence of Mobile Devices and Wireless Sensor Networks

Frank H.P. Fitzek[1] and Stephan Rein[2]

[1] Aalborg University `ff@es.aau.dk`
[2] Technical University Berlin `stephan.rein@tu-berlin.de`

Summary. Pervasive computing with mobile devices and wireless sensor networks has been identified as a future area of research. Wireless sensors have the possibility to sense the surrounding world, while mobile devices offer a sophisticated user interface together with the well-known cellular services to the symbiotic partnership. The strength of the envisioned convergence is shown throughout the next chapters and here possible architecture forms are classified.

19.1 Introduction

In a world of ubiquitous computing the mobile device is used in many different situations. It is not only a mobile phone anymore, but offers many additional services. These services are mostly services carried out by and used directly on the mobile device. While it is well known that such services add to the complexity of the mobile device, the need for such services is sometimes questionable (camera, blog, etc). Customers request context-aware services that are cognitive to their environment. To gather context-aware information, a centralized approach such as the one offered by cellular networks is often not enough. This and the following chapters advocate enriching the mobile device by services provided by wireless sensors surrounding it. The term *wireless sensor* represents a physical hardware entity that has sensing, computing, and wireless communication capabilities. The convergence of mobile phones with wireless sensors combines the advantages and benefits of both of them. While the wireless sensors enrich the mobile device with information that results in new services, the mobile device can act as the main interface for human interactions. Nowadays, the wireless sensors have nearly no or limited interface for human interactions. The mobile device on the other side is able to provide the following advantages:

Human interface: The mobile device offers a large colored display and audio output in terms of alarm sounds, predefined sound files, and text-to-speech

F.H.P. Fitzek and F. Reichert (eds.), Mobile Phone Programming and its Application to Wireless Networking, 375–380.
© 2007 *Springer.*

capabilities. Especially the last one could end up in new services for handicapped people.

Network interface toward the cellular network: The mobile devices in the form of mobile phones grant network access. Informing centralized servers or remote people over services such as SMS, MMS, and GPRS.

A sensor is a very small low-complexity device with wireless communication capabilities, which is mostly battery driven. Most researchers foresee these sensors to be stupid with straightforward functionality such as providing sample values. But throughout this chapter we will show that the sensors can also act in a cooperative way to achieve a common goal using the mobile device only as a gateway to convey their information to a predefined source.

19.2 Classification of Different Convergence Forms

The convergence of mobile devices with sensors is already realized at a certain integration level by some mobile phone manufactures such as Nokia or Apple. The Nokia model 5500, for example, is equipped with a motion sensor. This sensor is able to detect real movements of the mobile phone. The application of such a sensor is manifold such as the integration into mobile games or the support of seniors. This form of convergence is referred to as integrated sensors as given in Figure 19.1. If the sensors are using the mobile devices as a gateway to convey their data, we refer to this as the *gateway approach*. If the mobile device is gathering information from the sensors, we refer to this

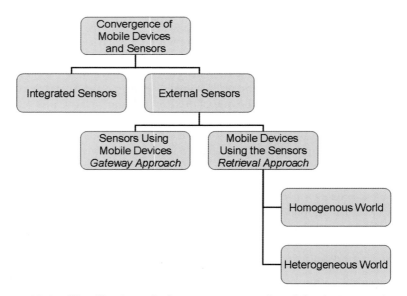

Figure 19.1. Classification of the convergence of mobile devices and sensor networks.

as the *information retrieval approach*. The latter approach can be separated into information retrieval from heterogeneous or homogeneous sensor clusters. In the following we introduce the different approaches in more detail.

The gateway approach allows sensors to contact the mobile device to use functionality that is not available on the sensors. To make low-complexity sensors, which are applied in large numbers, the sensor is equipped with rudimentary display, computational functionality, and wireless communication functionality for short-range links. This communication link is enough to reach the mobile device but not to perform 2G or 3G network compliant actions. Figure 19.2 and Figure 19.3 show the two different realization forms. In Figure 19.2 each sensor is capable to communicate with the mobile device. Such an architecture seems suitable for heterogeneous sensors, where each sensor performs a different task (temperature, distance, camera, surveillance, etc.), but for homogeneous sensor clusters there exists several drawbacks:

- numerous sensors will generate a lot of traffic resulting in high cost for the cellular link and larger energy consumption for the mobile device
- each sensor has to be identified individually and therefore the message to be conveyed becomes larger
- a direct communication link between mobile device and sensor results in a smaller coverage

Therefore, an architecture as illustrated in Figure 19.3 seems to be more promising. On requests or individual triggering, the sensors communicate among each other before a dedicated or selected sensor sends a request to the mobile device. This idea fits nicely with the overlay concept, where only some sensors can connect to the mobile device (using, e.g., Bluetooth

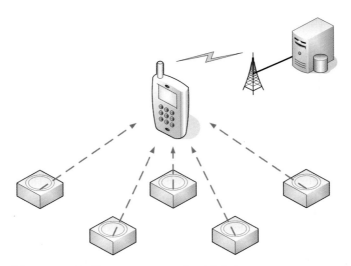

Figure 19.2. Gateway approach with heterogeneous sensors.

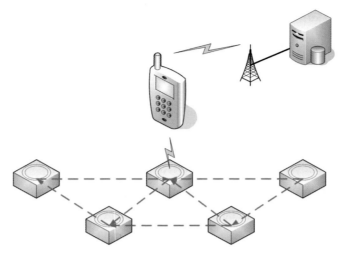

Figure 19.3. Gateway approach with homogeneous sensors.

communication) but all sensors can communicate among each other (low cost ISM band communication). This architecture form has its application especially for mean value extraction (mean temperature in a building) or node counting as explained in [1].

19.3 First Demonstrator

A first demonstrator for such an envisioned communication architecture has been implemented for Python S60. A simple Python client running on the mobile device and sensors (which can be emulated by any Bluetooth capable device and later be replaced by a real wireless sensor) is able to send text strings to it using the Bluetooth serial port with the following syntax:

- ACTION = DISPLAY, SPEECH, SMS, MMS, ALARM
- BODY = any
- KEYVALUE = any

This is only one possible communication syntax and it can easily be extended. The values for the ACTION field inform the mobile device what to do with the information sent by the wireless sensor. In case the value is set to DISPLAY, the information is just displayed on the mobile device's display. Some mobile devices even support the output in speech form. This is not limited to predefined sound files, but can be any text that is transformed into speech signals. It is one possibility to have a more sophisticated information output than the simple ALARM sound. Furthermore, a simple forwarding of the sensor information to a telephone number given by KEYVALUE is

Python for Series 60 –
software is Copyright (c)
2004 Nokia.

Server channel: 4
Connection from 00:0a:3a:51:
a7:1e
waiting for data
Waiting for data...

Options   **Exit**

(a) The mobile device is waiting for incoming data.

2004 Nokia.

Server channel: 4
Connection from 00:0a:3a:51:
a7:1e
waiting for data
Waiting for data...
received Text action
text: Temperature 25 C
Waiting for data...

Options   **Exit**

(b) First incoming data with un-critical temperature value.

waiting for data
Waiting for data...
received Text action
text: Temperature 25 C
Waiting for data...
received SMS action
text: Warning!!!Temperature
43 C
number: +393498523179
Sending SMS....

Options   **Exit**

(c) Incoming data from the sensor with critical temperature.

(d) Notification by the mobile device that SMS has been sent successfully.

Figure 19.4. Screenshots of the sensor gateway realization

provided by means of the actions SMS and MMS. If the sensor wants to send an SMS to the number +123 the text string looks as given in the following:

```
<ACTION>SMS<BODY>Hello  World!<KEYVALUE>+123
```

In the following we illustrate the communication between a sensor that provides temperature values and the mobile device by a first Python implementation. The following figures show the output:

In Figure 19.4(a) the screen shot of a simple Python screen offered on Series 60 Nokia mobile phones is given. The screen shot shows that the mobile device is waiting for incoming data in terms of the temperature from the external source. Later in time the external source conveys temperature information. In Figure 19.4(b) the screen shot shows that a temperature of 25°C degree is initially reported. The mobile device knows that this value is uncritical. The knowledge has to be entered by the customer beforehand or is set by the application communicating with the external sensors. The information is listed on the mobile device's display but no further action is taken.

In Figure 19.4(c) and Figure 19.4(d) screenshots for a critical temperature value are given. Once again the information of the wireless sensor is read out and displayed by the mobile device. This time an SMS message is generated as the reported value is identified as critical. The number to convey the SMS is provided by the sensor, but could also be predefined in the mobile device. The example given is a very first example of the envisioned convergence, but already shows the beauty of it.

The retrieval approach is separated into two different scenarios. The first is referred to as homogeneous scenario, where all sensors in one cluster are performing the same tasks. The second one is related to the heterogeneous scenario.

Homogeneous World As an example we refer to an example from the car manufacturers. Their biggest problem is that the cars are overpriced and with each additional feature (navigation systems, TV, etc.) in the car the prices increase and it makes it more attractive to car thieves. To both problems the use of the mobile device, which is always in the user's possession, can take over most of these services as they are on the device anyway. We have one chapter dedicated to this scenario.

Heterogeneous World To illustrate this, we refer to the example of the intelligent house. Already today, standard washing machines, dish washers, and ovens can report their status back to a centralized concentrator (this is often done over power line communications). The concentrator sends out the information wirelessly to a handheld device with a small display to the user. The whole technology as such is very expensive and could easily be replaced by small sensors in each domestic appliance and the existing mobile device, where a small application needs to be installed. Besides lowering the costs, the manufacturer can also update the software on the mobile phones (and maybe even on the appliances) over the cellular network.

The following chapters within this part give more examples of the convergence of mobile devices and sensor networks.

References

1. P. Popovski, F. Fitzek, and R. Prasad. A Class of Algorithms for Batch Conflict Resolution Algorithms with Multiplicity Estimation. *Algorithmica,* Springer, October 2005.

Using In-built RFID/NFC, Cameras, and 3D Accelerometers as Mobile Phone Sensors

Paul Coulton, Will Bamford, Fadi Chehimi, and Paul Gilberstson, Omer Rashid

Infolab21, Lancaster University, Lancaster, LA1 4WA, UK

Summary. One of the limiting factors in mobile applications development is the restrictions imposed on the user interface through reliance on the standard ITU-T keyboard which is only really optimal for dialling phone numbers. However, with the convergence of the mobile phone with other sensing technologies, such as Radio Frequency Identification (RFID) and the associated Near Field Communications (NFC), Cameras, and 3D motion sensors, we have the opportunity to use new relational interfaces [12] based on touch, vision, and movement to create new and exciting experiences for mobile application users. In this chapter we present the technologies and the software Application Program Interfaces (APIs) associated with these sensors together with methodologies for their use.

20.1 Using RFID/NFC on Mobile Phones

20.1.1 Introduction

In simple terms RFID is a wireless extension of bar code technology allowing identification without line-of-sight. Although it has become most well known through the recent wide-scale adoption of the technology by the likes of Wal-Mart and Tesco as a replacement for barcodes [13], it is also seeing more innovative uses. Another area that has greatly benefited from RFID is micro payments which has seen huge take-up in places such as Japan and Korea and has driven the convergence of this technology with the pervasive mobile phone.

RFID tags, a simple microchip and antenna, interact with radio waves from a receiver to transfer the information held on the microchip. These RFID tags are classified as either active or passive, with active RFID tags having their own transmitter and associated power supply. Passive RFID tags do not have their own transmitter; they reflect energy from the radio wave sent from the reader. Active RFID tags can be read from a range of 20–100 m, whilst passive RFID tags range from a few centimeters to around 5 m (dependent on operating frequency range).

F.H.P. Fitzek and F. Reichert (eds.), Mobile Phone Programming and its Application to Wireless Networking, 381–396.

NFC is an interface and protocol built on top of RFID and is targeted in particular at consumer electronic devices, providing them with a secure means of communicating without having to exert any intellectual effort in configuring the network [3]. To connect two devices together, one simply brings them very close together, a few centimeters, or makes them touch. The NFC protocol then automatically configures them for communication in a peer-to-peer network. Once the configuration data has been exchanged using NFC, the devices can be set up to continue communication over a longer range technology. The other advantage with NFC comes in terms of power saving, and it achieves this by having an Active Mode (AM) and Passive Mode (PM) for communication. In AM both devices generate an RF field over which they can transmit the data. In PM, only one device generates the RF field, the other device uses load modulation to transfer the data. The data rates available are relatively low, 106, 212, or 424 kbits/s, although for the applications envisaged this should be more than sufficient [3].

Nokia was the first to combine mobile phones with RFID/NFC when it introduced clip-on RFID and NFC shells (Nokia Xpress-on Mobile RFID/NFC Kits) for the 5140 and 5140i Series 40 phones, respectively. The RFID/NFC shells can be accessed via J2ME applications running on the phone to trigger defined actions within the application. The particular operating range for mobile phones is generally 13.5 MHz which limits the range to approximately 3 cm or touch.

These phones were but the first of a growing trend and the Japanese telecommunications giant NTT DoCoMo has reported to have shipped more than 5 million RFID-enabled mobile phones for use instead of printed tickets in the National Rail Network [1]. Further Sony has started to ship all of its laptops with RFID/NFC technology so that users can download straight to RFID cards or RFID mobile phones [1].

20.1.2 Overview of JSR-257 (Contactless Communication)

The contactless communication API, or JSR-257 [7], allows applications to access information on contactless targets, such as RFID tags and visual codes such as QR codes and semacodes [10]. The 2D bar codes are similar to RFID tags in that they contain data often in the form of a URL. However, they hold much smaller amounts of information than the newer RFID tags and they are generally slower and more difficult to read [1].

The primary objective of JSR-257 is to provide easy access to various contactless targets and transfer information to or from them. Before we provide details of programming RFID/NFC applications we shall first consider the mandatory and optional parts of the API, which are divided into five packages as shown in Figure 20.1 [7].

Only `javax.microedition.contactless` is the mandatory package whilst the rest are optional and can therefore be left unimplemented. A reference implementation must provide a list of target types it supports and it is this

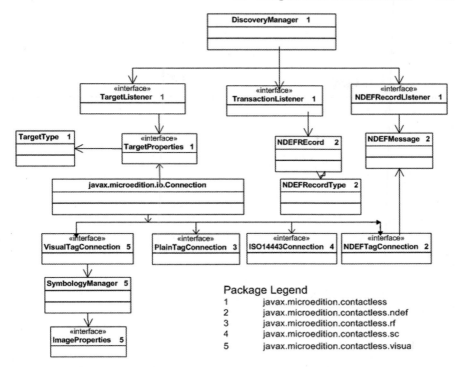

Figure 20.1. JSR 257 reference specification.

list of targets that dictates the packages that must be implemented. If a target type is listed in supported target types an implementation must be provided. All the targets supported by this API implementation are defined in the `TargetType` class. To aid understanding we shall briefly consider the functionality provided by each package.

`Javax.microedition.contactless:` This package provides functionality common to all contactless targets supported by this API. The `Discovery-Manager` Class sits on top of this API and is the starting point of contactless communication. In order for an application to receive notifications about targets in close vicinity it must obtain an instance of the `DiscoveryManager` class which in return provides the notifications to the application about contactless targets. An application can then use subpackages to communicate with different targets depending upon their type. Connections to contactless targets are built on top of Generic Connection Framework (GCF). Each target defines a new protocol to GCF, although only the visual code protocol is visible and all RFID-related protocols are hidden due to the nature of communication involved.

The `TargetListener` interface provides notifications back to `Discovery-Manager` instance once a target is discovered by the device hardware. In essence a Java Virtual Machine (J1VM) must provide a `TargetListener` for each `TargetType` obtained by calling `getSupportedTargetTypes()` function of `DiscoveryManager` class. Only one `TargetListener` can be registered at a time as RFID hardware can only handle one physical connection and failing to do so will result in a registration failure and an exception will be thrown.

Class `TargetType` provides all the contactless targets supported by the API whilst the `TargetProperties` interface provides the properties common to all supported contactless targets. Since the API can support card emulation mode, the interface `TransactionListener` provides information to the application relating activity between secure elements on the device and an external reader.

`Javax.microedition.contactless.ndef`: NDEF records have been defined by the NFC Forum to exchange data between two NFC devices or an NFC device and a tag. Since these records contain type, type format, identifier, and payload, a connection can be established with any physical target that supports the NDEF data format. To utilize this feature the application must register with an NDEF Listener to receive notifications about tags, or devices, detected. The data from the tag is passed on to the applications registered to receive NDEF notifications.

`NDEFMessage` class represents messages which can consist of a number of NDEF records. This class provides necessary functionality to manipulate these records, i.e., add, delete, or update data payload within the records in the message. Each individual record within the message is represented by the `NDEFRecord` class stored in a byte array. Class `NDEFRecordType` stores the name and format of the record type. Record types are stored as constants within the class and must abide by the rules of NFC Forum RTDs and RFCs. For example, MIME types and URL. Whilst other classes handle definition and storage of NDEF messages the `NDEFTagConnection` interface provides the basic connection for exchanging this NDEF data between NFC devices or NFC tag. This interface does not concern itself with physical type of the contactless target; it simply reads and writes the data. The protocol to do achieve this read and write process is extended from GCF.

`Javax.microedition.contactless.rf`: Although this API implementation aims to use NDEF records as standard, it also provides access to physical RF targets that do not support NDEF. This package contains one interface; `PlainTagConnection`; which encloses functionality to detect RFID tags. This is done to enable comprehensive support for various types of RFID tags since it would be undesirable to have an API support all different types of tags.

`Javax.microedition.contactless.sc`: This package is responsible for communication with external smart cards. Although JSR-177 covers communication with smart cards, through an APDU connection, JSR-257 defines an ISO14443 connection interface to access smart cards. Unlike and APDU connection, an ISO4443connection can read both resident and external smart cards and provides access at a much lower level.

`Javax.microedition.contactless.visual`: This package provides support for visual tag targets which are detected as contactless targets. It contains classes and interfaces to read information stored on the visual codes.

20.1.3 Programming with Nokia NFC and RFID SDK

Although Nokia were the first to launch the phones equipped with RFID/NFC readers and reference implementation for JSR-257 is available there are no mobile devices in the market that support it. However, the Nokia NFC and RFID API can be used under commercial license to develop MIDlets that utilise Nokia NFC shells or xpress on RFID shells on the 5140 and 3320 phones. Nokia provides an SDK to develop and test the MIDlets. This SDK consists of the following components;

- Nokia NFC and RFID API: enables developers to create MIDlets able to connect to NFC or RFID shells.
- LI Client API: enables MIDlets to employ the LI Server, which is also used in Nokia's Field Force Solution. The LI Server is web-based and not only supports reporting, but also manages users, mobile phones, tags, and events. It can communicate with a MIDlet using HTTPS over GPRS or via SMS.
- Nokia NFC and RFID Cover Emulator: enables development on a PC. It emulates tag touch and can not only emulate both NFC and RFID shells but can also be used to manage and store tags to provide a thorough test and development environment.

20.1.4 Using Nokia NFC and RFID API

Class `ContactlessConnection` sits on the top of this API and provides a connection to NFC or RFID shell and `InterfaceFeature` represents all the capabilities and functionalities of this underlying hardware. `Target` class defines physical target types supported by each shell. Table 20.1 lists different features supported by NFC and RFID shells, respectively.

In order to receive notifications about contactless targets a registration is required with a contactless listener which provides information about contactless events. Contactless events can also be registered with the MIDlet

Table 20.1. Nokia RFID and NFC shell feature support.

Feature Name	NFC Shell	RFID Shell
ntip_container	Yes	Yes
nfc_device	Yes	No
iso14443	Yes	No
mifare_4k	Yes	No
mifare_1k	Yes	No
mifare_ul	Yes	Yes

push registry which means that the applications can be auto-launched once a certain contactless event occurs, e.g., scanning a tag.

There are certain considerations that need to be taken into account when programming NFC/RFID applications. Due to the nature of the communication taking place it is always a good programming practice to have the listener registered with a separate thread so that the user interface stays responsive. If authentication of the data being read from the tag is to be done from a server, that information should be passed on to the network handler which should operate under a separate thread. This multithreading is important when programming applications or games such as PAC-LAN [9] and Mobspray [4].

Since all the information about the contactless targets is reported by the listener as events, it is of utmost importance that the application must handle all the events that may arise in. Further, our experience shows [1] that informative feedback must be given to the user, through visual alerts, vibration, and sounds, relating to the contactless event taking place.

20.2 Using Cameras on Mobile Phones

20.2.1 Introduction

Cameras are now a common feature of even the most basic mobile phone and indeed a reported 295.5 million [8] were shipped in 2005 which represented nearly 40 per cent of all phones shipped. There is thus a real opportunity for their use as a relational interface through both the use of the various forms of 2D bar codes, as previously discussed, as well as algorithms for the detection of objects or movement. In general, movement is detected by utilizing relatively simple algorithms that detect changes in the background whilst tracking generally requires more complex procedures. However, both can be used to provide novel interfaces, for example, there have been a number of very innovative games that have used the camera to detect movements of the phone and relating these to movements within the game. Probably the best known are from game developer Ojom (www.ojom.com) with its games "Attack of the Killer Virus" and "Mosquitoes" for Symbian S60 mobile phones. In both games, the enemy characters are superimposed on top of a live video stream from the mobile phone's camera and are 'shot' by pressing the defined key when the cross hairs, at the center of the phone screen, are over the target. There are other games that utilize visual codes in conjunction with the camera to detect movement but these appear to offer no great advantage to those that use the video input or the surroundings directly.

20.2.2 Programming the Camera in J2ME

The ability to access a mobile phone's integrated camera/s from within J2ME is not currently part of the standard platform implementation (MIDP 1/2).

This standard covers only the base level of functionality so that it is applicable to the lowest feature handsets on the market and in fact some phones do not have cameras!. However, most of the newer J2ME-enabled handsets now support camera access through the Mobile Media API (JSR-135) [5], which is a standard extension to the J2ME platform (MIDP), allowing Midlets to access a phone's native multimedia resources, such as audio/video recording/playback.

Programming Guide

In the first instance the `javax.microedition.media` package should be imported into the code. There are many different classes within this package that support media functionality. These are comprehensively described in the MMAPI Javadocs, but a brief summary of the most important classes for cameras follows:

- `Manager` – The `Manager` class is used to access system-dependant resources such as a `Player` for multimedia content. In the case of the camera, the `Manager` returns a `Player` capable of interacting with the handset's primary camera.
- `Player` – An instance of the `Player` class will know how to interpret (control and render) certain time-based media formats. For instance, playing back a sampled audio format or displaying a streaming video clip. A media-specific `Player` object can be obtained from the `Manager` by passing in either a media locator (URI), or an `InputStream` as an argument to the `Manager`'s `createPlayer()` method. In the case of the camera, the following URI is used; "capture://video." Note: a J2ME handset's support of the Mobile Media API does not guarantee that you will also be able to access the camera.
- `Control` – Controls are used to modify the behaviour of the `Player`. For example a `VolumeControl` can be used to alter the playback volume for a particular media. Controls are created by Player instances and in the case of camera applications, a `VideoControl` is needed in order to access the viewfinder and subsequently capture a frame (i.e., create a snapshot) from the camera.

The video frames from the camera can be displayed either as an Item in a Form (high-level user interface component) or as part of a `Canvas` (low-level user interface component). However, the viewfinder does not need to be visible in order for a snapshot to be obtained.

Considerations

Mobile applications that use the phone's camera to track objects or detect motion require real-time image processing. Unfortunately, this type of operation is processor-intensive and will consume large amounts of processor power and consequently drain the battery. Java is usually considered slightly slower than native level programming languages [2] because it works on top of a Java Virtual Machine (JVM) and, as such, is abstracted to a layer above native code execution. This facilitates cross-platform compatibility, but usually comes with some kind of performance penalty. With certain operations, this performance hit can be compensated through just-in-time (JIT) compilation and other optimisation techniques. Access to the camera is essentially a native operation, so this in itself should not be that much slower than using, for instance, Symbian OS. If speed is an issue, selecting a relatively small resolution (e.g., 160×120 pixels) and uncompressed encoding scheme (e.g., BMP if it is supported), will speed up captures and subsequent processing times. The biggest drawback to using Java to access the phone's camera is that you have very limited control over advanced functionality such as exposure, macro mode, colour/white balance, zoom, focus, second camera access, etc.

When taking a photo from within a MIDlet, the user will be asked if they will allow the application permission. If the application is unsigned, the application will also ask for the user's permission for each subsequent capture. However, by signing the application with a digital certificate and adding the following into the list of requested API permissions – `javax.microedition. media.control. VideoControl.getSnapshot`, subsequent captures can be made without repeated prompts to the user. This behaviour is common across all platforms, as required by J2ME's security architecture.

Testing

The latest versions of Sun's Java Wireless Toolkit (WTK) [11] provide J2ME emulators which can be used to test your camera midlets before deploying to real handsets. However, it is still important to frequently test applications on real devices due to differences between the emulator and real hardware.

There are also differences between the various J2ME devices currently on the market. For instance, whilst most modern handsets feature the Mobile Media API, the encoding schemes supported vary considerably. Some allow JPEG encoding of image data, others only supporting PNG. There are further differences in the image resolutions supported; although 160×120 and 640×480 pixels are quite common formats. Trying to use an encoding scheme that is unsupported will generate a `MediaException`. To find out which encodings are supported by a device, you can query the `System.getProperty`("property_name") with "`video.snapshot.encodings`" as the argument. This method will return a string containing all the supported formats delimited by a space.

There are efforts to standardize J2ME across platforms which, due to the current fragmentation of the handset market, would be a great help to mobile developers. JSR-248, the Mobile Service Architecture for CLDC defines a base set of requirements that all supporting handsets will need to adhere to. For instance, it will be required that "Implementations MUST support the same resolutions for image capture as supported by the system camera applications of the phone" [6].

20.2.3 Programming the Camera in S60

Introduction

Programming an application that uses the camera for either the Nokia S60 or Sony Ericsson UIQ platforms requires the camera API `CCamera` in the Symbian OS. Note that this API was significantly altered in version 7.0 of the Symbian OS for simplicity we only consider its implementation from this point. Because of difficulties often experienced using the Symbian OS we will illustrate the use of the camera through an application that enables a S60 phone to identify specific colour in a scene and track it. This example could be used for various applications for example, as an input method for a game. Further we will highlight some optimization techniques for camera-based Symbian applications to improve performance.

The example assumes a basic knowledge of the architecture of a Symbian application and readers should familiarize themselves with this if they want to get the most out of the example.

Introduction to `CCamera` API

`CCamera` provides several asynchronous functions that enable applications to control cameras on mobile phones. These functions return immediately once entered leaving their code to be completed at a later stage, thus preventing an application blocking the phones operation whilst a portion of the code is awaiting completion.

The first step in using a camera is to reserve it, which ensures that an application will be granted a handle to the camera either directly or once it is free in cases where it is used by another application(s). To perform this task the asynchronous `CCamera::Reserve()` function must be called on the `CCamera` instance object in the client application.

Once the camera is reserved for the client it has to be powered on by calling the asynchronous function `CCamera::PowerOn()`. The camera is then ready to receive data from the camera viewfinder in bitmap buffers via either of the activation functions:

```
CCamera :: Start View Finder Bitmaps L ( . . . )
CCamera :: Start View Finder Direct L ( . . . )
```

The client application now needs to inherit from the camera interface in Symbian OS `MCameraObserver` in order to actively interact with the camera API. This observer class provides five pure virtual functions that must be implemented:

```
   void  ReserveComplete(TInt);
   void  PowerOnComplete(TInt);
 3 void  ViewFinderFrameReady(CFbsBitmap&);
   void  ImageReady(CFbsBitmap*, HBufC8*, TInt);
   void  FrameBufferReady(MFrameBuffer*, TInt);
```

ReserveComplete handles the events of completing camera reservation. Similarly, PowerOnComplete indicates that power-on of the camera has been completed, while ViewFinderFrameReady receives the data from the camera viewfinder in continuous bitmaps. ImageReady and FrameBufferReady are responsible of handling image and video-capturing events, respectively.

Note that to use the camera API the client application must be linked against the ecam.lib and fbscli.lib, for bitmap API, libraries, and include ecam.h and fbs.h in the source files.

Color Tracker Application

The architecture of any Symbian-based application is based on four component classes that instantiate each other hierarchically until the entry-point class, usually called the "container" or "view," is reached. In the ColourTracker example a container class is used. In the following code we will implement its header file ColourTrackerContainer.h to include ecam.h and fbs.h and inherit from MCameraObserver and declare its pure virtual functions. The code then declares the tracking function and instantiates iCamera, iViewFinderBitmap (which is a bitmap holding individual frames from the camera) and a Boolean value indicating completion of transfer from the camera to the viewfinder bitmap.

```
   ...
   #include <ecam.h> // Camera API
   #include <fbs.h> // Bitmap API
   ...
 5 class  CColourTrackerContainer  :
     public CCoeControl,
            MCoeControlObserver,
            MCameraObserver
     {
10 public:
       // Class functions declaration
       ...
       // Pure virtual functions
       void  ReserveComplete(TInt);
15     void  PowerOnComplete(TInt);
       void  ViewFinderFrameReady(CFbsBitmap&);
       void  ImageReady(CFbsBitmap*, HBufC8*, TInt);
       void  FrameBufferReady(MFrameBuffer*, TInt);

20     // The function of action in our example
       // @arg1: The colour to track
       // @arg2: Reference to the graphics context
       void  TrackColor(TRgb&, CWindowGc&);

25   private:
       // Members declarations
       CCamera *iCamera;
       CFbsBitmap* iViewFinderBitmap;
```

```
          T Bool  i A l l o w D r a w ;
30        . . .
          } ;
```

The next implementation stage is contained within `ColourTrackerContainer.cpp` and performs the construction of the `iCamera` object which is reserved it in `ConstructL()`.

The next action is to construct the viewfinder bitmap `iViewFinderBitmap` and initialize the Boolean value to false before adding `iViewFinderBitmap` and `iCamera` to the destructor

```
     . . .
     void  C Colour Tracker Container :: Construct L (const  TRect&  aRect )
          {
4         // Basic Symbian-window  initializations
          CreateWindowL ( ) ;
          SetRect ( aRect ) ;
          ActivateL ( ) ;

9         // Constructing iCamera :
          // @arg1 :  reference  to  McameraObserver  interface
          // @arg2 :  the  index  of  the  device  camera  to  use .  It
          //   starts  with  0  and  increments  for  each  other  camera
          //   available  on  the  phone .
14        iCamera = CCamera :: NewL (* this , 0 ) ;
          iCamera–>Reserve ( ) ;  // Reserving
          // Construct  the  bitmap
          iViewFinderBitmap  new ( ELeave )  CFbsBitmap ( ) ;
          iAllowDraw = EFalse ;
19        }
          // Destructor
          C Colour Tracker Container :: ~ C Colour Tracker Container ( )
          {
          delete  iCamera ;
24        delete  iViewFinderBitmap ;
          }
     . . .
```

`Reserve()`, `PowerOn()`, and `StartViewFinderBitmapsL()` are all asynchronous functions that run in parallel with other operations either at the application or OS level. Thus, it is not possible to call these functions sequentially as shown in the following code snippet as running the later functions depends on the states of the previous ones, which may have not completed execution.

```
iCamera–>Reserve ( ) ;
iCamera–>PowerOn ( ) ;
iCamera–>StartViewFinderBitmapsL ( ) ;
```

For example, when `PowerOn()` is called `Reserve()`may have not completed reserving the camera. This results in `PowerOn()`considering the camera as not being available and will terminate its usage in the application.

To solve this dependency issue, we implement the observing functions of the interface `McameraObserver` which provides a framework to monitor the completion of each camera operation. Therefore, when the camera reservation process is completed an event is caught by the OS which calls `McameraObserver::ReserveComplete()` to handle it.

This is implemented in our example code as:

```
   void CColourTrackerContainer :: ReserveComplete ( TInt )
 2   {
        iCamera−>PowerOn ( ) ;
     }

     void CColourTrackerContainer :: PowerOnComplete ( TInt )
 7   {
         // @arg: the size of the viewfinder on the screen
         iCamera−>StartViewFinderBitmapsL ( TSize ( 176 , 208 ) ) ;
     }

12   void CColourTrackerContainer :: ViewFinderFrameReady (
            CFbsBitmap& aFrame )
     {
         iAllowDraw = ETrue ;
         iViewFinderBitmap−>Duplicate ( aFrame . Handle ( ) ) ;
17       DrawNow ( ) ;
     }
     // This function is already implemented in the class
     // but needs its content be modified .
     void CColourTrackerContainer :: Draw ( const TRect& aRect ) const
22   {
         // Painting a white background
         CWindowGc& gc = SystemGc ( ) ;
         gc . SetPenStyle ( CGraphicsContext :: ENullPen ) ;
         gc . SetBrushStyle ( CGraphicsContext :: ESolidBrush ) ;
27       gc . DrawRect ( aRect ) ;
         // Allow bitmap drawing and manipulation
         if ( iAllowDraw )
            {
               gc . BitBlt ( TPoint ( 0 ,0 ) , iViewFinderBitmap ) ;
32          }
     }

     void CColourTrackerContainer :: ImageReady (
            CFbsBitmap∗ aBitmap , HBufC8∗ aData , TInt aError )
37   {
     }

     void CColourTrackerContainer :: FrameBufferReady (
            MFrameBuffer∗ aFrameBuffer , TInt aError )
42   {
     }
```

Since the camera reservation is running in the background asynchronously it is logical to call CCamera::PowerOn() within the handler function
ReserveComplete() which will be called once the reservation process is finished.
The same applies to CCamera::StartViewFinderBitmapsL() which will start
acquiring bitmaps from the camera viewfinder once powering-on is completed.
When the first bitmap frame is ready, the ViewFinderFrameReady() function will be called by the observer class passing as a parameter that bitmap
(aFrame) to be copied by iViewFinderBitmap for later manipulation. This will
be performed continuously as long as the camera is powered-on.

DrawNow() will call Draw() to initiate the on-screen drawing and we cannot
call Draw() directly as there are several steps the OS takes before and after
Draw() which DrawNow()deals with.

The code thus far is the basic structure for any application utilizing the
camera. To complete the example the following codes implement a function

that will perform the actual tracking of a colour (in this case red) and fetch it to `ColourTrackerContainer.cpp`. The function is called from `Draw()` which will draw on screen white dots in place of red pixels.

`TrackColor()` calculates a threshold for the colour to track in a hemisphere shape surrounding the true possibilities of RGB values and centred at the target colour. The function is self-explanatory with the comments included.

```
void CColourTrackerContainer :: TrackColor (
        TRgb &aColour , CWindowGc &aGc )
    {
        TUint r, g, b, distance ;
        TRgb pixelColour ;
        //The colour of the dots to draw
        aGc . SetBrushColor ( KRgbWhite ) ;
        //Traversing the 176x134 viewfinder bitmap
        for (TUint x=0; x<176; x+=2)
            {
                for (TUint y=0; y<134; y+=2)
                    {
                        iViewFinder−>GetPixel ( pixelColour , Point (x,y)) ;
                        r = pixelColour . Red () ;
                        g = pixelColour . Green () ;
                        b = pixelColour . Blue () ;

                        //Calculating threshold as a sphere radius
                        distance =
                            (r−aColour . Red ()) * (r−aColour . Red ()) +
                            (g−aColour . Green ()) * (g−aColour . Green ()) +
                            (b−aColour . Blue ()) * (b−aColour . Blue ()) ;

                        //If target colour is within 125 radius
                        if ( distance <= 15625) // 125^2
                        aGc . DrawRect ( TRect ( TPoint (x,y) ,
                            TPoint (x+2, y+2))) ;
                    }
            }
    }
```

20.3 Motion Interfaces using 3D Sensors

20.3.1 Introduction

3D motion sensors provide an intuitive and novel way to interact with games and services. These sensors provide an output when the user tilts the mobile around any axis by using on-board accelerometers. Movement of the mobile in the form of gestures and tilts can be used to control menu handling, player input, scrolling, and a variety of other systems. Motion sensing can provide an innovative alternative to keyboard input particularly in games. For instance a driving game could use the rotation of the phone as a form of intuitive steering wheel control, whilst vertical tilting could be used for acceleration and braking.

The first 3D motions sensors have only just been integrated into mobile phones, and these phones are yet to become widespread. The Nokia 5500 is one of the first such equipped phones and forms the basis of the following discussion. Nokia have targeted the phone at sports users, utilizing the built-in accelerometer as both a pedometer and speed/distance tracker for various exercising purposes. The utility of the motion sensors in gaming cannot be overestimated, and the mobile comes with one motion sensing game called GrooveLabyrinth.

Other phones with motion sensors include Samsung's SCH-S310 (the world's first 3D motion sensing mobile), and NTT's N702.

20.3.2 Programming 3D Motion Sensors in Symbian

Accessing the 3D motion sensors requires the use of the Symbian Sensor API. Based off the Java Sensors API (JSR-256) this API provides access to a wide range of sensors. These sensors can include accelerometers, thermometers, barometers, and humidity monitors, in fact, any type of sensor designed to be incorporated in a mobile phone, or be accessible via Bluetooth. Sensors need only be supported by the API library to be usable. The Symbian Sensor API is available from Nokia and requires the use of Symbian S60 3rd Edition SDK. The Sensor SDK includes a header file and GCCE compiler-based lib file.

The API contains a number of key classes for retrieving sensor information. These classes and types provide the main method for reading sensor data by callback function. They are defined in **RRSensorApi.h** and the application needs to link against **RRSensorApi.lib**. Both these files are provided in the SDK.

- **CRRSensorApi** is the sensor manager class. It provides a **FindSensorL** static function for querying available sensors. This returns an array of **TRRSensorInfo** objects. Instances of **CRRSensorApi** class are created by passing a particular sensor's **TRRSensorInfo** object to the constructor. The class then allows for the application to send commands to the sensor. It also allows for the addition and removal of data listener classes.
- **MRRSensorDataListener** interface should be implemented by the class which is to receive and interpret sensor data. It contains one function **HandleDataEventL**. This function has two arguments; **TRRSensorInfo** contains the identifier for the sensor that raised the event and **TRRSensorEvent** that contains the sensor information.
- **TRRSensorInfo** contains a human readable sensor name, a unique identifier for the sensor, and a category that identifies if the sensor is internal to the device or external over Bluetooth.
- **TRRSensorEvent** contains three TInt fields that contain sensor data. Each sensor will provide different data in these fields; some may provide absolute values, whilst others may provide values relative to the previous event.

The accelerometers in a Nokia 5500 appear as a single sensor with ID 0x10273024, and Category 0x10010FFF which corresponds to an internal sensor. The type **TRRSensorEvent** is filled with absolute values of the acceleration detected around X, Y, and Z-axis respectively.

The best method for receiving and using sensor data from the Nokia 5500 is to create a class to handle the sensors. It should inherit `MRRSensorDataList-ener` and implement `HandleDataEventL` to receive events. Within the constructor is the best place to retrieve the needed `TRRSensorInfo` object and register itself with `CRRSensorApi` to receive data.

Whilst the sensors provide spot acceleration data, a form of position is more valuable for most applications. The easiest way to achieve this is to maintain three TInts initialized to zero, one for each axis. Then sum the raw acceleration data received to each TInt. By making these variables accessible via get methods.

Testing can only be done on the device at the moment. There is currently, no Symbian emulator for the sensors. Also, there is no WinSCW library provided for use with the standard 3rd Edition emulators. This currently makes development difficult and time-consuming.

The main issue with the sensors at the moment is that they appear to be fairly noisy. The sensors are extremely sensitive and constantly fire events even when the device is static. The position data appears to wobble around a mean value. When using the data for games this would cause undesired movement and triggers.

Various methods can be used to alleviate this. Firstly only acceleration events over a certain magnitude could be used. This would eliminate the "wobble", however it would also cause the system to ignore subtle movements. This is fine for menu selections and other binary events where only the motion but not the actual position is needed. The other method is to average the apparent position over a number of samples. This is better for applications that are using the motion intensity or device position. There is a small lag from movement to update, and short sharp "twitches" are missed.

20.4 Conclusions

In this chapter we have explored the utilization of some of the expanding feature set of mobile phones which can be implemented in a variety of sensor applications. Whilst it is difficult to imagine a standard feature set on a device that purposefully allows consumers to select a very personal choice there is no doubt that mobile phones will increasingly be used as readily deployable sensors using inbuilt components.

Acknowledgements

The authors wish to acknowledge the support of Nokia for the provision of phones, software, and Nokia Forum Pro access to the Mobile Radicals Research Group at Lancaster University for the development of the projects featured in this chapter.

References

1. P. Coulton, O. Rashid, and W. Bamford. Experiencing 'touch' in mobile mixed reality games. *The Fourth Annual International Conference in Computer Game Design and Technology*, Liverpool, 15–16 November 2006.

2. Paul Coulton, Omer Rashid, Reuben Edwards, and Robert Thompson. Creating entertainment applications for cellular phones. *Comput. Entertain.*, 3(3):3–3, 2005.

3. ECMA. Near field communication: White paper. Technical Report Ecma/TC32–TG19/2004/1, ECMA, 2004.

4. P. Garner, O. Rashid, P. Coulton, and R. Edwards. The mobile phone as a digital spraycan. *Proceedings of ACM SIGCHI International Conference On Advances In Computer Entertainment Technology*, Hollywood, 14–16 June 2006.

5. JSR 135 Expert Group. Jsr 135: Mobile media api, final release 3. Technical report, Sun Microsystems, Inc., 22 June 2006.

6. JSR 248 Expert Group. Jsr 248: Mobile service architecture, proposed final draft. Technical report, Sun Microsystems, Inc., 17 May 2006.

7. JSR 257 Expert Group. Jsr 257: Contactless communication api, final release. Technical report, Sun Microsystems, Inc., 17 October 2006.

8. Nokia. The mobile device market. `http://www.nokia.com/nokia/0,,73210,00.htm`, August 2005.

9. Omer Rashid, Will Bamford, Paul Coulton, Reuben Edwards, and Jurgen Scheible. Pac-lan: mixed-reality gaming with rfid-enabled mobile phones. *Comput. Entertain.*, 4(4):4, 2006.

10. Omer Rashid, Ian Mullins, Paul Coulton, and Reuben Edwards. Extending cyberspace: location based games using cellular phones. *Comput. Entertain.*, 4(1):4, 2006.

11. Sun. Java wireless toolkit for cldc. `http://java.sun.com/products/sjwtoolkit/`.

12. B. Ullmer and H. Ishii. *Human-Computer Interaction in the New Millenium*, chapter Emerging Frameworks for Tangible User Interfaces, pp. 579–601. Addison-Wesley, August 2001.

13. R. Want. An introduction to rfid technology. *Pervasive Computing, IEEE*, 5(1):25–33, 2006.

21

Sensor Networks for Distributed Computing
openSensor for everybody!

Stephan Rein[1], Clemens Gühmann[1], and Frank H.P. Fitzek[2]

[1] Technical University Berlin {stephan.rein|clemens.guehmann}@tu-berlin.de
[2] Aalborg University ff@es.aau.dk

Summary. This chapter describes a sensor network for signal processing and distributed computing. In such a network, the sensors do not necessarily transfer the collected data to the user right away but cooperate to inspect the data in a shorter time for the desired information conveying the result to the user. As an example, the computation of a system of linear equations is given performed on the newly introduced *openSensor*. The hardware design and the developed software are made freely available.

21.1 Introduction

Designing a wireless sensor ought to be simple: Take a microcontroller, a sensing circuitry, a transceiver chip, and put them together with the appropriate software – these may be the basic steps. If network communication protocols are included in the software architecture, a low-cost and very flexible sensor network is established, which can lead to many creative applications, including personal assistance, medical care, or surveillance. This may be one of the reasons why sensor networks have aroused so much interest in the communication society, and more recently also in the data and signal processing community. There exist ready-to-use sensor network solutions like the family of *Mica modes* from Berkeley [3], for which the operating system *TinyOS* [1] is available. Such solutions allow for immediate start of sophisticated research investigations beyond of basic hardware and software problems that have already been solved by others. However, to our best knowledge, there exist no solutions in the field of sensor networks for distributed computing. Therefore, this chapter presents the design of a new sensor board referred to as *openSensor*. The newly introduced sensor board is more advisable, especially when complex computations for immediate data evaluations have to be performed within the sensor network. Most of the commercial solutions are designed for the communications engineer, who needs a platform for introducing novel protocols or related algorithms to improve energy efficiency or medium usage. If algorithms that work on the data are proposed

F.H.P. Fitzek and F. Reichert (eds.), Mobile Phone Programming and its Application to Wireless Networking, 397–410.

in this field, they are generally less complex, i.e., kinds of data aggregation procedures detecting redundancy in order to reduce the overall stream of data. In contrast to this, the data processing engineer concentrates on the development of complex algorithms to inspect the data, and is not so much interested in the communication aspects. Hopefully, those algorithms can run on top of the communication architectures. From a methodic point of view, a data processing sensor network is the ideal tool to combine these two disciplines – the design of data processing algorithms and communication protocols – thus inspiring more effective solutions for future applications. This chapter describes a platform for distributed and embedded signal processing algorithms. In this field, the single sensors take advantage of their different knowledge and abilities, and also cooperate to solve a problem in a shorter time. The general idea is well known as *distributive computing*. A related idea that can be applied to the sensors is *distributive storage*. Applied algorithms can lead to complex pattern detection or signal separating techniques for different kinds of data, including sensor measurements, audio data, and pictures. Another aspect that demands an own sensor network are the joint chances of this in education. Class work in wireless communication is often performed by the study of the underlying principles, i.e., network architectures and protocols. A large set of low-cost sensors can easily be distributed to the students for conducting own experiments and even programming some basic algorithms, thus getting a more thorough understanding in communications. The authors have experienced that such sensors are welcomed as a nice playground to develop novel applications. Furthermore, as small and mobile devices have already become part of our daily lives, experience in embedded programming is a worthwhile educational investigation.

Sensor networks impose a broad field of disciplines including sensor network architecture, energy saving communication protocols, and security. A very thorough description of these is given in [4]. This chapter focuses on the signal processing aspects of a small sensor network and only scarcely considers the communication aspects – which include medium access protocols (MAC), time synchronization, localization, and routing, to name a few. The application of sensor networks in military environments is widely believed to be typical, because a large number of cheap sensors is employed, thus requiring complex protocols and topology control. However, the authors do not think so as already a small number of smart sensors can lead to interesting applications. In such a scenario the communication aspects are of less importance. This chapter gives an overview of the basic components of the proposed sensor network, including the software and hardware modules, which both are set up for ease-of-use as a modular system. The hardware consists of the main board with the processor and a transceiver board, which are described in Section 21.2. The appropriate software for the sensor is given in Section 21.3. A linear algebra example that applies the whole system is detailed in Section 21.4. Technical details and descriptions of possible sensor extension boards can be found on http://mobiledevices.kom.aau.dk/sensor.

21.2 Hardware

In this section the processor and the transceiver chip of the openSensor platform is described. As illustrated in Figure 21.1, it is possible to plug own sensor extension modules onto the system. A schematic of the sensor is depicted in Figure 21.2. Extension modules can concern audio, picture, ultrasonic, and global positioning system (GPS) data, to name a few examples. A memory card extension may allow for building a data logging system or be useful when the collected data cannot be processed in time. An ethernet module may be helpful for remote access allowing for distant surveillance (software for an Internet protocol stack is freely available). The extensions are not described in this chapter to keep it succinct and general.

Figure 21.1. The openSensor platform. Two layers – one containing the main board with the processor and USB-interface to be applied with a free terminal software, and another with ISM-band and optional Bluetooth module – build the core of the system.

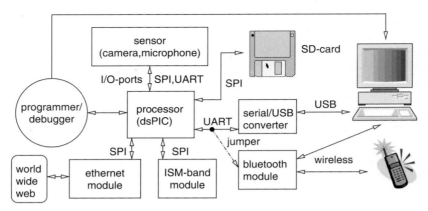

Figure 21.2. The hardware design of the openSensor.

21.2.1 Processor

The first step when constructing a wireless sensor is to choose an appropriate processing unit. Especially in the field of sensor networks, some fundamental requirements are

1. Low cost
2. Low energy
3. Interface for sensors
4. Embedded design (highly integrated circuitry)
5. High-level math abilities
6. Developer's ease of use.

The *low energy* feature allows for mobile and battery-driven units with a long lifetime. A *sensor interface* can be a set of serial ports on the processor, like serial peripheral interface bus (SPI) or universal asynchronous receiver/-transmitter bus (UART). Many processors support analog signal input for a limited voltage range using an integrated analog/digital converter (AD), however, a sensor with a digital output is preferable as it allows for a minimum of hardware. Thus, the idea of an *embedded design* is supported. For the processor this means that it should be a highly integrated device that can start operation just when power is supplied. More precisely, this processor contains all electronic components except the sensors for an ordinary operation. This includes a well-sized random access memory (RAM) and a flash memory for the program code. Finally, the processor should be *easy* to use for the developer. For instance, this can be an in-circuit serial programming feature (ICSP) that allows the programmer to control the program on the processor in operation, or a high-level language like C including a set of libraries that simplify peripheral and math activities. For the math abilities there should be hardware support to relieve the compiler from severe translation issues, thus allowing for fast signal processing operations.

There exist two main classes of architectures, the *microcontrollers* and the *digital signal processors* (DSP). The microcontrollers fulfill most of the requirements despite the math abilities. The DSPs are explicitly optimized for high-level math – more specifically, vector and matrix operations – excluding most of the other required features. In the literature, a typical sensor is generally believed to perform communication and only few data processing tasks [4]. By contrast, this chapter describes a wireless sensor suitable for complex signal processing tasks. In the recent years, the borderline between the architectures became blurred as mixed ones were introduced – the so-called *digital signal controllers* or *mixed controllers* – to combine the individual benefits. Such a processor is employed in this chapter, the *dsPIC* from Microchip with a RAM of 30 KByte. This unit is an exemplary choice as other companies offer useful solutions as well. It is a 16-bit processor with a math coprocessor that is available with different specifications, see www.microchip.com for details.

The general way to get started with such a processor is to buy a developer board, install the software, and program an example. A good alternative toward designing an own sensor is to skip buying the board and to use a very simple but own circuitry instead. A more demanding problem is to program the processor. To do so the authors employed a commercial C compiler and the ICD2 in-circuit debugger/programmer, both from Microchip. There also exist freeware solutions for the programming device, i.e., clones of the ICD2 or boot-loaders that allow to program the processor with a software via UART.

21.2.2 Transceiver

For transmission of data packets, a wireless sensor needs a radio frequency transceiver circuitry. Due to the here required embedded design it makes sense to select a transceiver chip that almost needs an antenna but not many more components to start operation. These transceivers are available for different commercial standards, which exist for wireless local area networks (WLAN) like IEEE 802.11 and wireless personal area networks (WPAN) like IEEE 802.15.4. Two examples for WPANs are Bluetooth specified in IEEE 802.15.1 and Zigbee as an extension of IEEE 802.15.4, both operating in the industrial, scientific, and medical band (ISM). The standards include physical layer and medium access protocol (MAC) specifications.

One reason why IEEE 802.11 and Bluetooth are considered as less suitable for wireless sensor networks are the higher energy consumption, which is caused by the high bit rates of IEEE 802.11 and in case of Bluetooth by the master node activity to poll the slaves [4]. Even if there exist more suitable standards, the here detailed flexibility requirement requests to use none of them:

- Using a standard does not allow for the design of own and completely novel protocols.
- It might give a too complex environment for testing novel signal processing algorithms and for learning the basics in wireless communications in the classroom.
- The costs for a transceiver with a commercial standard are higher.

In this chapter, the single chip radio transceiver nRF905 from Nordic is employed. Its features are given as follows:

- 433/868/915 MHz ISM band
- Channel resolution of 100/200 kHz
- Gaussian frequency shift keying modulation (GFSK)
- Maximum data rate of 50 kbits per second
- Internal Manchester coding
- SPI interface
- Hardware cyclic redundancy check (CRC)
- Carrier detect function
- Manchester coding of the packets
- Packet payload from 1–32 bytes and address size from 1–4 bytes

Similar to the selected processor unit, the nRF905 is an exemplary choice. As packet size and transmit time can be set by the user it is possible to design own MAC protocols with the restrictions given by the GFSK modulation, which for instance does not allow for code division multiple access (CDMA) schemes. However, a schedule-based protocol like Time Division Multiple Access (TDMA) or a contention-based protocol like Carrier Sense Multiple Access (CSMA) are possible. A simple variant of CSMA to be employed in this chapter is presented in Section 21.3. The exploration of more complex MAC protocols is left to the interested reader. TDMA requires precise time synchronization and imposes more signaling overhead. A frequency division multiple access (FDMA) protocol can be employed as well as the transceiver supports multiple channels. Space division multiple access (SDMA) is not regarded as a candidate for sensor networks in [4] because of the required sophisticated arrays of antennas and signal processing techniques. However, for the specific sensor network described in this chapter it may be an interesting field of research.

To exploit the full flexibility each *openSensor* is equipped with the nRF905 unit. Furthermore the *openSensor* can also host a Bluetooth connection. As the Bluetooth technology is less flexible in terms of reprogramming, quite expensive compared to the nRF90 and using more energy, this technology will not be carried by all sensors in a sensor networks. Still, some sensors may offer a Bluetooth entry point to allow standard mobile phones to gather information from the sensor network.

21.3 Software Design

The software modules of the openSensor are illustrated in Figure 21.3. Of major importance is the MAC module which is implemented in *trans.c* and the *stack* module, both described in this section.

21.3.1 Medium Access Protocol

The MAC module is one of the basic components to establish communication. In case all sensors share the same channel, a protocol like ALOHA or CSMA may be considered. In pure ALOHA each node can transmit whenever it has data. Thus, a channel utilization up to 18% can be achieved [7]. Slotted ALOHA can give a throughput of 37% [7], however, synchronization between the nodes is necessary. Carrier sense protocols where the node first listens to the channel can give better performance. In *1-persistent* CSMA a packet is sent immediately when an idle channel is detected. In *nonpersistent* CSMA the nodes wait for a random time when the channel is in use and then sense the channel again. This prevents a collision from two nodes that wait until the end of a transmission of a third node before they both immediately transmit. A very simple variant of a nonpersistent CSMA protocol is employed here, see Figure 21.4. It is similar to the protocol presented in [8] except that the RTS/CTS handshake is only done once for the complete array of packets. Thus signaling overhead is reduced as the packets have a relatively small payload.

For streaming applications a TDMA protocol may be more suitable. The packet layout is given in Figure 21.5 for a data packet and in Figure 21.6 for

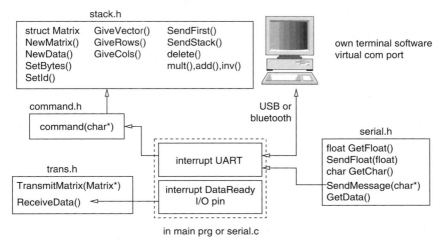

Figure 21.3. The software design of the openSensor. It consists of a command wrapper for user interaction, a serial interface module with help functions to allow data exchange between the user and the sensor, a module for wireless transmission of data via the ISM-band, and a stack module for data processing.

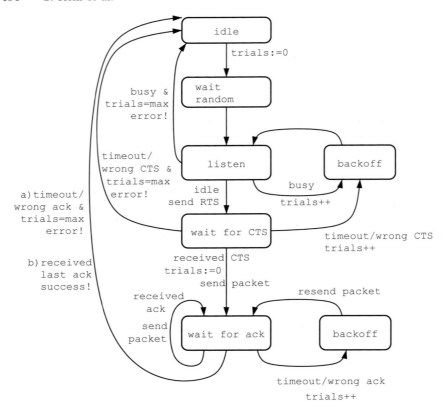

Figure 21.4. CSMA protocol as it is employed on the openSensor. It allows for a reliable transmission where each packet gets an acknowledgment. RTS/CTS - handshake is only done once when the connection is established, and not for each packet.

Figure 21.5. Fields of a data packet.

the signaling packets. There are fixed fields for the preamble, the receiver's address (ADDR) with four Bytes at most, and the CRC checksum. The user payload can be 32 Bytes at most. Each packet has a *type Byte* to specify the type of packet, which can be data, RTS, CTS, or Ack. The RTS packet has a Byte to specify the data type such as float, integer, or character, an integer field for the total number of packets, and one Byte to give the length in Bytes for the last packet. This information is stored by the receiver when a connection is established and is not contained in the data packets themselves.

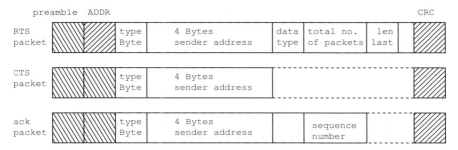

preamble ADDR   CRC

Figure 21.6. Fields of the three types of signaling packets.

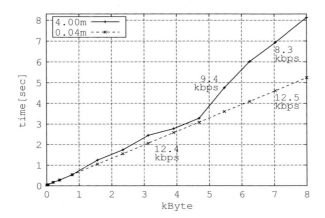

Figure 21.7. Time that is needed to transfer up to 8 kBytes achieved by the employed transceiver chip using a CSMA protocol. Measurements were conducted in a small office room for the distances 4 cm and 4 m. For the larger distance, the sensors had no line of sight and were placed at different heights from ground.

The achieved data rates for two distances are given in Figure 21.7. In ideal case, a transmission rate of 12.5 kBits per second is achieved.

21.3.2 Research in Signal Processing: the Problem

A very convenient method for developing novel signal processing algorithms is to use a high-level language like Matlab or a similar software. A tremendous part of research investigations is performed this way and the work is believed to be completed with the appropriate results and plots attained by such computer simulations. Relatively little work is devoted to demonstrate the algorithm applicability to small and low-cost devices in the real world – which of course can be realized by embedded systems. This is first, due to the difficulties in code porting from high-level to a lower level language. Even if there already exist tools for this purpose they suffer from code overhead or limitations on the hardware platform choice. Second, the proposed algorithms themselves can be too complex to cope with the embedded hardware design concerning speed and memory. A solution to construct a link between signal processing and embedded systems is to directly start programming on the embedded device, as it is proposed in [6] for a set of well-known algorithms. For the development and test of novel signal processing algorithms on embedded systems this section gives a methodology as a trade-off between a PC simulation and directly programming on a microprocessor. This trade-off is achieved by the recent performance improvements of embedded systems, and realized by the so-called *stack*, a system to allow for script-language-based measurements within the sensor network. It will even make more sense in the future as the development goes ahead.

21.3.3 The Stack: An Embedded Calculator

The *stack* is the core of the presented system and is an embedded calculator related to the reverse polish notation system (RPN). In RPN the operands precede the operator. For example, $(2+5)/3$ is done by entering 3, 2, and 5 into the stack and then doing an *add* and a *divide* operation. In this chapter the stack is twofold, it is the data structure of a linked list, and second, it contains a set of functions that can be performed on the stack elements, i.e., matrix multiply, add, or convolution, which is a typical signal processing task. A stack element can not only be a matrix with sensor data but also a script with stack commands, a data file, or a text message for the user. The stack elements are defined as *matrix objects* as follows:

```
typedef struct {
   char type;
   char *id;
   Matrix *next;
} Matrix;
```

The definition is kept very general to allow for a large number of object types. The object's data is contained in a character array that is allocated in time. If an object needs more data fields, i.e., number of data elements or type of number format (integer/float), they are hidden in the first characters of the character array. Data can be put on the stack with the supplied terminal program. By default the stack is dynamically allocated in the RAM memory. However, the user can outsource stack elements on a flash memory card.

21.4 Distributed Computing

Distributed computing is a technique where a program is divided into subparts to run on many computers for reducing the absolute computation time. Very popular examples are projects where PC users are invited to help in solving a large computing problem in a selected discipline, as for instance in biology, medicine, climate modeling, or physics. A platform for setting up such an environment is the Berkeley Open Infrastructure for Network Computing (BOINC), which allows for 475,000 active computers that are connected within the Internet. For distributed signal processing on PCs there exist commercial toolboxes, like the Matlab distributed Computing toolbox. A small extension to the set of stack commands can give a similar tool but for the wireless sensor network. This extension concerns a remote terminal function that allows to send data and stack commands to other sensors. Thus, the following example can be performed.

The scenario is now that a master sensor has to perform a set of independent tasks. Other sensors are nearby waiting for computational load to help the master sensor. One of the tasks is to solve a system of N equations. Generally, such a system can be written as $\mathbf{A} \cdot \mathbf{x} = \mathbf{b}$. The solution can be computed by matrix inversion as

$$\mathbf{x} = \mathbf{A}^{-1} \cdot \mathbf{b}. \tag{21.1}$$

The idea of the master sensor is now to outsource this task to another sensor in order to minimize its own computational load. The operational times for this procedure are given as

1. $t_{snd(A)}$ for sending the matrix A to the slave sensor,
2. $t_{snd(b)}$ for sending the vector b to the slave,
3. $t_{A^{-1}}$ for computing A^{-1} on the slave,
4. $t_{A^{-1}b}$ for computing the product $A^{-1}b$ on the slave, and
5. $t_{rcv(A^{-1}b)}$ for receiving the result vector $x = A^{-1}b$ on the master sensor.

The operational time gain t_{gain} for the master sensor is given by

$$t_{gain} = t_{A^{-1}} + t_{A^{-1}b} - (t_{snd(A)} + t_{snd(b)} + t_{rcv(A^{-1}b)}) \tag{21.2}$$

Figure 21.8 shows the operational times and the time gain for systems with $N = 5 \ldots 70$ equations. As a frequency reference a crystal oscillator with a frequency of 22.118 MHz together with a phase-locked loop (PLL) that divides the reference by four is employed, thus giving $5.53 \cdot 10^6$ operational cycles per second. Matrix $\mathbf{A}$ and vector $\mathbf{b}$ are given by floating point numbers, each number allocating four bytes. The idea of outsourcing starts to pay of for $N > 40$. Note that this idea only makes sense if the result is not needed urgently, because the overall time to solve the system of equations becomes longer with the additional time falling off for transmission of data. However, the operational time of the master sensor is reduced for $N > 40$, and thus the master sensor may solve a set of tasks in a shorter time.

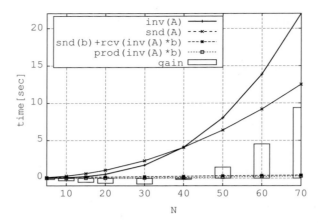

Figure 21.8. Gain achieved when outsourcing the task of solving a system of N equations. The most time is spent for calculating the matrix inverse, indicated as *inv(A)*, and for transmission of the matrix data, indicated as *snd(A)*. Outsourcing of such a task makes sense for systems with more than 40 unknown variables.

21.5 Conclusion

In this chapter the *openSensor* platform for distributed computing especially in the field of signal processing is described. The *openSensor* is realized by a mixed controller that is easy to handle yet including hardware support for matrix operations. A standard transceiver chip with a CSMA software module lets the sensors exchange data with rates of 12.5 kBits per second. Although this seems to be a fairly low rate, it may be sufficient for transmitting the result of data processing algorithms. The simple communication protocol fulfills the low energy constraints that are typical for sensor networks. The *openSensor* is provided together with software that allows to conduct measurements in a script-language-like environment. Own functions can be included and tested using the developed PC terminal program. The linear algebra example gives the conclusion that transmission of data is timely very expensive, similar than it was verified in [2,5] for mobile phones. Outsourcing a task only makes sense when a more complex computation has to be performed. The data output of the computation should be much smaller than the input. In data processing, this can be achieved with statistical summarization tools. When difficult patterns have to be recognized, the result can just be a single number indicating the detected object.

As we envision the convergence of sensor networks with mobile phones, the *openSensor* may offer Bluetooth as the main communication entry point for the phones. In Figure 21.9 the interaction of the user, the mobile phone, the *openSensor*, and the real world is shown. As each human being has a limited number of senses, the wireless sensor may support him to increase this number of senses and even to increase the range of sensibility. The number of senses

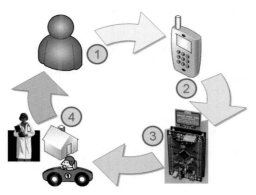

Figure 21.9. Interaction between the user, the mobile phone, the wireless sensor, and the real world.

is increased, e.g., by measuring radioactivity, which cannot be sensed by a human being, but could be performed by the *openSensor*. On the other side human beings are able to sense temperature, but by the use of one or multiple *openSensors* it is also possible to sense the temperature in all the rooms of the house without actually going there. Having said this, the sensors are very poor in the capability to convey their measurements to the user. For this task the mobile phone comes into the picture. The mobile phone is the main enabler in this communication chain due to its capability with a huge colored display, text2speech functionality, sound, and others. In Chapter 22 we will give one possible example for this convergence of sensors and mobile phones.

The idea of the *openSensor* is to encourage developers, students, and researchers to design their own distributed algorithms. A lot of open fields are left to be explored by the interested reader, including distributed audio processing, image processing, distributed coding, streaming, low-complexity data compression, security, routing, and localization. We hope that the *openSensor* can serve as a good starting point for own future investigations and are grateful for any comments and suggestions.

References

1. J. Hill et al. System architecture directions for networked sensors. *Proceedings of the 9th International Conference on Architectural Support for Programming Languages and Operating Systems*, Cambridge, MA, 2000.
2. F.H.P. Fitzek, S. Rein, M.V. Pedersen, G. Perucci, T. Schneider, and C. Gühmann. Low Complex and Power Efficient Text Compressor for Cellular and Sensor Networks. *IST Mobile Summit 2006*, Greece, 2006.
3. J. Hill and D. Culler. Mica: A wireless platform for deeply embedded networks. *IEEE Micro*, 22(6): 12–24, 2002.

4. H. Karl and A. Willig, editors. *Protocols and Architectures for Wireless Sensor Networks*. Springer, 2005.

5. S. Rein, C. Gühmann, and Frank Fitzek. Compression of short text on embedded systems. *Journal of Computers (JCP)*, 6 (Sept), 2006.

6. S.W. Smith, editor. *The Scientist and Engineer's Guide to Digital Signal Processing*. California Technical Pub, 1997.

7. A.S. Tanenbaum, editor. *Computer Networks*. Prentice-Hall, 1996.

8. A. Woo and D. Culler. A transmission control scheme for media access in sensor networks. *Proceedings of the 7th Annual International Conference on Mobile Computing and Networking (MobiCom)*, 2001.

22

Parking Assistant Application

Janne Dahl Rasmussen, Peter Østergaard, Jeppe Jensen, Anders Grauballe, Gian Paolo Perrucci, Ben Krøyer, and Frank H.P. Fitzek

Aalborg University {jannedr|petero|jeppe85|agraubal|gpp|bk|ff}@es.aau.dk

Summary. This chapter describes the first practical example for the wireless sensors described before. The idea is to equip a car with distance measurement sensors. The result of the distance measurements are displayed on mobile phone. The presented application is realized in JAVA and Python for S60.

22.1 Motivation

After introducing the wireless sensors, we give an example how mobile phones can collaborate with wireless sensors. One example is the parking assistant application. Often high class cars have a system where the driver gets a rudimentary information when the car is close to nearby objects. Medium and low class cars do not have this functionality as the price of the parking assistant is too high. By our approach, we enable even the drivers of those cars to be assisted in parking their car. But the envisioned application can even do more than any onboard parking assistant system. Nowadays the parking assistant only gives the current distance to the nearest object. Due to the sophisticated display of the mobile device, not only the actual distance measurements are displayed. Furthermore, it is possible to compose a measurement series to an understanding of the surrounding obstacles.

22.2 Mounting the Sensor

The parking assistant is realized by mounting a wireless sensor on each corner of the car. These sensors need to send the measurements to the mobile phone and this can be done by establishing a connection between each sensor and the mobile phone using a Bluetooth communication. Bluetooth has been chosen for this prototype as nearly all phones have such technology onboard. In Figure 22.1 the usage of the mobile application with the wireless sensors is displayed.

F.H.P. Fitzek and F. Reichert (eds.), Mobile Phone Programming and its Application to Wireless Networking, 411–417.
© 2007 *Springer.*

Figure 22.1. This picture shows the sensors plugged on the car and the screen of the phone during the test phase of the application.

22.3 JAVA Realization

The JAVA-based parking assistant application was developed as a part of a fifth semester project at Aalborg University (Denmark). It was tested on a Nokia N70, and the sensors were done in collaboration with the Technical University of Berlin (Germany). The application was developed using J2ME and the source code can be found on the DVD. To use the parking assistant application, the user starts the application on the mobile phone, and it will connect to the sensors that are located on the car. The mobile phone can then be placed in a retainer, so both hands are free to park the car. The driver starts to turn into the parking bay and the display on the mobile phone shows the distance to any nearby objects. When the car gets too close to an object, the application notifies the driver that he or she must stop the car. The different steps in the parking process are shown in Figure 22.2. When the application is started, the screen will show a drawing of a car seen from the top as in the Figure 22.3. The main menu, which can be accessed by pressing the left soft key, contains the following entries:

- Install Sensors
- Settings
- Start
- Exit

22.3.1 Install Sensors

If this option is chosen a submenu will appear and it is possible to connect with new sensors and bind them to the application. The submenu contains the following entries:

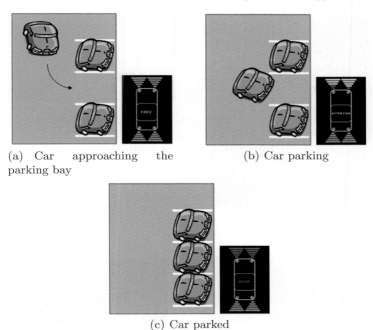

(a) Car approaching the parking bay

(b) Car parking

(c) Car parked

Figure 22.2. Parking Assistant Application testing.

Search The application will start to search for devices. Because sensors may have the same Bluetooth-name, the MAC-addresses of the found devices are shown on the display. When the application has completed the search, any found device can be chosen and a new screen will appear where it is possible to choose the position for this device (front-right, front-left, back-right, or back-left). When a position is chosen, the screen with the found devices will appear again. Now it is possible to choose other devices and assign a position for it.

Save When all the sensors are chosen, the settings can be saved using this button.

Load It can be used to load some sensor settings previously saved.

View It can be used to view the sensor settings saved.

22.3.2 Start

When the sensors are installed, the 'Start' will enable the parking assistant. The application starts the connection with the installed sensors and the distance bars are updated on the screen of the phone in real time. When any of the four sensors measure a distance below 100 cm the application will show a message: *Attention*; when it gets below 50 cm: *Careful* and below 10 cm: *Stop*.

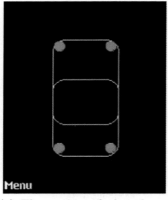

(a) The screen of the phone when the application is launched.

(b) Main menu.

(c) Start submenu

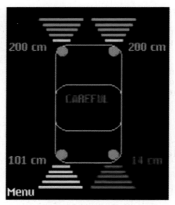

(d) Application running. On the screen the values of the distance are updated every time a new value is received from the sensors

Figure 22.3. Screenshots of the J2ME realization of the Parking Assistant Application

22.3.3 Exit

Choosing this option the application will exit.

22.4 Python Realization

The same application can also be realized with Python for S60. One possible script is given in the Listing 22.1. To make it easy only one sensor is connected to the mobile phone and we leave it up to the developer to make the needed changes for multiple sensors. The script opens a Bluetooth connection for retrieving data from the wireless sensor. The values of the distance are sent by the sensor and shown on the screen. A Nokia N70 has been used for testing the script and the source code can be found in the DVD.

When the application starts, the screen of the phone looks like the one in Figure 22.4(a). Pressing the "Options" button and then "Connect" (see Figure 22.4(b)), the phone starts to search for Bluetooth devices in its range. Selecting the "Serial Port Device" in the list (see Figure 22.4(c)), the Bluetooth connection is established. The sensor is continuously sending the values of the distance from an object using the RFCOMM protocol which emulates a RS232 serial port. The flow of data is parsed and the actual value of the distance is updated on the screen as shown in Figure 22.4(d).

Listing 22.1. Distance Sensor Gateway

```
   import appuifw,e32,socket
   from graphics import *

   appuifw.app.screen='large'
 5 appuifw.app.body=c=appuifw.Canvas()
   c.clear(0x000000)
   c.text((30,20),u'Distance_sensor',0xffffff,font='title')
   running =1
   sock=socket.socket(socket.AF_BT,socket.SOCK_STREAM)
10 def connect():
       global sock
       addr,services=socket.bt_discover()
       port=services[services.keys()[0]]
       sock.connect((addr,port))
15
   def getDist():
       global sock,running
       io=''
       dist=''
20     line=[]
       while running:
           ch=sock.recv(1)
           if(ch=='\n'):
               temp=''.join(line)
25             if '.' in temp:
                   dist=temp[1:temp.find('.')]
               else:
                   dist='0'
               print dist
30             update(int(dist))
               line=[]
           line.append(ch)

   def update(measure):
35     c.clear(0x000000)
       c.text((30,20),u'Distance_sensor',0xffffff,font='title')
       c.text((25,70),u"Distance:_"+str(measure)+u'_cm',
               0xffffff,font='title')

40 def start():
       connect()
       appuifw.app.menu=[]
       getDist()

45 def Quit():
```

```
      global running , sock
      running =0
      sock . close ()
      appuifw . app . set_exit ()
50
   appuifw . app . menu = [( u"Connect" , start )]
   appuifw . app . exit_key_handler=Quit
   app_lock = e32 . Ao_lock ()
   app_lock . wait ()
```

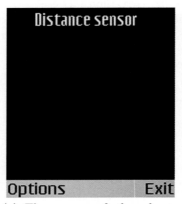

(a) The screen of the phone when the application is launched.

(b) Using the "Connect" entry in the menu for starting the Bluetooth device discovery.

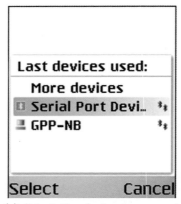

(c) Select the device to connect to. In this case the sensor is represented by the "Serial Port Device" entry in the list of Bluetooth devices found.

(d) Application running. On the screen the values of the distance are updated every time a new value is received from the sensor

Figure 22.4. Screenshots of the Python for S60 realization.

22.5 Outlook

The parking assistant is only one possible combination of wireless sensors and mobile phones. A lot of more applications can be found. In the field of car manufacturing, another example is to substitute the onboard navigation system, which has become very popular these days. The onboard navigation system will increase the overall price of the car and is often a reason for thieves to break into the car. A solution could be to use a navigation application on the mobile phones, which makes the onboard unit obsolete. This would not increase the price of the car and as the mobile phone is always with the customer it is not prone to criminal activities.

Part VIII

Power Consumption in Mobile Devices

Energy Efficiency of Video Decoder Implementations

Olli Silvén[1] and Tero Rintaluoma[2]

[1] University of Oulu, Olli.Silven@ee.oulu.fi
[2] Hantro Products Oy, Tero.Rintaluoma@hantro.com

Summary. High-end mobile communication devices integrate video cameras, color displays, high-speed data modems, net browsers, media players, and phones into small battery-powered packages. The physical size limits the heat dissipation, while the battery capacity needs to be used conservatively to provide for satisfactory untethered active use time. Together with the required versatile capabilities of the devices, the physical size and battery capacity are essential constraints that must be taken into account from hardware to application software design. In video decoding additional constraints come from the need to support multiple digital video coding standards, and the platform-oriented design regimes of the device manufacturers. Along these lines, we consider the implementations of video capabilities for mobile devices in a top-down manner starting from typical applications, progressing to energy efficiency analysis via device architectures to codec implementations, and software platforms.

23.1 Introduction

Wireless multimedia applications typically use content provided via the web or broadcast services such as DVB-H [26], or play back locally stored music and movies. In addition, users can create content and stream it to the network for redistribution or make video calls that require real-time streaming. The popularity of laptop PCs as DVD players and as a means to access multimedia content via public WiFi networks give a prediction of the future uses of wireless terminals.

The technical requirements for wireless terminals are tough, especially when considered from the energy efficiency point of view. At the same time high demands are placed on the usability that includes not only the intuitiveness of the user interface, but also the length of active usage time between charging the batteries.

A typical laptop PC user carries a charger and connects to the mains whenever possible, and uses the device while sitting down. In contrast, a handheld device is expected to provide for a longer active use time as they

421

F.H.P. Fitzek and F. Reichert (eds.), Mobile Phone Programming and its Application to Wireless Networking, 421–439.

Table 23.1. Characteristics of typical portable multimedia devices.

	Portable Laptop PC	Handheld Multimedia Terminal	Typical ratio
Display size (inches)	12–15	2–4	5x (area 20x)
Display resolution (pixels)	1024x768– 1600x1200	176x208– 640x240	15x
Processor DRAM (MB)	256–1024	16–64	16x
Processor clock (GHz)	1–3	0.1–0.3	10x
Max. power dissipation (W)	60	3	20x
Surface area (cm^2)	1500	150	10x
Heat dissipation (mW/cm^2)	40	20	2x
Video resolution	720x576/25 Hz	640x480/30Hz	1x
Battery capacity	4000 mAh/14.4V	1000 mAh/3.6V	15x

are used anywhere in an untethered manner, and are charged only at night. Another energy efficiency related aspect is heat dissipation. This is mostly the concern of the handhelds, as most of the time the laptop devices are desktop operated.

The characteristics of typical wireless handheld and laptop multimedia devices are compared in Table 23.1. The application requirements are almost the same, but the handheld devices provide the services using around one-tenth of the size, energy, and processor speed, while the maximum heat dissipation via surfaces can be only half of the laptop level. The power consumption of the larger display of the laptop explains less than 10W of the difference.

The usability of the handheld devices is critically dependent on their active use times; these in turn depend on the energy efficiency. Table 23.2 presents the power consumption breakdown of an early 3G phone in 384 kbit/s video streaming mode [16], clearly showing the limitations of the multimedia implementation. Only 600 mW is available for application processing, in this case decoding of video bit stream into sequences of image frames. This is a very hard requirement for software solutions. With a 1000 mAh battery the active time is limited to around an hour. The application power needs of the PDA device [21] are in the same region, while the larger display and frame buffer memory explain most of the higher power consumption. The hypothetical power budget for a future device that provides for three hours of active use time has been estimated based on the data of the most power-efficient system components available today.

The increasing bandwidth needs and the increasing complexity of air interfaces make it very difficult to save in RF and baseband signal processing, while essential efficiency improvements can be expected from the display technologies, e.g., by switching from TFT LCDs to OLEDs. Still, the application processing is under very severe pressure to cut down power.

Table 23.2. Power consumption breakdown examples of pocket-sized devices.

System component	Power consumption (mW)		
	3G phone in video streaming mode [16]	PDA device in MPEG-4 playback [21]	Expected future mobile devices
Application processor and memories	600	840	100
Display, audio, keyboard and backlights (UI)	1000	2510	400
Misc. memories	200	670	100
RF and cellular modem	1200	N/A	1200
Total	3000	4020	1800
Battery capacity mAh/usage time	1000/1 h	N/A	1500/3 h

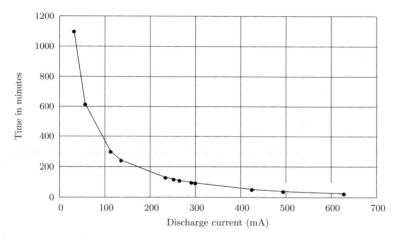

Figure 23.1. Discharge time of 640 mAh LiON battery under constant load.

A standard way to estimate the capacity that can be obtained from a battery at a given rate of discharge is to use the well-known Peukert's law [19], but it can be inaccurate for mobile devices. This is explained by the dependency of the battery capacity on temperature that in the absence of active cooling strongly depends on the load current [6]. Figure 23.1 below shows the actual behaviour of a 640 mAh LiON battery in a PDA device under constant load based on the experiments in [19]. As the battery life is a non-linear function of the load current, improved power efficiency in the *knee* region of the curve will result in superlinear improvements.

Hints on how to lower application power consumption can be seen, e.g., in the estimates provided by a multimedia processor supplier [18] for a hypothetical mobile device with a *smart* energy-efficient display shown below in

Table 23.3. Multimedia use cases for a processor platform on 3.6V 800 mAh battery.

Use case	Target bitrate (Mb/s)	Power consumption (mW)	Usage time (h)
Video capture	1–4	350	8
Movie playback	1	500	6

Table 23.3. The video standard in question is MPEG-4 SP [13] at 30 frames/s that is close to DVD quality. The power results are for a full system including all components such as camera, display, speakers, etc., but this device lacks wireless connectivity. The Microsoft Zune player based on this particular processor technology achieves 4 h video playback in QVGA format on 800 mAh 3.7V battery. In other words, the device dissipates around 700 mW.

Interestingly, in Table 23.3 the computationally expensive video encoding consumes less power than decoding. This is explained by hardware acceleration provided for encoding, while decoding is performed in software. In this case hardware acceleration has been used as an encoding application enabler, rather than to save power. We may conclude that hardware for video decoding would cut the consumption figures of playback below the currently estimated ones. The assumed processing platform implementation technology in this case is 90 nm CMOS.

In the following sections we review the energy efficiency issues of mobile video decoders to better understand the implementation alternatives. Comparative evaluations are presented when there is available data.

23.2 Mobile Video Applications

The user experience of portable devices can be significantly harmed by the battery running out prematurely. These kinds of multimedia applications include the use of the mobile devices as camcorders, video phones or mobile digital TVs, and have an impact on system designs and power budgets. The more versatile devices tend to be less energy efficient due to the added software and hardware complexity, and platform technologies needed to support rapid development.

The usual organization of a portable multimedia device is shown in Figure 23.2 in which two interconnects are present for transfers between subsystems. The *Interconnect 2* is designed to provide high transfer bandwidths between the camera, memory, and the video and image processing units. *Interconnect 1* is the system bus that interfaces the essential units with the master processor.

Camcorder use requires real-time encoding and preview capabilities, while during playback decoding and display are needed as illustrated in Figure 23.3. Encoding in consumer use will be mostly limited to short D1 (720*576@25frames/s) sequences within the memory capacity of the device,

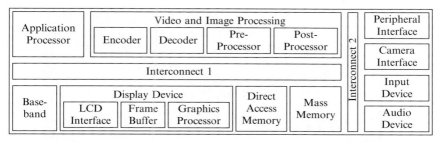

Figure 23.2. Organization of portable multimedia device.

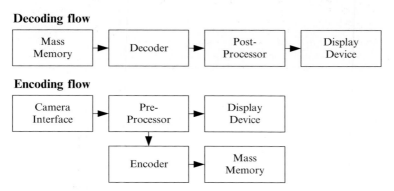

Figure 23.3. Decoding and encoding data flow.

but requires significantly more computing and electrical power than decoding. Interestingly, typical consumer camcorders need around 8–9 W of power in encoding mode, while their displays are in the same size class as multimedia capable mobile phones. Part of the 6–7 W disparity, around 1–2 W [5], is explained by the electro-mechanics of the DVD drive and additional electrical interfaces of the camcorders, but a significant portion comes from the computing platform, including the display interface.

Video calls on a mobile device require the simultaneous operation of video encoder, previewer, and decoder together with both uplink and downlink streaming via the baseband. Clearly, video calls are the most power consuming application, and are likely to be bound by the heat dissipation limit even in the future, although the durations of such calls are usually moderate. The current video call resolutions are typically limited to QCIF (176*144@15frames/s) and the bit rates to 64–128 kbits/s in cellular networks, but the user needs are leading to the level of current fixed line video call resolutions, CIF to D1, and 1–2 Mb/s bandwidths.

The emergence of mobile TV is expected to result in the introduction of terminals that support several simultaneously decoded program streams. This is necessary to provide for living thumbnails, as shown in Figure 23.4,

Figure 23.4. Picture-In-Picture mobile television demonstration.

and seamless channel surfing despite the energy saving time-slicing technique used in the air-interface of DVB-H. However, thumbnails or split displays effectively multiply the power and memory bandwidth needs of the decoding task. In practice, either at least two decoders, as in desktop digital TV set-top boxes, or the shared use of the decoding resources are needed.

An interesting opportunity is to employ several instances of software-based decoders on a fast processor that supports frequency and voltage scaling. For instance, only about 10–15% of the maximum performance of Core Duo processor is needed for the H.264/AVC [9] decoding task at CIF and QVGA resolutions [8]. This choice results in an energy efficiency and cost compromise that can occasionally be justifiable when all the factors influencing the system design are added up.

23.3 Mobile Video Codec Implementations/Platforms

The two basic alternatives to implement video coding in mobile devices are the completely software-based approach and hardware acceleration. The energy efficiency of software is lower, whereas hardware acceleration, when implemented in a monolithic manner, can result in inflexibility. The actual design problem is to find a cost-efficient way to deliver the desired capabilities.

23.3.1 Software-based Decoders

The power needs of mobile and general purpose processor-based platforms in multimedia applications are approaching each other and in the future they may not differ significantly. To illustrate these developments, Table 23.4 shows the nominal energy per instruction data (EPI) of Intel PC processors normalized to the same 65 nm 1.33V silicon process [8]. The table also contains the required cycles/s for MPEG-4 and H.264 decoding [7], and the approximate power consumptions of the decoders when the processors are clocked at their nominal rates without dynamic power management. All the video figures have been scaled from the data given for Pentium 4 [7]. In power consumption calculations 2.59 cycles per instruction (CPI) has been assumed based on the findings in [14].

For comparison, Table 23.5 shows the power consumptions of selected ARM architecture processors in decoding of 30 frames/s VGA baseline H.264 bitstreams. DVD/MPEG-2 [1] decoding at 3Mbit/s and MPEG-4 at 512kbit/s rate need 30–50% of these figures. The costs of accessing external memories are not included. In practice, they may more than double the power needs. The processor implementations have been roughly scaled to the same technology as above (65 nm, 1.33 V) by assuming that the power consumption is proportional to supply voltage squared and the design rule (line width). The CPIs (cycles per instruction) with this application are 1.2 and 1.4 for ARM10 and ARM11, respectively.

Although the energy efficiency disparity between the processor categories in this application is significant, it is narrowing, and can have an impact on platform designs. At the moment, the mobile processors essentially offer low-power technology and lower cycle rates, but at the cost of lower peak

Table 23.4. Processor cycles/s and power needs of MPEG-4 and H.264 decoders (VGA 30 frames/s, 470 kbit/s) on three Intel processors.

Processor	EPI (nJ)	MPEG-4		H.264	
		Cycle rate (MHz)	Power needs (mW)	Cycle Rate (MHz)	Power needs (mW)
i486	10	N/A	N/A	N/A	N/A
Pentium 4 (Cedar Mill)	48	273	5060	725	13440
Pentium M (Dothan)	15	400	2320	1060	6140
Core Duo (Yonah)	11	280	1190	744	3160

Table 23.5. Power needs of H.264 decoders on ARM processors (VGA 30 frames/s, 512 kbit/s).

Processor	Cycle rate (MHz)	Power consumption (mW)
ARM10 (1022E)	384	222
ARM11 (1136J-S)	434	332

Table 23.6. Energy characteristics and silicon areas of ARM processors.

Processor	Clock frequency (MHz)	Silicon area (mm^2)	Power consumption (mW/MHz)	EPI (nJ)
ARM7 (7EJ-S)	260	0.5	0.1	0.22
ARM9 (926EJ-S)	500	1.55	0.29	0.46
ARM10 (1026EJ-S)	540	2.45	0.45	0.54
ARM11 (1136J-S)	620	2.5	0.45	0.63

Table 23.7. Instruction counts in millions for decoding 1s video bit stream (VGA 30 frames/s, 512 kbit/s).

Processor	MPEG-4	H.264
Pentium 4	105	280
Pentium M	154	410
Core Duo	108	288
ARM9	129	322
ARM10	129	322
ARM11	99	311

performance when compared to PC processors. Table 23.6 shows the power needs of ARM series cores implemented using a 90 nm 1V CMOS process, demonstrating the impacts from pursuing higher computing throughput via architectural changes. The EPI values have been calculated from H.264 video decoding, in which the CPI for ARM7 and ARM 9 are 2.2 and 1.6, respectively.

The instruction counts presented in Table 23.7 of the above highly optimized MPEG-4 and H.264 decoders are similar on ARM and Pentium processors. We may conclude that the big differences in energy efficiency are at least partly explained by architectural features that are not used at this application level, such as memory management, multitasking support, and superscalar execution.

We can estimate that a DVD playback application consumes about 150–200 mW on mobile processors implemented using the 65 nm 1.33V CMOS process. This can be compared to a software implementation on Core Duo CPU [5] that through exploitation of dynamic voltage and frequency scaling techniques consumes approximately 700 mW, saving at least 500 mW from the nominal level.

Now, returning to the expected future devices in Table 23.2, we realize that use of a modern PC processor for video streaming applications would increase the system power consumption by around only 30%. This can be an acceptable trade-off, although the higher load currents could drop the battery lifetime more than proportionally. However, light devices with long active use times still need mobile processors and hardware acceleration.

23.3.2 Hardware-based Decoders

Hardware acceleration has the potential for high energy efficiency when compared to software implementations because of the lack of instruction fetch and decoding cycles. In addition, a substantial share of the data accesses can be eliminated. Figure 23.5 shows the pipelined internal functional blocks of the monolithic MPEG-4 decoder, in which energy efficiency is achieved by clock-gating and by avoiding the use of external and shared memories. The data buffers needed between the blocks are implemented either as registers or memories, depending on the capacity requirements and silicon area consumed.

Table 23.8 shows the relative energy consumptions of two MPEG-4 decoder implementations that have very similar input data access characteristics, but the hardware accelerated solution uses only 5% of the total energy needed by the software solution. Furthermore, the software solution, although optimized, needs 30 times more energy for memory accesses than the accelerated application. The explanation is in the accesses needed to store intermediate results.

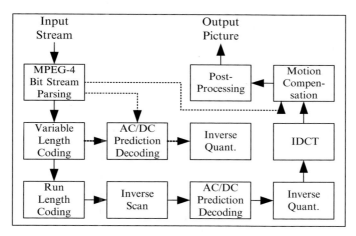

Figure 23.5. The internal pipelined organization of a monolithic MPEG-4 accelerator.

Table 23.8. Estimated energy shares of system components in MPEG-4 video decoding.

Platform component	Hardware decoder [11]	Software decoder [22]	Relative energy (HW:SW)
CPU	1% (ARM9)	20% (ARM7)	1:10
Memories and memory interfaces	54%	80%	1:30
Decoder HW	45%	N/A	N/A
Relative total energy	1	20	1:20

Table 23.9. Gate count and memory needs of video decoder implementations for (VGA, 30 frames/s).

Implementation	Gate count (kgates)	Memory footprint (kB)	External memory (kB)	Internal memory (kB)	Relative energy consumption
Software H.264 on ARM11 processor	N/A	65	1164	N/A	2
Software MPEG-4 on ARM11 processor	N/A	45	1061	N/A	1
Monolithic hardware multiformat decoder (MPEG-4, H.264, JPEG)	221	105	947	18	0.25 (H.264) 0.06 (MPEG-4)
Monolithic hardware MPEG-4 decoder	96	60	938	3.7	0.05

If support for multiple coding standards is needed in the same device, software is obviously the more flexible implementation technique. To illustrate the rapid complexity growth of monolithic hardware accelerator designs, Table 23.9 [27] shows four decoders, two software and two hardware implementations. The accelerators are a single and a multistandard decoder. The software codes for MPEG-4 and H.264 are essentially rewrites, and reuse only the instruction set, while the hardware implementation reuses the functional blocks and suffers from increasing amount of control logic with each added decoder. We can also notice the growth of *internal memory* that is the buffers between the accelerator blocks, however, the silicon area consumed by buffering is still small in comparison to the total.

The complexity growth is continuing with new standards and extensions such as VC-1 [2] and the main profile of H.264. These add new functional blocks that complicate the routing of data in multistandard accelerators, making the silicon area gap to single standard-based designs narrower. The consumers may witness this development as high-end dedicated devices that employ hardware acceleration, and as all-in-one mostly software-based compromise solutions.

A way to curb the complexity growth of hardware solutions is to resort to finer grained acceleration. Then, more of the control complexity could be allocated to software. Unfortunately, fine grained hardware acceleration is difficult to implement in an energy-efficient manner, because of the overheads in software/hardware interfacing, and the obviously needed shared memories.

To illuminate the overhead issues we can consider the costs of interrupts that are the usual mechanism to synchronize hardware accelerators to software. A typical interrupt overhead in a platform with an operating system is around 300 CPU cycles. On the other hand, the 8x8 IDCT (Inverse Discrete Cosine Transform) that is common in video decoders needs around 300 CPU cycles when implemented in mobile microcontroller software, and less when

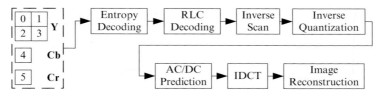

Figure 23.6. Decoding of a MPEG-4 intra macroblock.

a suitable DSP or media processor is employed. Although a straightforward hardware implementation can execute in a few tens of clock cycles, such a solution is more expensive than software after the interrupt overhead and the actual interrupt service cycles are added.

The next step towards coarser granularity is to employ hardware acceleration for decoding macroblocks; Figure 23.6 shows the idea for the *intra* case. The top level of the finite state machine control of the decoding pipeline is now replaced by software. With MPEG-4 each macroblock may comprise 0–6 8 × 8 blocks that each need to be variable-length decoded, AC/DC predicted, and IDCT'd. In addition, motion compensation that employs bilinear interpolation of image data is required for *inter*-coded macro-blocks.

Depending on the degree of parallelism in hardware, and the number of coded blocks and the type of the macroblock, a typical macroblock accelerator needs around 800–1200 cycles. So the interrupts needed for software/hardware synchronization add 20–30% in overheads. As decoding a macroblock in software consumes 5,000–6,000 cycles, the accelerator enables significant savings, but with a VGA-sized bitstream, 36,000 interrupts are generated each second, wasting a significant portion of the processor resources in overheads, extra administration, and control. Furthermore, the gate counts of macroblock decoders are not much lower and the flexibility is not any better than with monolithic accelerators, so the system designers tend to choose the latter approach.

An ideal fine grain accelerator solution should work as shown in Figure 23.7 where an *inter*-macroblock in a MPEG-4 stream is being decoded. VLD and MV denote to Variable Length Decoding and Motion Vector, respectively. The pipeline of the monolithic hardware solution has been replaced by cooperatively scheduled threads that switch between software and hardware execution. The latencies of the accelerator blocks are deterministic and known when the threads are scheduled based on the contents of the input bit stream. Interrupts are not needed, as each hardware execution is guaranteed to have finished before the respective next step in software continues.

Based on our experiments significant energy efficiency improvements can be achieved through the cooperative thread scheduling as software/hardware synchronization overheads are low in comparison to the interrupt driven scheme. Early digital mobile phone baseband implementations built in this manner [23] show the efficiency of the approach when compared to the current

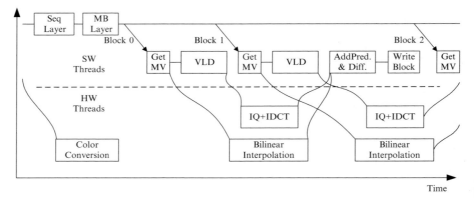

Figure 23.7. A use-case for fine-grained hardware accelerators.

interrupt driven implementations. Finally, from the hardware development point of view the fine-grained accelerators are faster and easier to verify than the monolithic ones, that alone make the approach attractive.

23.4 Software Interfacing Issues

As mobile devices must satisfy a multitude of requirements, the result is often a complex hardware and software system. For instance, support for several standards is needed, such as JPEG, H.264, MPEG-4, and VC-1, in the same execution environment. To deal with the complexity the manufacturers have created platforms for their product families, and define stable interfaces for software developers to facilitate the integration of applications to the devices. The interfacing solutions that include multitasking operating systems, APIs, and middleware, can have big impacts on system performance. The interface overheads have had a tendency to increase with platform generations, and have influenced system design principles.

23.4.1 Operating System Costs

Practically every multimedia application runs in an operating system environment and shares the CPU resources including clock cycles and cache memories. As a result the applications run slower on operating systems than on naked processors, even if a single task is running and the only operating system activity are the clock/scheduling interrupts. Park et al. (2004) found that an MPEG-4 encoder ran 20% and the decoder 27% slower [17] on ARM926 processor using embedded Linux operating system. Cache effects due to the more fragmented memory access pattern were pinpointed as the key reason for the degradation, while the executed kernel code increased the instruction count only by around 6%.

When several applications are active, the overheads increase due to context switching. Based on overhead measurements by Mogul and Borg [15] in 1991 and Sebek [20] in 2002 the context switch latencies appear to have remained the same for more than a decade despite processors becoming much faster. This is explained by the low cache hit ratios during the context switches.

Mobile operating systems, such as Symbian [24], have been designed to operate at low overheads, although the application start-up costs can be high. However, few applications interface to the plain operating system, instead, they are usually built to run on supposedly stable interfacing frameworks. Much of the software overheads originate from these layers that provide for compatibility between platform and operating system versions.

23.4.2 The Cost of a Multimedia Software Framework

Symbian Multimedia Framework (MMF, Figure 23.8) is a multithreaded app-roach for handling multimedia data, and provides audio and video streaming functionalities. It is intended to standardize the integration of software and hardware based video coding solutions and provides mechanisms that enable building multimedia applications that are portable between platform genera-tions.

With MMF, regardless whether the codecs are implemented in software or hardware, they are interfaced as *plug-ins* to the Multimedia Device Framework (MDF). With actual hardware codecs the plug-ins hide the vendor-specific device drivers. The MDF with plug-ins is middleware that can be used to hide an underlying multiprocessor implementation, for example, a decoder plug-in may hide a decoder running on a Texas Instruments DSP processor behind XDAIS [4] interface. The codec vendors implement the MDF plug-ins with specified interfaces, and the MMF controller plug-ins that take care of, e.g., synchronization between audio and video [25].

The application builders use the Client API that handles requests such as *record, play, pause*. At a minimum, these activations of requests go through five software interface layers before reaching the codec, but the performance depends greatly on the vendor provided controller plug-ins.

In Symbian operating system version 7 of 2003 the MDF was the whole framework that increased with two new abstraction layers, Client API and Controller Framework, in version 9 released in 2005. We can expect additional layers in the future to support more versatile multimedia applications, based on, e.g., the emerging MPEG-21 standard.

The proprietary solutions from mobile video codec manufacturers app-roach the portability issue in a different manner. For instance, in [10] thin software wrapper layers are used to mould hardware and software codecs into *multimedia engines* that implement video recording, playback, and other functionalities in a tightly integrated manner. The controller functionalities are internal and as such not accessible to application builders.

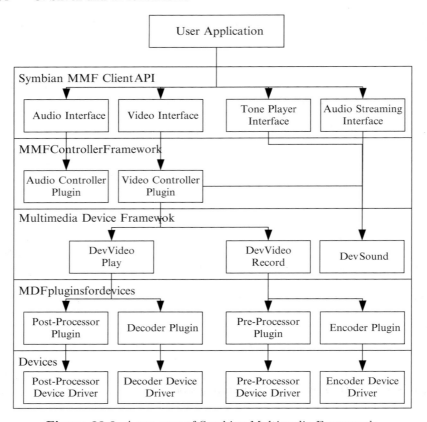

Figure 23.8. A use case of Symbian Multimedia Framework.

Table 23.10. The costs of multimedia APIs.

	Decoder software interfaces		
	Proprietary API	Symbian MMF	Difference
Total cycles	220890240	225165845	2.14 MHz
D-cache misses	1599992	1633425	33433
I-cache misses	250821	322635	71813
D-cache hit ratio	94.7%	94.6%	0.1%
I-cache hit ratio	99.7%	99.6%	0.1%

Table 23.10 compares the costs of accessing the video decoder functionality directly via its own proprietary API, and through Symbian MDF level. The experiments were made on an actual ARM11 platform without SIMD optimizations and with a system supporting a single video coding standard. With more codecs the overheads of using any of them are slightly higher. In power consumption the difference between these multimedia application interfacing solutions is around 1mW based on Table 23.6. The cache hit ratios are

Table 23.11. Typical software interface costs in an embedded system environment (Symbian 9).

Mechanism	Overhead/cycles
Procedure call	3–7
System call (user-kernel)	1000–2500
Interrupt latency	300–600
Context switch	400
Middleware	60000

essentially the same, highlighting the small memory footprint of the mobile operating system. The costs are approximately the same for both software and hardware decoders.

In the above Table 23.10 the bulk of the overhead cycles originates from the middleware. As an interfacing mechanism it must be exploited sparingly, primarily with rare long latency services. Previously, Verhoeven et al. (2001) found that the performance of different middleware solutions varied between 260 and 7,500 calls per second [28]. This translates into at least tens of thousands of cycles per call and is in line with our observations. Table 23.11 shows the typical overheads of interface mechanisms as ARM11 processor cycles on a Symbian operating system.

As the older mobile processor architectures appear more energy-efficient than the more recent ones (see Table 23.6), a system designer is tempted to employ multiple processors. This kind of solution, is employed at least on Agere's Vision x115 platform that employs both ARM7 and ARM9 cores [3].

23.4.3 Internal Interface Overheads

The internal organization of the decoder requires at least function interfaces, and perhaps simple APIs to provide for configurability. As typical video decoders consist of 10–20 algorithms that are in total invoked around 1–2 million times each second for VGA sequences, the costs of invocation mechanisms are important, regardless of whether the implementation is in the software or hardware.

Figure 23.9 shows an indicative organization of a MPEG-4 decoder [12] that consists of layers that each provide decoding functions for the upper layer. This is the structure designed already into the standards and regularly imitated in software implementations. The sequence layer is executed once for each frame or video packet, and extracts information on the employed coding tools and parameters from the input stream. The macroblock layer in turn controls the block layer decoding functions. For a VGA bit stream, the macroblock layer is invoked at most 1,200 times per frame, while the block layer is run at most 7,200 times.

Table 23.12 demonstrates the costs of internal software interfaces, when the APIs enabling reusability of functionalities are placed on the sequence,

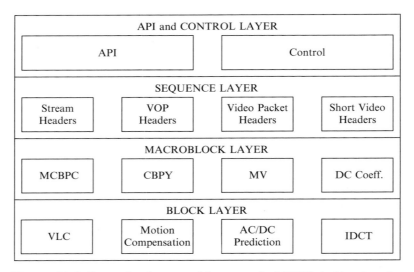

Figure 23.9. Layered software architecture of a MPEG-4 video decoder.

Table 23.12. The power used into internal overheads with three API layer options (MPEG-4 decoder on ARM11 implemented at 90 nm CMOS, VGA 30 frames/s, 512 kb/s).

APIs	Overhead (MHz)	Power consumption (mW)
Sequence layer only	0	0
Sequence and macroblock layers	2.7	1.2
Sequence, macroblock, and block layers	11.4	5.1

macroblock, and block layers, and the assumed call overhead is 7 cycles. These overheads were measured with a pure software implementation, and are an approximate lower bound of fine-grained accelerator solutions. The figures do not contain the costs of any functionality in the layers, and the experiments were run without an operating system for maximum efficiency.

The internal overheads of the decoder are about 5.4% of the total decoder effort. Energywise the overheads are about the same as the total needs of a monolithic MPEG-4 decoder accelerator implemented using the same silicon technology.

23.5 Summary

Due to its computing power, user interface, and data communication needs, video is among the most demanding services in mobile devices that are increasingly used for entertainment applications. These require at least 3-4 h of battery life from small devices, and are a big challenge for energy-efficient design.

The decision between software and hardware implementation of the video functionality sets the baseline for power consumption in the active state. With software implementations the energy efficiency gap between naked mobile and desktop processors is narrowing, while at the same time the application diversity of mobile devices is growing. This can be expected to influence the platform designs.

The energy efficiency of hardware accelerated video decoding is very good, and higher resolutions than with software implementations can be supported. On the other hand, the monolithic hardware solutions suffer from increasing complexity, and silicon area, due to the frequently required multistandard support. As a result, finer-grained video accelerators are becoming attractive.

Regardless of the implementation technology of video decoding capability, it needs to be integrated with applications. This is done via operating systems and software frameworks that hide the implementation details, and provide standardized service interfaces. From the energy efficiency point of view the interfacing overheads can be significant. Mobile software platform designs with small footprints perform well in that respect.

Combining energy-efficient system components does not automatically lead to efficiency. Meticulous analysis of the coding flow from bit stream to display together with solving intermingled hardware, software, and application problems is needed.

23.6 Acknowledgments

Numerous people have directly and indirectly, sometimes even unknowingly, contributed to this paper by providing us their observations, comments, and questions. In particular, we wish to thank Messrs. Juuso Raekallio, Jani Huoponen, and Juha Valtavaara from Hantro Products Oy, Mr. Kari Jyrkkä from the Nokia Corporation, Dr. Mark Barnard and Mr. Jani Boutellier, both from the University of Oulu.

References

1. ISO/IEC 13818-2:2000. *Information technology - Generic coding of moving pictures and associated audio information: Video.* ISO/IEC, 2nd edition, December 2000.
2. SMPTE 421M-2006. *VC-1 Compressed Video Bitstream Format and Decoding Process.* SMPTE, February 2006.
3. Agere. Agere Systems Announces Chipset Solution for Mainstream Mobile Phones Capable of Cinema-Quality Video and CD-Quality Sound. *NEWS Release.* Agere Systems, October 2005.

4. S. Blonstein. The TMS320 DSP Algorithm Standard. *Application Report SPRA581C.* Texas Instruments, May 2002.
5. R. Chabukswar. DVD Playback Power Consumption Analysis. Intel, 2007.
6. D. Doerffel and A. Sharkh. A critical review of using the Peukert equation for determining the remaining capacity of lead-acid and lithium-ion batteries. *Journal of Power Sources,* 155: 395–400, April 2006.
7. Fraunhofer. Fraunhofer IIS MPEG-4 Video Software. Fraunhofer Institut Integrierte Schaltungen IIS, May 2006.
8. E. Grochowski and M. Annavaram. Energy per Instruction Trends in Intel Microprocessors. *Technology@Intel Magazine.* Intel, March 2006.
9. ISO/IEC 14496-10:2005; Recommendation ITU-T H.264. *SERIES H: AUDIO-VISUAL AND MULTIMEDIA SYSTEMS Infrastructure of audiovisual services – Coding of moving: Advanced video coding for generic audiovisual services video.* ITU-T, November 2005.
10. Hantro. 8300 Multimedia Application Development Platform. 2006.
11. Hantro. Hardware Video Codecs for Wireless IC's. Hantro Products Oy, 2006.
12. Hantro. Software Video Codecs for Wireless IC's and Handsets. Hantro Products Oy, 2006.
13. ISO/IEC. *Information technology – Coding of audio-visual objects – Part 2: Visual,* 3rd edition, June 2004.
14. C. Martinez, M. Pinnamaneni, and E.B. John. Multimedia Workloads versus SPEC CPU2000. *2006 SPEC Benchmark Workshop,* The University of Texas, January 2006.
15. J.C. Mogul and A. Borg. The effect of context switches on cache performance. *Proceedings of the 4th Int. Conf. on Architectural Support for Programming Languages and Operating Systems,* pp. 75–84, Santa Clara, CA. 1991.
16. Y. Neuvo. Cellular phones as embedded systems. In *Solid-State Circuits Conference,* 1: 32–37, 2004.
17. S. Park, Y. Lee, and H. Shin. An experimental analysis of the effect of the operating system on memory performance in embedded multimedia computing. *Proceedings of the 4th Int. Conf. on Architectural Support for Programming Languages and Operating Systems,* pages 26–33. EMSOFT 04, ACM 1-58113-860-1/04/0009, 2004.
18. O. Pelc. Multimedia Support in the i.MX31 and i.MX31L Applications Processors. Freescale Semiconductor, February 2006. Document Number: IMX31MULTIWP, Rev. 0.
19. D. Rakhmatov, S. Vrudhula, and D.A. Wallach. A model for battery lifetime analysis for organizing applications on a pocket computer. *Very Large Scale Integration (VLSI) Systems, IEEE Transactions,* 11: 1019–1030, December 2003.
20. F. Sebek. Instruction cache memory issues in real-time systems. Master's thesis, Department of Computer Science and Engineering, Mälardalen University, Västeroas, Sweden, 2002.
21. H. Shim. System-Level Power Reduction Techniques for Color TFT Liquid Crystal Displays. Master's thesis, School of Computer Science and Engineering, Seoul National University, Korea, 2006.
22. D. Shin, H. Shim, Y. Joo, H-S. Yun, J. Kim, and N. Chang. Energy-Monitoring Tool for Low-Power Embedded Programs. *IEEE Design & Test,* 19: 7–17. IEEE Computer Society Press, July 2002.

23. O. Silven and K. Jyrkkä. Observations on power-efficiency trends in mobile communication devices. *Proceedings of the 5th Int. Workshop on Embedded Computer Systems: Architectures, Modeling, and Simulation*, 3553/2005: 142–151. Springer Berlin/Heidelberg, July 2005.

24. Symbian. Introduction to the ECOM Architecture. 2006.

25. Symbian. Multimedia framework client overview. 2007.

26. ETSI EN 302 304 V1.1.1. *Digital Video Broadcasting (DVB); Transmission System for Handheld Terminals (DVB-H)*. European Telecommunications Standards Institute, November 2004.

27. K. Vainio. Video Decoder and Post-processing in Portable Device. Master's thesis, Department of Electrical and Information Engineering, University of Oulu, Oulu, Finland, 2006.

28. P.H.F.M. Verhoeven, J. Huang, and J.J. Lukkien. Network middleware and mobility. *PROGRESS workshop*, 2001.

External Energy Consumption Measurements on Mobile Phones

A Black Box Approach for Energy Consumption Measurements!

Frank H.P. Fitzek

Aalborg University `ff@es.aau.dk`

Summary. This chapter shows how to perform energy consumption measurements on mobile phones. We refer to this method as *black box* measurement approach as we are interested only in the overall consumption of the mobile phone. With this approach it is not possible to give a detailed description of the energy consumption per entity such as display, memory, or others, but will be explained in the following chapter. On the other side this approach is platform-independent and can therefore be used for all battery-driven devices.

24.1 Introduction

From the mobile manufacturer's perspective the energy consumption problem is critical, not only technically but also taking into account the market expectations from a newly introduced technology. The long operational time capability of terminals is both satisfying and vital for users; it gives them a truly wireless experience. This feature has been put at the top of the wish list by consumers as shown recently in [5], and therefore it must be taken seriously by the industry, and indirectly, by the developer community. In [1] it is claimed that the battery capacity has only increased by 80% within the last ten years, while the processor performance double every 18 months following *Moore's Law*. As the developer gets more and more processing power, the way applications are implemented will have an impact on how much energy the application is consuming.

As a concrete example, Figure 24.1 shows the power requirements of different wireless terminal generations, including an approximate power consumption breakdown. In terms of power consumption we have moved from a relatively low 1–2 W range in the first generations to around twice of that in 3G. The prospective for the future does not look encouraging in this aspect, as one could easily expect again another doubling in the power consumption figure. Many of the above listed factors having a direct impact on

F.H.P. Fitzek and F. Reichert (eds.), Mobile Phone Programming and its Application to Wireless Networking, 441–447.

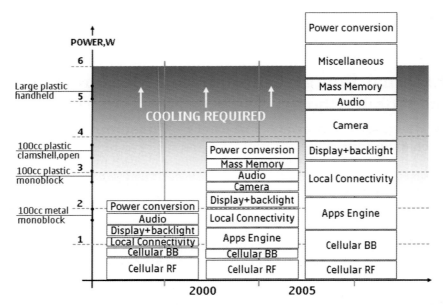

Figure 24.1. Power consumption breakdown of wireless terminals. (Courtesy of Nokia).

power consumption are directly related to the basic communications and signal processing capabilities of the terminal and, as shown in Figure 24.1, they account for roughly 50% of the power budget. Therefore, any reduction in the power consumption in these functionalities will have a substantial impact on the overall power demand.

Energy consumption in future terminals is undoubtedly going to increase considerably due to the following facts as given in [2]:

- Higher data rates: As energy per bit decreases, in order to operate with acceptable signal-to-noise ratios, the transmitted power needs to be increased.
- The ability to provide users with a continuous connection, or as it is typically referred to as *always being connected*. From the battery standpoint, this can also be interpreted as *always being drained*.
- The emergence of multimodality, multifunction terminals. Multimodality particularly means the support of multiple air interfaces which can, in principle, be used simultaneously.
- Increased DSP power is needed to process faster, wider bandwidth data. The power consumption of the processing unit increases as higher clock rates are needed to support greater processing power. Video signal processing is particularly costly in this regard, not only for the high data rate required but also because of the on-board image processing requirements.

- The inherent advanced still and video imaging capabilities of 4G terminals mean displays with higher resolution, higher contrast, and higher frame displaying rates. These all have an adverse effect on power consumption.
- Audio capabilities, particularly high-fidelity applications, sometimes incorporating stereo loudspeakers on the handheld device.
- Terminals contain large amounts of mass memory, and one is witnessing that this trend is on the rise. The use of semiconductor memory and particularly the recently introduced hard-disk equipped terminals increase power consumption significantly.
- New services may also unfavorably impact power consumption. For instance, they may exploit user location information obtained by using either an onboard satellite receiver (GPS/Galileo positioning system), or based on processing specific system signalling for that purpose (e.g., time-of arrival and triangularization.)

Therefore the developer of new applications has to take into account the energy consumption of the application as well.

24.2 Measurement Setup

Having said this, the question arises as to how much energy my application consumes. There are mainly two ways of answering that, namely external measurements and onboard measurements. This chapter deals with the external measurements.

Even for the external measurements there exist two ways of doing the measurements. Here we refer to the NOKIA N70 as reference platform, but the solution discussed here can be applied to any mobile device. The first approach is given in Figure 24.2. After finding out where the mobile phone gets its energy from, one contact is split up in such a way that we get two cables out of the phone. Those cables are then used to measure the current that is taken out of the battery. Whether you want to use a simple multimeter or a more advanced oscilloscope depends on the measurements you need to perform. The more dynamic the energy consumption of the application is, the better suited is the oscilloscope. On the other hand the oscilloscope is not really mobile and limits the measurements to the lab.

If you have a battery emulator such as the Agilent 66319D your life becomes much easier. Instead of using the mobile phone's battery you use the battery emulator as energy supply. The emulator will not only power the mobile phone up, it can also measure the energy used with high granularity. An advantage of the external power supply is that the voltage supplied is stable over the time, which may not be the case with the battery. The electrical energy E (in Joule) is calculated by multiplying the voltage U (in Volt), the current I (in Ampere), and the time t (in seconds)

$$E = U \cdot I \cdot t. \tag{24.1}$$

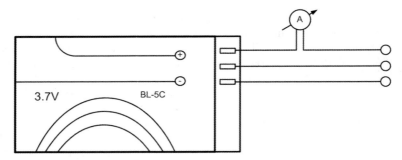

Figure 24.2. External Approach to measure the energy consumption with a battery.

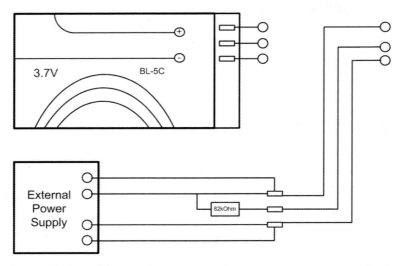

Figure 24.3. External Approach to measure the energy consumption with a battery emulator.

As given in Figure 24.3 the approach using a battery emulator is shown. But the external power supply is not that easy. Just connecting the mobile phone to a power supply may not work. As given in the figure we used a $82k\Omega$ resistor for the NOKIA N70. This was done to cheat the mobile phone, which checks whether a battery is attached. Without the resistor it would not work. In case you have a different mobile phone, you need to find out the appropriate work-arounds. But using a multimeter to check the internal resistance of the battery between the contacts should help. Note, that the cables used are also contributing to the resistance with

$$R = \rho \cdot \frac{l}{A},\qquad(24.2)$$

where ρ is the electrical resistivity of the material, l the length of the cable, and A the cross section of the cable.

In Figure 24.4 one realization is shown. The battery is actually not used and is totally isolated, but still the battery is used as the main body to hold our external contacts.

In Figure 24.5 a screenshot is shown for a PC-based application working together with the Agilent machine. The curve gives the current coming out of

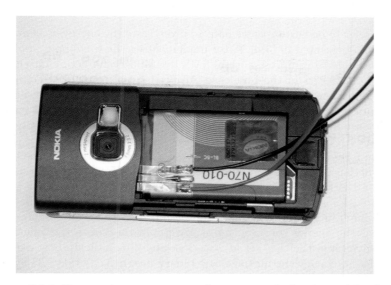

Figure 24.4. Preparation to use external power supply for the mobile phone.

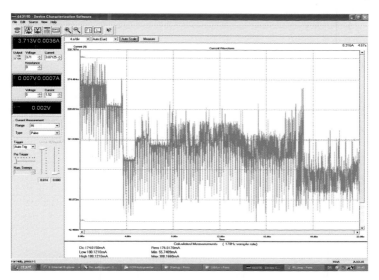

Figure 24.5. Screenshot of the Agilent tool on the PC showing energy used at the boot process.

the battery into the phone while the boot process of an N70. As can be seen in the figure, the advantage is the visibility even for highly dynamic changes in the energy consumption.

24.3 Example of Energy Measurements

Now we refer to one example we used to investigate the energy consumed. The full work is presented in [3,4]. After implementing the smsZipper application, which was able to compress normal SMS text of 500 characters into one SMS with 160 characters in the best case, we needed to answer how complex the whole application is and how much energy is consumed. We split the energy consumption into two parts. The first part had to answer how much energy we spend by sending SMS text with a given length. The energy is mostly related to the RF/BB of the mobile phone. The second step was to look into the costs of the compression. Obviously the compression costs energy, but the question arises whether the whole communication will pay off in the end by using less energy on the RF/BB.

Figure 24.6 gives the energy consumption for the sender side (the receiver looks similar) including the compression applied and the sending process. On the y-axis the energy consumption is given. On the x-axis the length of the SMS is given and the tics, set at 133, 256, and 384, indicate how many SMS are needed to convey the message as a binary message can take 133 bytes. We

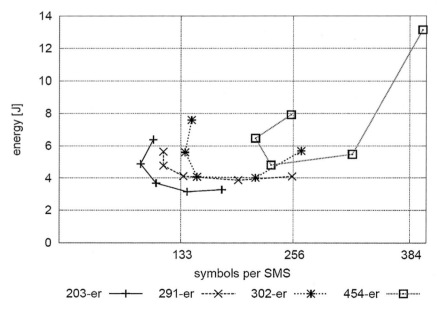

Figure 24.6. Energy trade-off plot for the sender side with model loading.

have four messages under investigation and therefore the plot contains four curves. Each curve is composed out of five points. The rightmost point represents the uncompressed sending process. The second, third, fourth, and fifth point (moving from the right to the left on the curve) represents the higher compression. Each point indicates the energy needed to compress and send the SMS. If we have a look at the message with 302 characters, we would reduce the needed SMS message from three to two with some compression reducing also the needed energy by nearly 2 Joule. Further compression increases the energy by 2 Joule compared to the uncompressed one, but on the other side we need only one SMS reducing the cost by a third. The user would probably go for the lower cost than the saved energy in this case. But in general we can find an optimal working point to save both money and energy.

This small example illustrated the way the external energy measurements can be done.

References

1. C. Andersson, D. Freeman, I. James, A. Johnston, and S. Ljung. *Mobile Media and Applications, From Concept to Cash: Successful Service Creation and Launch.* Wiley, 2006.
2. F.H.P. Fitzek and M. Katz, editors. *Cooperation in Wireless Networks: Principles and Applications – Real Egoistic Behavior is to Cooperate!* ISBN 1-4020-4710-X. Springer, April 2006.
3. F.H.P. Fitzek, S. Rein, M.V. Pedersen, G.. Perucci, T. Schneider, and C. Gühmann. Low Complex and Power Efficient Text Compressor for Cellular and Sensor Networks. *IST Mobile Summit 2006*, Greece, 2006.
4. S. Rein, C. Gühmann, and F.H.P. Fitzek. Low Complexity Compression of Short Messages. *IEEE Data Compression Conference (DCC), IEEE Computer Society Press*, Snowbird, UT, March 2006.
5. TNS. Two-day battery life tops wish list for future all-in-one phone device. Technical report, Taylor Nelson Sofres, September 2005.

Optimizing Mobile Software with Built-in Power Profiling

Everybody can make the Battery Last Longer

Gerard Bosch Creus[1] and Mika Kuulusa[2]

[1] Nokia Research Center `gerard.bosch@nokia.com`
[2] Nokia Technology Platforms `mika.kuulusa@nokia.com`

Summary. Software development for battery-powered mobile devices is tricky business because they are constrained platforms. Mobile applications are typically developed using development kits and emulators and later tested on a real device. Generally, developers are mainly concerned about correct application behavior. Badly designed software can easily impact the standby and active use-time of mobile phones, but power consumption profiling needs a costly and cumbersome measurement setup. In this chapter we describe S60 software profiling tools that allow every developer to measure power consumption without any external equipment. Measurement analysis is carried out either on the mobile device or a PC. In addition, we present a set of guidelines and good practices that energy-conscious developers should follow to maximize application use-time.

25.1 Motivation

The exploding number of features is rapidly adding to the amount of processing power and related hardware needed for their implementation. Consumers desire more performance, more impressive multimedia, faster data connections, and better usability. As a result, devices are getting more power-hungry to the point where power consumption and thermal issues have become seriously limiting factors.

Power management is required because mobile phones are battery-operated devices and run on a limited power supply. Additionally phones are becoming smaller in physical size which can make excessive power consumption heat them up more easily. Battery technology improves at a steady rate, but is not able to keep pace with the continuous upscaling of processing performance and resource usage. Current battery technology cannot offer the energy densities required to make the power consumption problem disappear. Given that battery technology does not seem to provide the necessary improvements regarding energy and power management, the next solution is to try improving

F.H.P. Fitzek and F. Reichert (eds.), Mobile Phone Programming and its Application to
Wireless Networking, 449–462.
©2007 Springer.

the phone platforms so that the desired features can be implemented at a much lower cost in energy consumption.

Possible solutions can be addressed using both hardware and software approaches. The hardware approach is often emphasized since hardware is the part physically draining energy from the battery. From this viewpoint, it makes sense to focus on hardware optimizations. However, since the only mission of hardware is actually to fulfill software needs, one can argue that software is the ultimate consumer of energy and therefore the focus should be on software optimizations [6].

Extensive research has been produced on both hardware and software sides. The approaches should not be considered mutually exclusive, but rather synergistic in nature. Hardware should ideally provide an optimal trade-off between energy and other nonfunctional attributes such as performance. On the other side, software should strive to use those hardware pathways offering the optimal trade-offs for the application at hand. Most of the software is constructed with the help of supporting software development tools like compilers that may prioritize attributes such as speed or memory footprint at the expense of energy efficiency [9]. However there seems to be a growing understanding that the application code itself holds the solution to the energy consumption problem [2, 3, 8, 10].

While hardware optimizations are the domain of hardware vendors and platform integrators, a significant value offered to customers comes from the software integrated into the devices. The software comes from a variety of providers: smartphone manufacturers, open-source contributors, tool vendors, and other third parties. Together, these individuals represent a vast community with high impact on the energy expenditure of the software they create.

Knowledge of the starting point is necessary in any optimization process before deciding on the actions necessary to reach the desired objectives. Once this knowledge exists, the optimization process must proceed until some initial requirements are met. However, an average software developer knows very little about the energy consumption of the developed applications. Whereas an electrical engineer smoothly works with current and voltage numbers, these may be hard to grasp for a software engineer more familiar with measures such as CPU load, bandwidth, or memory footprint. Software engineers would prefer to have support for power consumption measurements in their advanced software development tools. Occasionally, neglecting application energy consumption from the software design process can lead to unacceptable situations. For example, imagine a game having the greatest 3D graphics and effects being soon ready to ship to the shops. During testing it is observed that the game play time is less than 1 hour. Consumers may consider this as a fairly unacceptable trade-off for a battery-operated device.

Without the necessary knowledge about the most power-hungry parts of an application, or so-called power hogs, trying to solve the power problem can be rather difficult. Earlier research recognizes that one of the main areas for action in order to improve energy efficiency are measurement tools and methods to evaluate power consumption [1]. Naturally, before creating any solutions, one needs first to see the problem. This leads to the conclusion that some kind of power consumption profiler is needed for mobile software development.

25.2 S60 Power Profiling Application

We have created a power consumption profiling application to assist developers in creating energy-efficient applications and squeeze more use time from the battery. Juice, the power consumption profiler, is an application for S60 3rd Edition smartphones. Its main objective is to graphically show the total power consumption of the device on which it is executing. In this sense, Juice is very similar to a digital oscilloscope displaying voltage, but since it is a computer program it can have much more advanced features. The rationale behind the Juice concept is that if developers can pinpoint the problematic parts of their applications, they may be able to take corrective action. Detecting basic mistakes early in the development process can bring quick benefits without large programming efforts. A key advantage is that it works on the latest Nokia S60 phones and no additional hardware is needed. Juice exploits existing hardware that is responsible for measuring the remaining energy levels for the battery level bars on phone displays. Typically, most battery-operated devices have this kind of hardware although they may have varying degrees of achievable measurement speed and accuracy. Nokia S60 smartphones support power consumption measurements at up to 4 Hz frequency.

According to our analysis, the built-in measurements are quite accurate. We used a calibrated digital multimeter as the external reference and found that the Juice built-in measurements are comparable to those obtained with the multimeter. Built-in measurements may miss some short-lived power spikes, but from a long-term energy consumption perspective the measurements remain very accurate with respect to the external reference.

Power consumption measurements are implemented by observing battery current and voltage values. Battery current values are truly average values because the hardware integrates current consumption in the analog domain over the entire measurement period. An important characteristic for any kind of instrumentation is that it should introduce minimal disturbance to the actual measurements. This characteristic must apply particularly for an energy profiling application. In other words, Juice processing and sleep periods have to be heavily optimized for low-power operation. This is important because any kind of processing adds to the total energy consumption, so Juice must be as lightweight as possible.

Graphical power measurement analysis on a small phone display must be very easy to use. This requires careful user interface (UI) design on a device with restricted display and input capabilities. Especially measurement analysis is something that most software developers would like to perform on a desktop PC. Still, having the measurement analysis available on-device adds to what makes Juice attractive for developers. There is no need to transfer files between phone and PC, making quick analysis and testing very fast. Figure 25.1 presents the user interface for Juice.

The main mode of operation for Juice is detailed in the list below. Figure 25.2 illustrates the same process in a graphical fashion.

1. Set the desired trace settings through the *Options → Settings* dialog. Parameters such as the measurement speed and the way battery capacity is calculated can be adjusted. For example, a developer may be interested only in current consumption while voltage is not relevant for the case at hand.
2. Start the measurement. Juice starts receiving power consumption values from hardware.
3. Send Juice to the background to profile the desired use case. The user can then freely experiment with S60 applications.
4. Start the developed application and execute through the use case. This could be previously identified problematic use cases or simply the use cases present in the test plan. Take screenshots manually if needed.
5. Bring Juice to the foreground.
6. Stop the measurement.
7. Analyze the power consumption profile and identify problematic areas. Juice has some features to assist developers in identifying these spots.

One question remaining is how developers can use power traces to optimize their applications. The main problem consists in relating the instantaneous power consumption to the profiled application. This power could very well be consumed by another application running concurrently or by some plat-

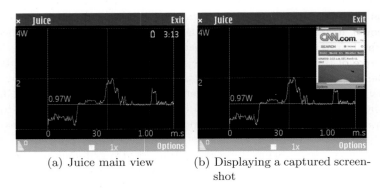

(a) Juice main view   (b) Displaying a captured screenshot

Figure 25.1. Juice showing the power consumption of web browsing over a 3G data connection. Average power is 0.97 W during the measurement, lasting 1 minute and 18 seconds. Repeating the use case forever would last 3 hours and 13 minutes for a fully charged 850 mAh battery, as shown in (a). In (b), the red vertical lines indicate that a screenshot was captured.

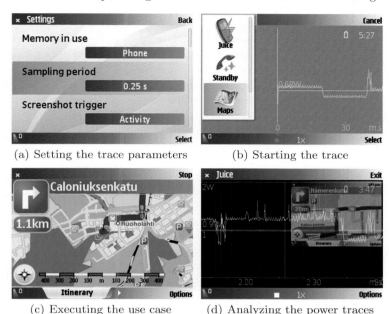

(a) Setting the trace parameters   (b) Starting the trace

(c) Executing the use case   (d) Analyzing the power traces

Figure 25.2. Juice measurements on Nokia N95. The Maps application acquires the GPS location and the user navigates to a location using the route search functionality. The average power consumption was 0.97 W and this use case could be repeated for 3 hours and 37 minutes with a full battery.

form service. This is a reasonable concern and as with any other kind of test, the testing environment should be as free as possible of external influences such as concurrent applications or unnecessary services. However, being able to account for every Joule spent even at the application level is something quite far from the current smartphone platform capabilities. Even with the knowledge of the threads and functions executing on the processor, the load on other hardware components may very well be due to requests by other applications, threads, or functions. Nevertheless, any kind of information provided in relation with the energy expenditure can potentially assist developers in pinpointing problematic threads or functions.

Presenting all the profiled data on a small screen can very easily lead to a case of information overload. Instead, Juice uses a different method to bind the power traces to specific parts of the application: visual screenshots of the phone display. By analyzing the power consumption in real time, Juice takes screenshots at those time instants in which there is a sudden peak or dip in the power consumption or when something significantly changes on the screen. Changes in the screen content may indicate some event-related processing. Additionally the user can manually take a screenshot through a specific key combination.

The purpose of screenshots is to allow developers to later browse through these screenshots in order to understand what was being done exactly at that time instant. As illustrated in Figure 25.3, the screenshots form a kind of visual history superimposed on the power graph. The capability to map application states to unusual power events can help to focus the developer's attention on a problematic construct or specific design decision. For example, a specific algorithm could be replaced with another with lower computational complexity or some buffering could be introduced to eliminate the problem. The solutions vary from application to application, we present some general guidelines for energy-efficient software later in this chapter.

A key feature in the Juice display is the time estimate (hours:minutes) that we refer to as *battery time*. This time describes how long a full phone battery would last for the measured use case. The calculation is based on nominal battery capacity (mAh) rating and real measured power consumption. For a developer, this number is very straightforward to understand since it reflects the design objective: the higher the number is, the longer use time your software has. It is interesting to note that battery time may open a whole new differentiation opportunity for S60 application developers. In a world where similar applications compete, better battery time performance can give application vendors a competitive edge over the rest. For example, two different web browser or e-mail client application offerings could have a similar set of features, but one allows for 2 hours more browsing or 5 days more push e-mail standby time. Battery life already plays a differentiating role in some embedded devices such as portable media players. Juice gives a new market opportunity for energy-conscious software developers.

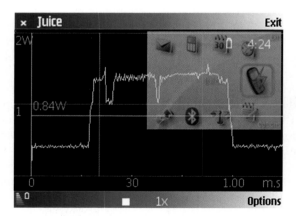

Figure 25.3. Juice graph viewing with screenshots. Visual history with screenshots helps to see what was done at a specific instant during application testing. Here, the user was browsing a 3D menu that uses hardware-accelerated graphics at the instant marked with a bright red vertical marker.

Although the main purpose of Juice is to measure active application use cases, file storage space is the only limitation for the length of a measurement. Figure 25.4 presents an example of an extended measurement spanning 75 hours.

We believe that every platform should provide developers with support for power consumption analysis. Even though Juice is currently available for S60 devices only, we are planning support for other product offerings, such as Maemo. Figure 25.5 shows a concept drawing of *MaemoJuice*.

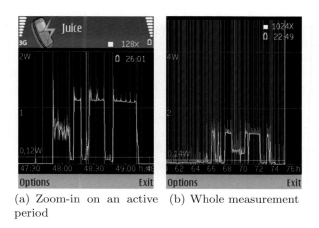

(a) Zoom-in on an active period   (b) Whole measurement

Figure 25.4. Juice graph examples. Figure (a) shows one GSM and three 3G calls in downtown Helsinki. Power consumption is around 0.7W for GSM and 1.2W for 3G. Maximum zoom-out from a 75 hour measurement with over 1,000 screenshots is shown on (b).

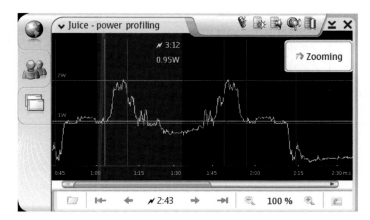

Figure 25.5. Concept drawing of Juice running on the Maemo platform. The bigger display area and pen input available in Internet tablets allow for an improved interface for power consumption analysis.

25.3 Carbide.c++ Power-Performance Profiling

A stand-alone S60 energy profiling application may become limited for more complicated applications or use cases where a deeper analysis is required on the development PC workstation. Many software developers are familiar with Carbide.c++, an integrated development environment (IDE) based on the open-source Eclipse IDE. Carbide.c++ contains a development tool called the Performance Investigator (PI). The PI integrates an on-device profiler together with a PC-based analyzer for S60 software performance optimization. In addition to CPU activity traces, the PI included in Carbide.c++ version 1.2 supports also power consumption measurements. The power measurement functionality in Carbide.c++ uses the same low-level interfaces as Juice and provides exactly the same measurement speeds at the same levels of accuracy. However, while Juice was designed to be a stand-alone application that developers can use on the device, the PI is aimed to be used in the IDE on the PC.

The PI is divided into two different components: an on-device profiler application and an analyzer plug-in running on the PC. The profiler is responsible for generating the runtime trace files that are subsequently imported to the analyzer. Figure 25.6 shows the user interface for the profiler application. Typically the profiler spends most of the time in the background while the profiled activity takes place on the foreground. The PI offers developers a complete picture of the CPU load and Symbian OS threads together with the power consumption measured from the battery. This kind of side-by-side view can be of significant importance in isolating problematic binary executables, threads, or even function calls. Figure 25.7 shows the PI analyzer with performance and power consumption traces.

Setting up the necessary components for the PI power tracing requires a few steps, detailed in the following.

1. Enable the desired traces on the profiler. Power traces can be enabled from the *Tracing options* settings dialog. The sampling resolution can be set through the *Power usage interval* setting in the *Advanced options* dialog. The minimum sampling period of 250 ms corresponds to the maximum sampling speed achievable in current devices.
2. Start the profiler and send it to the background. This can be achieved through the *Application* key, either by pressing it once and navigating to your application or keeping it pressed for about a second and selecting your application from the task list.
3. Execute the use case to profile.
4. Bring the profiler to the foreground (you can access it through the open tasks list) and stop tracing.
5. Transfer the trace files to the PC. The location of the files on the profiled device can be set up through the *Output settings* dialog in the PI Profiler.
6. Import the PI trace files to Carbide.c++ through the *File → Import* menu option.

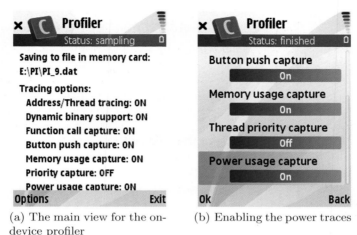

(a) The main view for the on-device profiler

(b) Enabling the power traces

Figure 25.6. User interface for the Performance Investigator on-device profiler. The *Options* dialog allows for the setup of the parameters to trace, including power consumption.

7. Analyze the traces through the PI plug-in on the PC. Try to determine if there are any execution constructs that trigger a particularly high power consumption.

The real advantage provided by the PI plug-in is that it allows developers to see a concurrent view of the CPU activity and the power consumption at the device level. Still, many hardware components work asynchronously of the main CPU and thus it may be hard to find valid correspondences between the power consumption and specific program threads in cases where the CPU is not the dominant power consumer. In any case, the PI is a very useful tool in identifying recurrent instances of particular function calls or threads that seem to have a significant contribution to the power consumption, for example, causing periodic power peaks over time. This may be due to the function activating a certain hardware component with a known power consumption pattern.

One interesting feature of power consumption analysis in conjunction with the CPU activity traces is that it can reveal problems at the system level. For example, the S60 web browser designers were investigating a slowdown when communicating with the 3G modem device driver. Power analysis revealed that the power consumption dropped dramatically after a certain function was executed. A quick look on the power levels confirmed that the 3G modem was erroneously disabled for some reason. A hidden bug in the driver interrupt handling routine was found and fixed. This programming error could have remained undetected for much longer had it not been uncovered through power profiling.

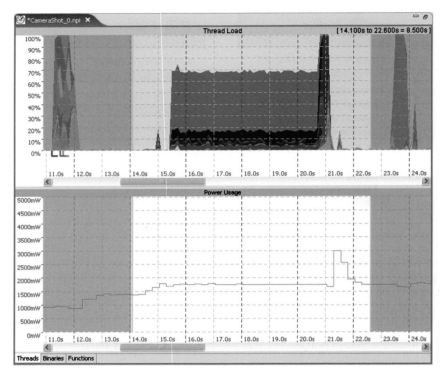

Figure 25.7. Power consumption analysis in the Carbide.c++ PI. The profiled use case involved the camera application, where 5s were spent viewfinding and then a picture was taken. The thread load on the CPU is shown on the top, with the power consumption graph on the bottom.

25.4 Energy-Efficient Design Guidelines

There are not many specific studies looking into the energy breakdown of mobile phone use cases. Such knowledge is important to understand which components draw most of the energy from the battery. Developers can then better figure out what are the actions to avoid. Figure 25.8 illustrates a power consumption breakdown in a mobile phone.

These figures are in agreement with previous findings on the power consumption of mobile computers [3, 5, 11, 12]. The major consumers are (in decreasing order): wireless modems (WLAN, 3G, GSM/EDGE), application processor, and display backlight. An important share of the total power is consumed by a mix of components with marginal power consumption by themselves, such as the radio modem control processor and different integrated circuits. Knowing that three components may account for roughly 65% of the total energy consumption, it becomes clear that reducing the load on those components will have a noticeable effect on the overall energy consumption.

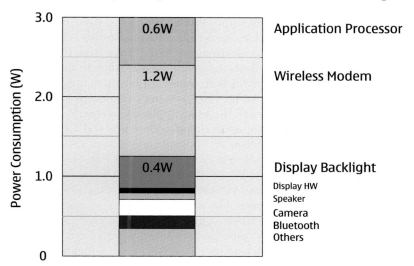

Figure 25.8. Power consumption for video conferencing in the Nokia 6630 [4, 7]. Constant average 3W can be tolerated without uncomfortable heating, but smaller products may have power budgets closer to 2W.

In practice, taking energy considerations into account will result in a more responsible use of the available resources. For example, since radio interfaces tend to be rather power-hungry, they should be disabled whenever possible.

Most mobile platforms provide some sort of automatic power saving modes and for this reason applications should do their best to support them. Applications may consider implementing some kind of buffering to avoid frequent use of the communications bearer. Similarly, applications should avoid keeping unnecessary networking sockets open if they serve no purpose, allowing the bearer to automatically enter a reduced power mode. In the following, we present a set of guidelines to support the development of energy-efficient software. While these guidelines do not necessarily guarantee energy efficiency, not respecting them will most likely result in wasted energy and lower battery times. Likewise, in some cases there may be other factors preventing their implementation such as hard performance requirements. This is better left to the judgment of the smart software developer.

1. Avoid unnecessary or fast timers. Reducing the amount of timer constructs in software spares resources. Timers may also prevent the device power management from automatically entering a low-power sleep mode or can cause the device to leave sleep modes early. Smartphones make extensive use of sleep modes to conserve energy and thus frequent timer wake-ups are not desirable. Make sure that you absolutely need that timer and, if it ticks periodically, select the lowest possible frequency. Frequent timers are a major source of energy consumption.

2. Avoid unnecessary computation. While this may sound obvious, many applications do not obey this simple rule. Most platforms support notifications for the cases where applications are moved to or from the background and foreground. Most applications should be idle when in the background, i.e., with no resource consumption. Some applications may require some form of background processing, such as e-mail clients, peer-to-peer sharing, or anti-virus applications. However, applications should definitely stop updating the user interface and related code when the UI is not visible to the user. This applies when an application is not in the foreground or when the display is not visible in clamshell phones, for example.

3. Make the UI energy-aware. User interfaces account for a significant share of the energy consumption in mobile computers. This is especially true since the speed gap between users and computers is considerable, and a fair amount of time is spent simply waiting for user input. Previous research looking into this topic has produced a set of recommendations to create more energy-efficient user interfaces [13], which we summarize below.

   • Accelerate user interaction. Make the frequently used functionality easily accessible.
   • Minimize screen changes. Avoid unnecessary animations or progress indicators for quickly completed tasks. Buffer multiple consecutive changes before redrawing the screen, so that a single update is needed.
   • Avoid or minimize slow user input. Text input is slow on smartphones and should therefore be minimized when possible. Different kind of lists can be used as alternative, faster input mechanisms.
   • Reduce redundancy. Replace progress bars with busy indicators when appropriate and avoid features slowing down the user interaction.
   • Speculate on the user input. Use text autocompletion, last entries, and pre-compute data for the most used functionality. Even though some of your predictions are going to be wrong, it's better than wasting energy simply waiting for the user input.

4. Take advantage of advanced platform features. Exploit the platform features allowing your application to hint or even directly control some platform resources. For example, S60 offers the *Lights API* to directly set the state of the display backlight.

5. Synchronize periodic actions. Instead of distributing periodic tasks in time, try to perform them in a batch run. This reduces the amount of timers and lengthens the idle periods which enables staying longer in sleep modes.

6. Consider energy-performance trade-offs. Application design presents many alternatives with different energy-performance ratios. Compare different implementation alternatives and select the most energy-efficient one that fulfills the desired performance needs. This applies to the whole development process, from the choice of architectural patterns to the selection of a particular algorithm or UI widget.

7. Optimize the server side. Applications using wireless data transfers are typically the most energy consuming. Applications typically connect to an Internet server using specific protocols and exchange application-specific data to and from the mobile phone. Consider optimizing the traffic pattern or compressing the data to a format that is lightweight to process on the device. Try to minimize the amount of data that needs to be transmitted.

8. Experiment with different devices. Mobile phones differ in features and hardware components. For example, the main CPU may have a different performance or even be a different model, such as ARM9, ARM11, multicore ARM11 (MPCore), or Cortex-A8. Some devices may have higher speed data connectivity and WLAN to complement the cellular modems. Also display sizes are different. An optimization that works on a high-end phone may not work that well on a low-end one.

25.5 Conclusions

This chapter has covered the field of software optimization from an energy perspective. We have presented two different approaches for getting a global view on the power consumption of software requiring no additional tracing hardware than that already available on smartphones. In addition, we have given a set of guidelines and recommendations to develop energy-efficient software.

There seems to be no silver bullet to eliminate or even alleviate the energy limitations that mobile devices face nowadays. Even device manufacturers, with a relatively high degree of control over the hardware and software components integrated into their product offerings, cannot alone solve the energy problem. Smartphone platforms are designed to be open to third-party software that adds to the value provided by the product. In fact, a fair share of the active use time on a mobile phone may be spent on this externally provided add-on software. If such software is not developed taking energy considerations into account, the problem is only going to get worse as the energy demands of the available features increase.

Mobile phone application development already exhibits a considerable degree of complexity. In order to create energy-efficient software, developers need tools to support them in such a challenge. The Juice power consumption profiler and the power tracing functionality included in the Performance Investigator of Carbide.c++ aim to fulfill this need. These tools do not require a basic set of electrical engineering skills to operate. Instead, they provide embedded power consumption measurements so that developers can easily access the required data. Juice provides an easy-to-use tool for basic power consumption analysis directly on the profiled device. For more detailed analysis, the Carbide.c++ PI plugin offers the possibility to link power consumption to specific software constructs running on the device.

The solution to the energy problem lies in the hands of every software developer contributing to enrich the mobile software ecosystem. Carefully designing software to be more energy-efficient can greatly enhance the battery life of mobile devices. Everybody can make the battery last longer.

References

1. C.Ellis, A. Lebeck, and A. Vahdat. System support for energy management in mobile and embedded workloads: A white paper. Technical report, Duke University, Department of Computer Science, October 1999.
2. C. Ellis. The case for higher level power management. *Proceedings of the 7th Workshop on Hot Topics in Operating Systems (HotOS'99)*, March 1999.
3. J. Flinn and M. Satyanarayanan. Energy-aware adaptation for mobile applications. *Proceedings of the 17th Symposium on Operating System Principles (SOSP'99)*, December 1999.
4. Mika Kuulusa. Multiprocessors in wireless multimedia terminals. In-depth presentations, Sixth International Forum on Application-Specific Multi-Processor SoC (MPSoC'06), August 2006. [Online] Available: http://tima.imag.fr/mpsoc/2006/slides/Kuulusa.pdf.
5. J. Lorch. A complete picture of the energy consumption of a portable computer. Master's thesis, University of California at Berkeley, 1995.
6. Yung-Hsiang Lu, Luca Benini, and Giovanni De Micheli. Requester-aware power reduction. *International Symposium on System Synthesis*, pp. 18–23. Stanford University, September 2000.
7. Yrjö Neuvo. Cellular phones as embedded systems. *IEEE Solid-State Circuits Conference (ISSCC'04). Digest of Technical Papers*, vol. 1, pages 32–37, February 2004.
8. B. Noble. System support for mobile, adaptive applications. *IEEE Personal Communications*, 7(1):44–49, February 2000.
9. T. Simunic, L. Benini, and G. De Micheli. Energy-efficient design of battery-powered embedded systems. *Proceedings of the International Symposium on Low-Power Electronics and Design (ISLPED'98)*, June 1998.
10. T.K. Tan, A. Raghunathan, and N. Jha. Software architectural transformations: A new approach to low energy embedded software. *Proceedings of the Conference on Design Automation and Test in Europe (DATE'03)*, 2003.
11. Sanjay Udani and Jonathan Smith. The power broker: Intelligent power management for mobile computers. Technical Report MS-CIS-96-12, Department of Computer Science, University of Pennsylvania, May 1996.
12. Sanjay Udani and Jonathan Smith. Power management in mobile computing (a survey). Technical Report MS-CIS-98-26, Department of Computer Science, University of Pennsylvania, August 1996.
13. Lin Zhong and Niraj K. Jha. Graphical user interface energy characterization for handheld computers. *Proceedings of the International Conference on Compilers, Architecture, and Synthesis for Embedded Systems (CASES'03)*, October 2003.

Acknowledgments

The present book is the result of the coordinated efforts of many people around the globe, and it would have not been possible without the key contributions of our invited authors. We thank you all for sharing with us your technical expertise, and for the professionalism and endless enthusiasm showed during the writing period.

We would also like to thank Harri Pennanen, Nina Tammelin, Karsten Vandrup, and Per Møller from NOKIA for the great help in supporting us in any activity. We are pleased by this collaboration and we are more than thankful that three years ago Harri gave us the possibility to join the academic activities of NOKIA. We want to express our thanks to Andreas Fasbender and Martin Gerdes from Ericsson for their great enthusiasm for our research and the active collaboration that made our work possible. We would also like to thank Knut Yrvin from Trolltech for his help and support regarding the *Greenphone* activities. Furthermore, Martin Østergaard from Sonofon supported our activities in the field of cooperative networking and many other projects and we therefore would like to thank him also. We thank those companies who financed our student helpers such as Gerrit Schulte from acticom and Olaf Kehrer from *o&o* software.

Parts of the book were partially financed by the Danish government on behalf of the FTP activities within the X3MP project.

We are particularly thankful to Mark de Jongh and Cindy Zitter, from Springer for their encouragement, patience, and flexibility during the whole edition process. We wish sincerely to thank Dr. Tim Brown for his invaluable help in proofreading several chapters of the book.

We would also like to thank all our colleagues in Aalborg University and Agder College for their encouragement and interest. Henrik Benner, Finn Hybjerg Hansen, Per Mejdal Rasmussen, Bo Nygaard Bai, Svend Erik Volsgaard, and Torben H. Knudsen, from the KOM Computer Workshop deserve our sincere thanks and appreciation for keeping the project website up and running day and night, and for always providing us with technical assistance and handy solutions. Special thanks also to the head of the *Electronic Systems* Børge

Lindberg for supporting us with several mobile phones to start the free-study activity *Mobile Phone Programming*. Furthermore, the technical assistance from Ben Krøyer, Peter Boie Lind Jensen, and Kristian Bank in the APNet group. Especially, Ben's help in building the wireless sensors has been highly appreciated and we look forward to getting more out of this activity. Also we thank Kasper Rodil for his enthusiasm in making the book cover.

There are more people we would like to thank: Christophe Maire/gate5, Prof. Dr. Martin Mauve/Heinrich-Heine Universität Düsseldorf, and Flemming Bjerge Frederiksen/Aalborg University. Last but not least, we have to thank those people running our research activities such as Morten V. Pedersen, Morten L. Jørgensen, Gian Paolo Perrucci, Thomas Arildsen, and Andreas Häber.

For the *SMARTEX* application we thank Karsten Noergaard, Jesper H. Pedersen, Steen S. Nord, Morten Videbaek Pedersen, and Martin H. Larsen. We would also like to thank Christophe Maire, Gate 5 for his motivation and availability to discuss the ideas.

The *Walkie Talkie* project was carried out as sixth semester project at AAU. We thank the project members: Kim Nygaard Gravesen, Morten Lisborg Joergensen, Peter Baagoe Nejsum, Morten Videbaek Pedersen and Sune Studstrup.

List of Abbreviations and Symbols

1G   First Generation
2G   Second Generation
3D   Three-dimensional
3G   Third Generation
4G   Fourth Generation
5G   Fifth Generation
8-PSK 8-Phase Shift Keying
ACL   Asynchronous Connection-Less
AF   Amplify-and-Forward
AMC   Adaptive Modulation and Coding
AMR   Adaptive Multi-Rate
API   Application Programming Interface
ARMI ARM-Interoperable
ARQ   Automatic Repeat-reQuest
ASIC Application-Specific Integrated Circuit
AWT   Abstract Window Toolkit
BSD   Berkeley Software Distribution
CAL   Channel Adaptation Layer
CA   Collision Avoidance
CDC   Connected Device Configuration
CDMA Code-Division Multiple Access
CD   Collision Detection
CLDC Connected Limited Device Configuration
CPU   Central Processing Unit
CRC   Cyclic Redundancy Check
CSD   Circuit Switched Data
CSI   Channel State Information
CSMA Carrier Sense Multiple Access
CS   Coding Schemes
CTS   Clear To Send
DCF   Distributed Coordination Function
DCP   Device and Service Descriptions
DF   Decode-and-Forward
DHCP Dynamic Host Configuration Protocol
DHT   Distributed Hash Table
DLL   Dynamic Link Library
DNS   Domain Name System
DRI   Decision Reliability Information
DRM   Digital Rights Management
DSP   Digital Signal Processing
DVD   Digital Versatile Disc
DWGPS Dynamic-Weight Generalized Processor Sharing

EC   Executive Committee
EDGE Enhanced Data Rates for GSM Evolution
EGPRS Enhanced General Packet Radio Service
EG   Expert Group
EPM   Energy and Power Management
FEC   Forward Error Correction
FM   Frequency Modulation
FP   Foundation Profile
GCF   Generic Connection Framework
GENA Generic Event and Notification Architecture
GERAN GSM EDGE Radio Access Network
GFSK Gaussian Frequency Shift Keying Modulation
GMSK Gaussian Minimum Shift Keying
GOP   Group-Of-Pictures
GPRS General Packet Radio Service
GPS   Global Positioning System
GSM   Global System for Mobile communications
GUI   Graphical User Interface
GWebCache Gnutella Web Cache
HSCSD High Speed Circuit Switched Data
HTML HyperText Markup Language
HTTPMU HTTP/Multicast
HTTPU HTTP/Unicast
HTTP HyperText Transfer Protocol
IAP   Internet Access Point
ICMPv6 Internet Control Message Protocol, version 6
IDE   Integrated Development Environment
IMEI International Mobile Equipment Identity
IP   Internet Protocol
IPv6 Internet Protocol, version 6
IR-UWB Impulse Radio – Ultra-WideBand
ISM   Industrial, Scientific, Medical band
ISO   International Organisation for Standardisation
ISP   Internet Service Provider
ITU   International Telecommunications Union
J2EE Java 2 Enterprise Edition
J2ME Java 2 Micro Edition
J2SE Java 2 Standard Edition
JAD   Java Application Descriptor
JAR   Java Archive
JCP   Java Community Process
JDK   Java Development Kit
JSR   Java Specification Request
JTWI Java Technology for the Wireless Industry
JVM   Java Virtual Machine

L2CAP Logical Link and Control Adaption Protocol
LDAP Lightweight Directory Access Protocol
MAC Medium Access Control
MAC Medium Access Control
MIDP Mobile Information Device Profile
MIDlet Java application for MIDP
MIMO Multiple Input Multiple Output
MIMO Multiple-Input Multiple-Output
MMAPI Mobile Multimedia API
MMS Multimedia Message Service
MSA Mobile Service Architecture
MTM Message Type Module
MTU Maximum Transmission Unit
MVC Mobile-View-Controller
NAL Network Adaptation Layer
NAT Network Access Translation
OBEX OBject EXchange protocol
OEM Original Equipment Manufacturer
OFDMA Orthogonal Frequency-Division Multiple Access
OFDM Orthogonal Frequency Division Multiplex
OLED Organic Light-Emitting Diode
OSI Open Systems Interconnection
OS Operating System
P2P Peer-to-Peer
PBP Personal Basis Profile
PCF Point Coordination Function
PC Personal Computer
PDA Personal Digital Assistant
PHY PHYsical layer
PIM Personal Information Management (or Manager)
PI Performance Investigator
PLL Phase-Locked Looped frequency divider
PP Personal Profile
PSTN Public Switched Telephone Network
QoS Quality of Service
RAM Random Access Memory
RFCOMM Radio Frequency Communication
RFC Request For Comments
RF Radio Frequency
RI Reference Implementation
RLC Radio Link Control
RLP Radio Link Protocol
RPN Reverse Polish Notation System
RSP Remote Socket Protocol

RSS Really Simple Syndication
RTCP Real Time Control Protocol
RTS Ready To Send
S60 Series 60
SAL Source Adaptation Layer
SAX Simple API for XML
SCO Synchronous Connection Oriented
SCTP Stream Control Transmission Protocol
SDK Software Development Kit
SDP Service Discovery Protocol
SDR Software Defined Radio
SLP Service Location Protocol
SMS Short Message Service
SMTP Simple Mail Transfer Protocol
SOAP SOAP
SP2 Service Pack 2
SP4 Service Pack 4
SPI Serial Peripheral Interface
SSDP Simple Search Discovery Protocol
SSI Source Significance Information
TCK Test Compatibility Kit
TCP Transmission Control Protocol
TCP Transport Control Protocol
TDMA Time Division Multiple Access
UART Universal Asynchronous Receiver Transmitter
UDA UPnP Device Architecture
UDDI Universal Description Discovery and Integration
UDN Unique Device Name
UDP User Datagram Protocol
UIC UPnP Implementers Corporation
UI User Interface
UMTS Universal Mobile Telecommunications System
UPnP Universal Plug & Play
URI Uniform Resource Identifier
URL Uniform Resource Locator
USB Universal Serial Bus
UTF-8 8-bit Unicode Transformation Format
UUID Universally Unique Identifier
UWB Ultra Wide-Band
VAA Virtual Antenna Array
VM Virtual Machine
VoIP Voice over Internet Protocol
WC Working Committee
WLAN Wireless Local Area Network

WMA Wireless Messaging API
WPAN Wireless Personal Area Network
WSDL Web Service Description Language
WWRF Wireless World Research Forum
XML Extensible Markup Language

Index